AF555933

BIOINSTRUMENTATION

BIOINSTRUMENTATION

L. VEERAKUMARI

Reader
Department of Zoology
Pachaiyappa's College
Chennai 600 030

MJP PUBLISHERS
Chennai 600 005

ISBN 978-81-8094-288-4 (HB)

MJP 020
Publisher : C. Janarthanan

MJP Publishers
No. 44, Nallathambi Street,
Triplicane,
Chennai 600 005

Preface

The growth of science and technology is quite astonishing. This success is due to the development of new tools and techniques. All scientific investigations require the use of modern tools and techniques. Knowledge of the basic principles and applications of tools and techniques is mandatory for students of life sciences before they start doing any experiment. Bioinstrumentation is defined as the instrumentation techniques and principles for the measurement of physical, physiological, biochemical and biological factors in man or other living organisms. There was a long-felt need among the students and teachers of biology for a textbook that will concentrate on the bioinstrumentation syllabus that has been introduced in the curriculum of life sciences of many Indian universities. The present work is an attempt to make available a book that will provide comprehensive knowledge about the basic principles and applications of the instruments generally used in biology. This book also encompasses the tools and techniques used in the growing field of molecular biology.

The book comprises eighteen chapters. The first two introductory chapters give an elementary overview of safety in laboratories and units of measurements, the knowledge of which is prerequisite for students of biology before conducting any experiment in biology. The third chapter on microscopy describes the basic principles and applications of different types of microscopes including the most advanced non-optical scanning probe microscopy that allows the scientists to visualize the regions of high electron density and infer the position of individual atoms.

The fourth, fifth and sixth chapters are on balance, centrifuge and pH meter which are commonly used in the laboratory. The next two chapters are on manometry and osmometry which are used to measure pressure and the osmotic strength of solutions respectively. The subsequent chapters (Chapters 9 and 10) deal with separation techniques—chromatography and electrophoresis. Chapter 11 on spectroscopy describes the principle and applications of an array of spectoscopic techniques involved in practical biochemistry. Radioisotope techniques, biochemical applications of radioisotopes, radioimmunoassay, detection and measurement of radioactivity are explained in chapter 12.

Furthermore, a chapter on biosensors has also been included. A biosensor is an analytical device consisting of a biocatalyst (enzyme, cell or tissue) and a transducer, which can convert a biological or biochemical signal or response into a quantifiable electrical signal. Equipment capable of acquiring data as well as processing it could find wide application in monitoring personal health, the food we eat and our environment. Biosensors have already gained major commercial importance, and their significance is likely to increase as the technology develops. Hence, a flavour of biosensors is given in chapter 13. The last few chapters (Chapters 14–18) are focused on the modern techniques in molecular biology—DNA sequencing, polymerase chain reaction (PCR), DNA microarray, protein sequencing and bioinformatics. A chapter on bioinformatics is included because most biological science students and researchers have started using it as a tool for the management of biological data. A large number of photographs and illustrations are included in each chapter. Review questions are added at the end of each chapter, which will help the students to review their learning. The last section of this book includes glossary and appendices. But I admit that this book is not a complete treatise of bioinstrumentation since the term bioinstrumentation also includes biomedical instrumentation and including topics on biomedical instrumentation would be beyond the scope of this book.

I am greatly indebted to my husband Mr.S.Ramesh Babu for his motivation, constant support and encouragement. My sincere thanks are due to all those who stimulated my interest and supported me in my efforts in this venture. Grateful acknowledgement is made to all eminent authors of related books, which have been consulted for the preparation of this book. I extend my thanks to the publisher and the editorial team for publishing the book promptly.

Valuable suggestions from teachers and students are most welcome.

L. Veerakumari

Contents

SAFETY IN LABORATORIES

Working safely in a laboratory requires properly maintained equipment and engineering controls, following appropriate personal protective measures, proper work practices, knowing safety information for the materials and equipment used, and following safety instructions and laboratory protocols. Knowledge of safe use of laboratory equipment and personal protection to work in the laboratory is imperative for laboratory personnel. Some of the important aspects to be considered in safe use of laboratory equipment and personal protection are discussed here.

SAFE USE OF LABORATORY EQUIPMENT

Laboratory apparatus must be used only for its designed purpose unless appropriate safety modifications are made.

Electrical Equipment

1. All electrical equipments used in the laboratory must be grounded. Ground fault circuit interrupters must be used whenever equipment is in a wet environment such as a cold room.
2. Electrical apparatus must be plugged into sockets, which can be reached safely, without exposure to hazards.
3. Electrical apparatus used in a fume hood must be plugged in outside the hood.

4. Electrical cords must be short and must be placed in such a way that the risk of tripping or spills is minimized.
5. Extension cords must be avoided. If unavoidable, it must be ascertained that the extension cord is appropriate.
6. Equipment, including electrical plugs and cords, must be kept in good condition. Electrical equipment must be unplugged before routine replacement of parts or before making internal adjustments.
7. A qualified electrician must attend to electrical repairs.
8. Non-sparking electrical switches and motors are desirable in laboratory equipment to prevent combustion of flammable vapours.

Heating Devices

1. Uncontrolled heat sources such as Bunsen burners and heat guns must not be used near flammable substances and must not be left unattended in the laboratory.
2. Heating devices (i.e., steam baths) which have an inherent cut-off point are safer than those which do not.
3. Hot plates, heating mantles, and other heaters must have enclosed elements and controls with a thermal shut-off safety device.

Cryogenic Liquids

Cryogenic liquids are gases that have been transformed into extremely cold refrigerated liquids, which are stored at temperatures below –130°F (–90°C). They are normally stored at low pressures in specially constructed, multi-walled, vacuum-insulated containers.

Hazards The potential hazards that accompany cryogenic liquids may result from:

1. Extreme cold which can freeze human tissue on contact, and which can also cause embitterment of carbon steel, plastics and rubber.

2. Extreme pressure, which can result from rapid vaporization of the refrigerated liquid due to rising temperature from leakage of heat into the cryogenic container or systems.
3. Asphyxiation due to displacement of air by the escaping liquid and the resultant rapidly expanding gas (in the case of inert gases).

Personnel safety Because of the potential hazards resulting from the extremely low temperatures of cryogenic liquids, all personnel handling them must be properly trained in the use of specialized equipment designed for the storage, transfer, and handling of these products.

Heavy leather protective gloves, safety shoes, aprons, and eye protection glasses must be worn to prevent possible contact with the extremely cold surfaces of uninsulated piping, transfer connections, valves, and other equipment, or from the cold liquid or boil-off vapours which may result from spilled or splashed liquid.

Any transfer operations involving open containers must be conducted slowly to minimize boiling and splashing of the cryogenic liquid, and such operations must be conducted only in well-ventilated areas to prevent the possible accumulation of inert gas which can replace the oxygen in the atmosphere and cause asphyxiation.

Centrifuges

1. Each centrifuge operator must be instructed on the proper operating procedures of the centrifuge including balancing loads, selection of proper rotor, head, cups and tubes and use of accessory equipment. The operating manual, should be referred to for information and/or assistance.
2. Centrifugation presents a physical hazard in the event of mechanical disruption. Aerosols and droplets may also be generated.
3. Operating procedures for each centrifuge must be established by the supervisor in accordance with the procedures outlined

in the operating manual, and guidelines for centrifugation of infectious agents, chemical hazards and/or radioactive materials, and for the location of centrifuge.

Centrifuge tubes

1. Plastic centrifuge tubes should be used whenever possible to minimize breakage.
2. Tubes to be used in angle-head centrifuges must never be filled to the point that liquid is in contact with the lip of the tube when it is placed in the rotor, even though the meniscus will be vertical during rotation. When the tube lip is wetted, high gravitational forces drive the liquid past the cap seal and over the outside of the tube.
3. Nitrocellulose tubes should only be used when clear, without discoloration, and flexible. It is advisable to purchase small lots several times a year rather than one large lot. Storage at 4°C extends life. Nitrocellulose tubes must not be used in angle-head centrifuges.
4. All centrifuge tubes should be inspected prior to use. Broken, cracked, or damaged tubes should be discarded.
5. Operating manual should be referred to for selection of appropriate tubes, carrier cups and rotors.
6. Capped centrifuge tubes should be used whenever possible.

Carrier cups and rotors

1. Operating manual should be referred to for proper selection and use of carrier cups and rotors. Recommended speeds should not be exceeded.
2. Centrifuge cups and rotors should be kept clean to prevent corrosion.

Ultraviolet (UV) Light

Ultraviolet radiation includes those portions of the radiant energy spectrum between visible light and X-rays (approximately 3900 to

136 Å units). Under certain conditions, including radiation intensity and exposure time, UV radiations may kill certain types of microorganisms, their greatest effectiveness being against vegetative forms. UV light is not a sterilizing agent except in certain exceptional circumstances. It is used only to reduce the number of microorganisms on the surfaces and in the air. Factors such as lamp age and dust accumulation will contribute to decreased efficiency.

Radiation exposure

1. The eyes and skin should not be exposed to direct or strongly reflected UV radiation. The effect of radiation over exposure is determined by such factors as dosage, wavelength, portion of the body exposed, and the sensitivity of the individual.
2. Overexposure of the eyes will result in a painful inflammation of the conjunctiva, cornea, and iris. Symptoms will develop 3–9 hours following exposure. There is an unpleasant foreign body sensation accompanied by lacrimation. The symptoms usually disappear in a day or two.
3. Exposure to the skin will produce erythema (reddening) 1–8 hours following exposure.

Rules and procedures

1. A hazard warning sign must be affixed on the doors of laboratories, animal rooms, etc. which have ultraviolet light installations.
2. Adequate eye and skin protection measures must be taken when working in an irradiated area. Safety glasses with side shields or goggles with solid sidepieces must be worn. Face shields, caps, gloves, gowns, etc. provide skin protection.
3. UV lamp surfaces should be cleaned as often as necessary to maximize output.
4. UV lamps used as space and surface sanitizers should be checked regularly and replaced according to the manufacturer's recommendations.

Microwave Ovens

When melting agar, the following precautions must be taken:

1. Explosions may occur when melting agar using a microwave oven.
2. Caps on screw-cap bottles must be completely loosened before the bottles are heated in the microwave oven.
3. A long-sleeve laboratory coat must be worn when heating agar in a microwave oven.
4. Heat-resistant gloves must be worn to prevent burns and to protect the hands in case of an explosion.
5. Face-shields must be used when handling microwave-heated materials.

Autoclaves

1. Autoclaves must be operated in accordance with the manufacturer's and laboratory safety manual's instructions.
2. Operating instructions and emergency shutdown procedures must be pasted on or immediately adjacent to the autoclave.
3. Responsibility for operation and routine care must be assigned to trained personnel.
4. Eye protection, heat-resistant gloves, and aprons must be worn when loading and unloading a hot autoclave. Opening doors soon after a run is finished may blow hot fluids and noxious vapours on the operator.
5. Autoclaves must be checked monthly to assure decontamination effectiveness.
6. Filters should be installed whenever needed.

PERSONAL PROTECTION

Personal hygiene is extremely important to persons working in a laboratory. Contamination of food, beverages, or smoking materials is a potential route of exposure to toxic chemicals or biological agents through ingestion. Thus, laboratory personnel should not prepare, store, or consume food or beverages in the laboratory. Hand washing is a prime necessity to safeguard against inadvertent exposure to toxic chemicals or biological agents and should be always done before leaving the laboratory, even though gloves are used. Long hair and loose clothing should be confined when in the laboratory to keep them away from catching fire, dipping into chemicals, or becoming entangled in moving machinery. Wearing finger rings and wristwatches must be avoided since they may become contaminated, react with chemicals, or may be caught in the moving parts of equipment.

Personal Protective Clothing

Personal protective clothing protects the laboratory personnel from injury due to absorbing, inhaling, or coming into physical contact with hazardous materials. Ordinary clothing and eyeglasses afford some protection. Substantial leather shoes should be worn in the laboratory to protect against chemical splashes or broken glass. Sandals, cloth sport shoes, perforated shoes, or open-toed shoes should be avoided. To clean up a spill from the floor, the added protection of rubber boots or plastic shoe covers may be needed. Steel-toed shoes are required for handling heavy items such as gas cylinders or heavy equipment components.

Aprons, laboratory coats, and other protective clothing, preferably made of chemically inert material, should be used. Laboratory coats protect the user from chemical splashes and infectious material. Cotton is a better material for a lab coat than nylon as it has a greater absorptive capacity and is generally more resistant to chemical splashes. The standard open-neck coat may be adequate for most chemical work but a high-necked gown is more sui for work with animals and potentially dangerous microorganisms. For work involving carcinogens, disposable coats may be preferred. For work with mineral acids, acid-resistant protective wear is desirable.

Goggles or Safety Spectacles

Eye protection is mandatory in laboratories because of the obvious hazards of flying objects, splashing chemicals, and corrosive vapours. Eyes are very vascular and can quickly absorb many chemicals. Eye protection will be required in all laboratories where chemicals are used or stored. Safety glasses with clear side shields are adequate protection for general laboratory use. Goggles are to be worn when there is danger of splashing chemicals or flying particles, such as when chemicals are poured or glassware is used under elevated or reduced pressure. A face shield with goggles offers maximum protection.

Gloves

Gloves are worn to prevent contact with toxic or biological agents, burns from hot or extremely cold surfaces or corrosives, or cuts from sharp objects. Skin contact is a source of exposure to infectious agents and toxic chemicals, including carcinogens. Many types of gloves are made for specific uses. For adequate protection, the correct glove for the hazard in question must be selected. A leather glove provides good protection for picking up broken glass, handling objects with sharp edges, and inserting glass tubing into stoppers. However, because they absorb liquid, leather gloves do not provide protection from chemicals, nor are they adequate for handling extremely hot surfaces. Gloves designed to provide insulation against hot surfaces and dry ice are not suitable for handling chemicals.

Face Masks

These are not always necessary but need to be worn when there is a risk of chemical, smoke, fire, etc.

HAZARDS IN THE LABORATORY

A number of accidents are encountered in the laboratory due to chemicals, toxic substances, flammable substances, explosives, radionuclides, etc. Hence, knowledge about the basic causes of accidents is mandatory to protect oneself from these hazards.

Chemicals

All chemicals should be considered as potentially dangerous and handled carefully. Contact with skin and clothing should be avoided and even if a chemical is considered harmless it should not be tasted or smelt.

Hazard warning symbols, which are black on an orange background, (Figure 1.1) should be pasted on the reagent bottles to warn of specific dangers. The technical staff should also mark solutions of reagents placed for class work and use coloured adhesive labels for this purpose.

Corrosive Irritant Substances

A corrosive substance is one that destroys the living tissue, e.g. strong acids and alkalis. An irritant on the other hand causes local inflammation but not destruction of the tissue. Occasional contact with the skin may suggest that the substance has no detectable effect. However, repeated exposure can suddenly give rise to irritation as in the case of some organic solvents.

Use of a chemical fume hood is an important safety measure for preventing exposure to hazardous materials. In conjunction with sound laboratory techniques, a chemical fume hood serves as an effective means for capturing toxic, carcinogenic, offensive, or flammable vapours or other airborne contaminants that would otherwise enter the general laboratory atmosphere. With the sash lowered, the chemical fume hood also forms a physical barrier to protect workers from hazards such as chemical splashes or sprays, fires, and minor explosions. Chemical fume hoods may also provide effective containment for accidental spills of chemicals, although this is not their primary purpose.

Toxic Compounds

Compounds are graded as toxic or highly toxic, depending on the dose required for killing 50% of a population of animals (LD_{50}). The inherent dangers of swallowing a toxic compound are obvious but the dangers of absorption through the skin or inhalation are not always appreciated.

Some compounds take a long time before their toxicity becomes evident and this is particularly true for carcinogens and teratogens. Some common biochemical reagents show this long-term toxicity; ninhydrin for example is carcinogenic and thyroxine is teratogenic. If possible, a substitute should always be used, but if none is available then extra care must be taken when using these substances.

Flammable Substances

In the case of fire, flammable substances with a low flash point and all naked flames in the laboratory should be extinguished immediately. Sparks from electrical equipment are less obvious than a Bunsen burner but can be just as dangerous. Hence, to avoid fire, organic solvents must not be stored in the refrigerator.

Oxidizing substances may not be flammable themselves but may cause a fire when brought into contact with combustible material. The best precaution if such compounds need to be used is to have only the minimum amount required on the bench and to keep the main bulk in steel cabinets, well away from the work area.

The major cause of death in fires is the inhalation of carbon monoxide and other toxic fumes in the smoke rather than burns, so if there is any difficulty get away from the place. If the fire is small then it should be tackled with a fire extinguisher or blanket, provided there is a safe exit. There are several types of extinguishers available in a laboratory which can be identified by their colour and the best one for tackling the blaze is selected according to the nature of the fire.

It is vital to recognize the type of fire and use the correct type of extinguisher. Indeed, selection of the wrong extinguisher may actually make the situation worse. For example, water-based extinguishers should not be used on solvent fires to avoid spreading the blaze and should never be used if electrical equipment is involved because of the risk of electrocution.

In the case of a fire involving oils or fats, then smothering the blaze with a fibre glass blanket is often more effective than using an extinguisher.

As with flammable compounds, only small quantities of the compound should be used in the work area and preferably behind a protective screen. Explosives can also result from the mixture of two compounds, which are individually harmless, and an awareness of this is necessary to avoid a laboratory disaster.

Explosives

Explosives as such are not handled in the normal biochemical laboratory but some general laboratory reagents such as picric acid are explosive and must be handled with extreme caution.

Low or high pressure can be hazardous although the former is easier to guard against. High pressure, especially in the form of a gas cylinder, is more of a potential hazard because of the added risk of fire or an explosive reaction.

Gas cylinders are used with complete safety in many laboratories but they can be dangerous objects. Accidents can happen but these nearly always occur through ignorance and incorrect use. A gas cylinder in a fire is a potential bomb and the best course of action is to evacuate the building immediately and warn the fire brigade. Such fires should always be left to the experts to tackle.

Oil and grease can explode at a high temperature in the presence of an oxidizing gas and so these materials are never used on gas cylinders.

Ionizing Radiation

Radionuclides can give rise to α, β and γ-rays as well as neutrons while some high-voltage electrical equipment can produce X-rays.

Alpha particles (α) These are doubly charged helium nuclei ($^{4}_{2}He^{2+}$) and are only emitted by radioactive elements with a very high atomic number. α-emitters do not represent an external radiation hazard since the α-particles are stopped by a layer of skin but they are extremely dangerous if ingested. The tissue immediately in the vicinity of the compound becomes intensely ionized and this leads to cellular injury and the development of cancer. Fortunately α-emitters are only rarely used in biochemical work.

Beta radiation (β) This consists of high-speed electrons (β) with a range of kinetic energies. Low-energy particles are absorbed by the outer layers of the skin but those with a high kinetic energy can penetrate up to 3 mm and cause unpleasant skin burns. Many of the isotopes commonly used in biochemical laboratories are β-emitters (^{3}H, ^{14}C, ^{32}P).

Gamma radiation (γ) This is an electromagnetic radiation of short wavelength and has no mass or charge. It is therefore extremely penetrating and creates areas of ionization following collision with atoms. Some of the isotopes used in biochemical work are α-emitters (^{131}I, ^{54}Fe).

Neutrons Neutrons carry no charge and so are highly penetrating and thereby dangerous. However, they are rarely used in the biochemical laboratory.

Warning signs The symbol for radiation is black on a yellow background (Figure 1.1) and must be clearly displayed on the door of a laboratory where radiation could be a hazard whatever be the source.

Exposure All radiation must be treated as dangerous and exposure kept to an absolute minimum and always below the maximum permitted dose. It is important to remember that there is no such thing as a 'safe' dose of radiation.

Personal monitoring Anyone working in a laboratory with a radiation hazard is required to wear a film badge continuously. This is worn on the trunk to monitor whole body radiation or on the wrist to monitor exposure of the hands. If high-energy β- and γ-emitters are used, a pocket dosemeter or alarm is useful along with the film badge.

Handling The precautions routinely used when working with toxic materials should be adopted with the added safety of shielding with perspex or lead as appropriate.

On completion of the experiment no radioactive waste must be put down the sink but must be disposed off according to the instruction given by the demonstrator.

Non-ionizing Radiation

Direct exposure to radiation can be avoided with precautions but reflections can be quite dangerous especially if the radiation is invisible.

Visible radiation Lasers are the biggest hazard from invisible radiation and protective goggles must be worn. Lasers are extremely dangerous and can cause blindness from less than one second's exposure.

Ultraviolet The ultraviolet region of the electromagnetic spectrum that is hazardous to living organisms is from 200 nm to 315 nm with a maximum at 270 nm. These wavelengths are readily absorbed by the nucleic acids of the chromosomes causing long-term damage and genetic mutations. The immediate acute danger is to the eyes where UV rays can cause conjunctivitis, and to the skin where it can give rise to sunburn. These effects can be avoided by wearing glass goggles and using a barrier cream.

Other radiation The dangers from other forms of radiation are not so well understood but infrared and microwave radiations are said to cause cataract in the eyes. High-intensity ultrasonic beams may also cause damage to living tissue but the exact mechanism of this is not yet clear.

In view of the uncertainties, it is best to treat all forms of radiation with care and to take steps to provide adequate protection for the eyes and skin.

Biological Hazards

Microorganisms Infection can occur from a cut or by ingestion. Less obvious is the risk of infection by inhaling the particles from an aerosol and this is probably the biggest danger when handling bacteria or viruses.

Aerosols are formed whenever a fluid surface is broken, as for example with careless pipetting. Aerosols containing microorganisms are highly dangerous if they have a diameter of 1–5 μm as this size readily penetrates the lung.

Personal protection The laboratory coat should be of the surgical type with a high neck and buttoned sleeves. Goggles, gloves and a facemask should also be worn to give maximum protection.

Safety cabinets Microorganisms should be handled in a safety cabinet or hood and not in the open laboratory. In these cabinets, air flows vertically away from the user and passes through a filter, which must be changed frequently and sterilized.

Flame loops Wire loops used to plate out cultures can give rise to aerosols if handled carelessly. They should never be overloaded since they can spatter and spread infected material when placed in the flame of a Bunsen.

Animals and body fluids Infection can also occur when working with tissues, body fluids and whole animals and the same precautions should be used when handling these materials as with the direct manipulation of microorganisms.

Animals There are a large number of diseases that affect animals and can be transmitted to man so even apparently healthy animals need to be handled with care. Animals can also cause allergies in some individuals but there is little that one can do about this except to avoid direct exposure as far as possible.

Body fluids Blood and plasma which are frequently used in biochemical experiments also need to be treated with caution, as they are a potential source of viral infections. A much greater risk to the laboratory worker is from Acquired immunodeficiency Syndrome (AIDS) and serum hepatitis. Infection is primarily through the blood.

Furthermore, the virus can be carried by apparently healthy individuals so all apparatus in contact with human blood or urine should be sterilized immediately after use. All human fluids should therefore be treated with the greatest care and strict precautions taken.

Genetic engineering Work with bacteria has always involved the risk of mutation occurring either spontaneously or following exposure to radiation or chemical agents. Most mutations are harmless but some may be dangerous especially those that are pathogenic or resistant to antibiotics. However, recent advances in molecular biology have raised the possibility of occurrence of new hazards, particularly those involving the manipulation of genetic information between bacteria.

Recombinant DNA The experiments causing most concern are those that involve recombinant DNA in which a fragment of nucleic acid is removed from a bacterium and incorporated into the nucleic acid of a plasmid or bacteriophage. These recombinant molecules are then cloned in bacteria, usually *Escherichia coli*.

The biological properties of these new hybrid molecules are unknown and this coupled with the possible transfer to man has given rise to some concern. Working parties were set up by several countries to examine the possible hazards of such experiments and they all concluded that two particular types of experiment could be potentially hazardous.

i. New pathogenic bacteria *Escherichia coli* are normally present in the human intestine and accidental infection with bacteria containing hybrid DNA could lead to the transfer of genetic information to the gut *E. coli* or to other bacteria pathogenic to man. This would be particularly serious in the case of the bacterial genes for the production of a toxin or antibiotic resistance.

ii. Cancer The other hazard considered a risk is the possibility of cancer when manipulating oncogenes or hybrid molecules containing animal DNA whose properties would be completely unknown. So far, there is no evidence that these fears are real but work with recombinant DNA should only be carried out after official approval and under very strict safety precautions.

Spillages

Chemicals Reagent bottles should be carried firmly and well supported while large containers must be transported in a wire basket. The work area needs to be kept as clear as possible by keeping bottles and flasks away from the edge of the bench and placing them on a shelf when not needed. Leaving pipettes in flasks or bottles, which are very susceptible to a knock should be avoided.

Spilled chemicals should be mopped up with an absorbent material and the area should be thoroughly washed with water. If corrosive substances are spilled then they must be neutralized before being cleaned up.

Radioisotopes If the reagent is radioactive, workers in the immediate vicinity must be warned and expert help should be obtained while dealing with it.

Infected material The area should be thoroughly washed with a solution of disinfectant and the incident must be reported.

Some of the warning symbols are shown in Figure 1.1.

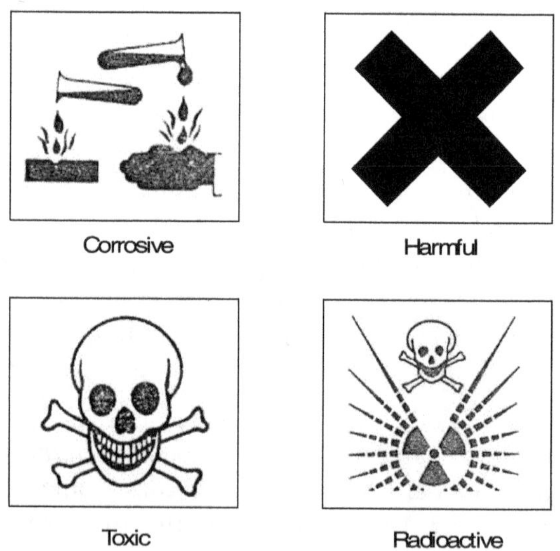

Figure 1.1 Warning symbols

WASTE DISPOSAL

Chemical Waste Disposal

Small quantities of many materials may be washed down the sink with plenty of water but some reagents must not be disposed of in this way. Chemicals such as acids and alkalis need to be neutralized first before washing down the sink while others can only be discarded by incineration. It is important that organic solvents are not poured down the sink but stored in metal drums. For safety reasons, the chlorinated solvents must be stored and not mixed with other organic solvents.

Radioactive Waste Disposal

The method of disposal depends on the isotope and may be via the sink, incineration or storage for later removal. Radioactive material should never be flushed down the sink unless it is safe to do so. It is advisable to consult the radiation officer about the method of disposal.

Radioactive waste is any waste that contains or is contaminated with radioactive material. This includes liquids, solids, and organisms used in scintillation counting liquids. Radioactive waste must never be placed in any non-radioactive waste container. No general (non-radioactive) waste may be disposed of in radioactive waste containers. Radioactive waste must never be placed in the corridor or any public area. All radioactive waste must be labelled with the appropriate label (Radioactive Waste Label) stating the radioisotope name, activity, date of disposal, and the radiation worker's full name and telephone number. All individual plastic containers, scintillation vials, bags and bottles of radioactive waste must be tagged with this label. Any information regarding other chemicals included in the radioactive waste must also be included on the label (e.g. strong acid). A Radioactive Waste Disposal Log should be used to compile a list of the radioisotopes disposed of in the waste cans. All of this information is necessary to correctly classify the waste for disposal (radiological, chemical, mixed, etc.). Each waste container will be used for disposal of one radioisotope only, except for dual labelled radioisotope experiments. Disposal procedures for these containers will be based on the longest half-life. The radioactive waste cans should be stored in an area within the laboratory where they will not be knocked over, used for other waste, or accidentally mistaken as cans for non-radioactive waste.

Creating multi-hazard waste that contains any combination of radioactive, biohazardous, and chemically-hazardous materials must be avoided. Disposal of multi-hazard waste is extremely costly and difficult.

Solid wastes include test tubes, beakers, absorbent papers, gloves, pipettes, and other dry items contaminated with radioactive material but not containing any liquid radioactive waste. These materials must be placed in plastic bags and sealed with tape. Hypodermic needles, capillary pipettes, and other sharp objects must be placed in puncture-proof containers before being put into the large waste cans. Containers bearing a radioactive label, but no longer containing radioactive material must be disposed of as ordinary waste only after the radioactive label is removed and after being decontaminated. Before any radioactive material contaminated with a microbiological organism (virus, fungus, or bacteria) is disposed of it must be chemically treated in a manner that destroys all living organisms (e.g. with fresh 10 percent bleach solution). Autoclaving or gamma cell irradiation should be used only when necessary. Care should be taken to protect autoclaves from any radioactive contamination, particularly, tritium, and radioiodines. Before beginning experiments on animals using radioisotopes, protocols must be approved by the Animal Use Committee so that proper arrangements can be made for disposal of radiologically contaminated or infectious carcasses. Animals that contain less than 0.05 microcurie of ^{3}H or ^{14}C per gram can be disposed of as biological waste. At concentrations higher than this or for other radioisotopes, the animal or tissues must be disposed of as radioactive waste.

Scintillation vials that contain less than 0.05 microcurie of ^{3}H or ^{14}C per gram of scintillation medium should be disposed of as chemical waste and not as radioactive waste. All scintillation vials containing radioactivity above these levels must be labelled as radioactive waste. Scintillation fluid and radioactive waste must be left in the original vials for disposal. Organic solvents that are insoluble, flammable, or toxic must be collected in inert, airtight plastic bottles and must never be disposed of in the sink.

No liquid radioactive waste should be disposed of in the sewage system unless the liquid is readily soluble or dispersible in water, and the material is diluted or flushed simultaneously with measured amounts of water sufficient to achieve those concentrations.

Infected Material

No infected material should ever leave the laboratory. Contaminated glassware must be placed immediately after use in a solution of a disinfectant. Other material should be sterilized in an autoclave kept specifically for that purpose and should then be disposed of.

FIRST AID

First aid as the name implies is the help given immediately to an injured person. It is not a substitute for medical or hospital attention and a doctor or ambulance. The most urgent medical emergencies, which require prompt action to save life, are severe bleeding, asphyxia, and poisoning. Shock may accompany any one of these conditions and is due to the failure of blood circulation. Some of the common accidents and few first aid methods are discussed here, which are useful to the people working in the laboratory.

Unconsciousness

In case of head injury, the brain is unable to function for more than a few minutes without oxygen. If the casualty has stopped breathing then artificial respiration must be started at once. It should be applied by someone trained in the technique. But until help arrives, something has to be done immediately. The following are some steps that can be taken until the arrival of the trained personnel.

1. Lay casualty on back.
2. Clear any obstruction of the mouth.
3. Place a coat under the shoulders so the head is tilted back.
4. Pinch the nostrils and apply mouth to mouth resuscitation 10 times a minute until breathing starts.

If the victim is breathing then lay him face down with his head on one side and the arm and leg of that side in a bent position. This posture makes breathing easier for the patient and provides a better circulation of blood to all parts of the body.

Shock

If conscious, keep the casualty lying down with his feet up and cover him with a light blanket to keep warm. Do not leave him alone but stay to reassure him that help is on the way. Do not give any drinks.

Burns

Severe burns If the casualty is on fire, act immediately to douse the flame by wrapping him in a fire blanket; then cool the affected area of the body under running water. Remove any contaminated clothing and wrap up the casualty to minimize shock.

Minor burns Do not touch the burn but wash the affected area under running cold water for at least 15 minutes, then cover with a dry sterilized dressing.

Alkali burns Flush the burn with water for about 15 minutes and apply a dry dressing.

Acid burns Treat the same as alkali burns but bathe the area gently with a solution of sodium bicarbonate before applying the dressing.

Electric burns Switch off the current or if this is not possible free the person using anything that is non-conducing. Check breathing and treat the burns as described.

Some Don'ts

- Do not remove clothing stuck to a burn.
- Do not burst any blisters.
- Do not apply any cream or ointment.
- Do not apply cotton wool or similar material.
- Do not use an adhesive dressing.

Eye Injuries

Foreign body A piece of dirt can often be removed by getting the person to blink rapidly. In most cases the movement of the eyelid coupled with an increased flow of tears is usually sufficient to remove the object. If this is unsuccessful, the dirt can sometimes be removed with the corner of a clean handkerchief. However, if these methods are unsuccessful do not persist but cover the eye with a pad and bandage flat to keep the eye shut until medical assistance is obtained.

Chemicals in the eye Flush the eye with cold water for 15 minutes. Lightly cover the eye with a sterile eyepad and take the casualty to hospital.

Bleeding

Severe bleeding Make the victim lie down and apply pressure to the wound with a sterilized pad and a firm bandage. If an arm or leg is badly cut but not broken, raise the limb above the rest of the body.

Minor cuts Thoroughly wash the cut with soap water or antiseptic, remove any foreign bodies, and then cover with a sterilized dressing.

Toxic Materials

Gassing Get the casualty out into the fresh air, keep warm and apply artificial respiration if his breathing has stopped.

Poisoning All poisons should be diluted with water but do not give an emetic if the poison is a corrosive liquid or organic solvent. Keep the patient warm and stay with him until an ambulance arrives. Try to find out what poison was ingested by asking the victim and by examining the area of his work. Such information could be of vital importance to provide proper treatment to the patient.

UNITS OF MEASUREMENTS

All quantitative measurements are expressed in some kind of unit. In order to communicate properly, it is important that all measurements are expressed by a universally accepted system. In this chapter the basic International System of units (SI)—the metric system, and its application in the preparation of laboratory solutions are introduced.

THE METRIC SYSTEM

The metric system is the system of measurement used in clinical laboratories. In India and other developing countries, the metric system is used in everyday life although other systems also exist.

Most common laboratory units are those used for the measurement of weight, length, time and concentration of solutions. As different units were used in different parts of the world, scientific communication became difficult. As a result, an International System of Units (*Systeme International D'unites*) was introduced in 1971 and was called SIU. It acccpts the metric system of measurement (Table 2.1).

Table 2.1 SI Basic units

Quantity	Name	Symbol
Length	metre	m
Weight	gram	g or gm
Volume	litre	L or l
Temperature	celsius	°C
Amount of substance	mole	Mol
Solution	molarity	M

The metric system is named because it is based on the fundamental unit of distance, the metre. In the metric system, the metre (m) is the basic unit used to measure distance. The gram (g) is the basic unit to compare mass or weight. And, the litre (l) is the basic unit to measure volume. Mole (Molecular weight in grams) is the weight measurement in chemical language. The concentration of a solution is expressed in molarity, which is the gram molecular weight (GMW) of a substance dissolved in a litre of solution.

For the measurement of temperature, Celsius (°C) is the recommended unit, while Fahrenheit (°F) is still in use but is getting obsolete. Zero degree of the centigrade thermometer is the freezing point of water and 100 degrees is the boiling point of water. These values correspond to 32 degrees and 212 degrees, respectively, on the Fahrenheit scale.

For the conversion of Centigrade (C) and Fahrenheit (F) scales, the nomogram given in Figure 2.1 is used.

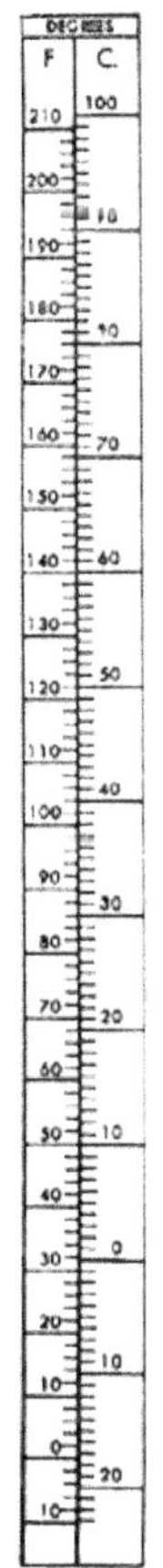

Figure 2.1 Conversion of Fahrenheit scale to Centigrade scale

Alternatively, the following formula is also used.

$$\frac{C}{5} = \frac{F - 32}{9}$$

where,

C = Centigrade,
F = Fahrenheit.

Examples

a) Convert 56°C to °F

$$\frac{56}{5} = \frac{(F-32)}{9} \text{ or } \mathrm{F} = 133.7^{\circ}$$

b) Convert 120°F to °C

$$\frac{C}{5} = \frac{(120-32)}{9} \text{ or } \mathrm{C} = 48.8^{\circ}$$

CONVERSION OF UNITS

The metric system uses decimal notations and the units are divided into increments of 10. This means that units larger or smaller than the basic units (metre, gram and litre) can be obtained by multiplying or dividing by increments of ten (or by a power of ten).

Prefixes may be added to the basic units to indicate larger or smaller units. For example, "kilo" means 1000, which is applied to express 1000 metres as one kilometre, 1000 grams as one kilogram, and 1000 litres as one kilolitre. The prefixes and their definitions are the same for the three basic units—metre, gram and litre.

In laboratory analyses, it is more common to measure units smaller than the basic units. Two common prefixes are "milli", which means one thousandth (0.001 or 10^{-3}), and "centi" which means one hundredth (0.01 or 10^{-2}); for example, a millilitre is 0.001 litre or 10^{-3} litre. Thus the interconversion of millilitre and litre can be expressed as follows:

1000 ml or 10^3 ml = 1 litre

1 ml = 0.001 litre or 10^{-3} litre

To indicate further smaller weights, volumes and lengths, the prefixes of micro-, nano-, pico- and femto- are used and each of these in sequence is one thousand times smaller than the previous one (Table 2.2). Thus 1 gram is 10^3 mg, 10^6 μg and so on.

Table 2.2 Metric weight measurements

Scale	Weight
unit (1)	1 gram (g)
milli (10^{-3})	1 milligram (mg) = 10^{-3} g (0.001 g)
micro (10^{-6})	1 microgram (μg) = 10^{-6} g (0.000001 g)
nano (10^{-9})	1 nanogram (ng) = 10^{-9} g (0.000000001 g)
pico (10^{-12})	1 picogram (pg) = 10^{-12} g (0.000000000001 g)
femto (10^{-15})	1 femtogram (fg) = 10^{-15} g (0.000000000000001 g)

Following the above relationship, we can express the weight, volume, and length as follows:

Weight

$1\ g = 10^3\ mg = 10^6\ \mu g = 10^9\ ng = 10^{12}\ pg = 10^{15}\ fg$

Volume

$1\ l = 10^3\ ml = 10^6\ \mu l = 10^9\ nl = 10^{12}\ pl = 10^{15}\ fl$

Length

$1\ m = 10^3\ mm = 10^6\ \mu m = 10^9\ nm = 10^{12}\ pm = 10^{15}\ fm$

Some SI equivalent of some older terms are given in the Table 2.3.

Table 2.3 New SI equivalent of some older terms

Older terms	New SI equivalent
micron (μ)	micrometre (μm; 10^{-6} m)
micro-micro-gram ($\mu\mu$g)	picogram (pg; 10^{-12} g)
microgram (mcg)	microgram (μg; 10^{-6} g)
millimicron (mμ)	nanometre (nm; 10^{-9} m)
cubic millimetre (mm^3)	microlitre (μL; 10^{-6} l)
per cent (%)	g/dl (w/v) or ml/dl (v/v)

A few other prefixes of the metric system commonly used are:

1 decilitre (dl)	=	100 ml
1 centimetre (cm)	=	10 mm
1 cubic centimetre (cc)	=	1 ml

In addition, the knowledge of the following conversion of English units to metric units may be necessary:

1 inch (in)	=	2.54 cm
1 pound (lb)	=	454 g
1 quart (qt)	=	0.95 l
1 fluid ounce (floz)	=	30 ml

The results of the chemical analysis of water and sewage are usually expressed in parts per million (ppm) by weight. One litre of water or sewage, with its contained impurities, is assumed to weigh 1 kilogram as it is used in the laboratory. It will be noticed from the weight relations given above that 1 ppm is the same as 1 milligram per litre. This is a convenient relation to have in mind when performing certain calculations. In special cases of waters containing large amounts of impurities, the results obtained should be divided by the specific gravity of the sample.

The results may also be expressed in grains per gallon. One grain per gallon (gpg) is equivalent to 17.12 ppm. Some of the convenient conversion factors are given below.

1 milligram per litre	=	1 part per million (ppm)
1 grain per gallon (gpg)	=	17.12 ppm
1 part per million	=	8.34 pounds per million gallons
1 litre	=	0.2642 gallons
1 kilogram	=	2.205 pounds
1 pound	=	453.6 grams

UNITS USED IN PREPARATION OF SOLUTIONS

When a substance is dissolved in a suitable solvent, it forms a solution. A solution is defined as a homogeneous mixture of two or more substances. A solution always consists of two components, a solvent and a solute. The solvent is usually a liquid and the solute may be a liquid, a solid or a gas.

In biochemical reactions, the concentration of solute present in a solution is expressed either based on the volume of the solvent or based on the weight of the solvent.

Expression of Concentration of Solutions

I. Solution concentration expressed based on the volume of the solvent

i. Molar (M) A solution is said to be one Molar, when it contains one gram molecular weight (1 mole) of a solute in one litre of solution (6.023×10^{23} molecules in one litre).

ii. Normal (N) A solution is said to be one Normal, when it contains one gram equivalent weight of a solute in one litre of solution.

iii. Weight/Volume per cent (W/V) A solution is said to be 1% W/V, when it contains one gram of solute in 100 ml solution.

iv. Osmolar The osmolar concentration is the molar concentration of a substance, multiplied by the number of particles or ions produced by that substance, when in solution, e.g. for sucrose or glucose, a non-dissociating solute, a 1M solution is one osmolar. But for sodium chloride, which dissociates into two particles, viz. Na^+ and Cl^-, a 1M solution is equal to 2 osmolar. Two solutions producing the same osmotic effect are called isoosmolar solutions, e.g. plasma and isotonic saline. The osmolar concentration of plasma is 0.308. Therefore, 0.308M glucose or 0.154M sodium chloride are isoosmolar to plasma.

$$\text{Osmolarity} = \text{Molarity} \times \text{No. of different ions present in that solution.}$$

II. Solution concentration expressed based on the weight of the solvent

i. Molal (m) A solution is said to be one Molal, when it contains one gram molecular weight (1 mole) of a solute plus 1000 g of solvent.

ii. Parts per million (ppm) For dilute solutions, it is convenient to express the concentration of the solute in terms of parts per million. 1 mg of any substance in one litre is 1 ppm. One μg of any substance in one litre is 1 ppb (parts per billion).

iii. Per cent by weight (W/W) A solution is said to be 1% W/W, when it contains one gram of solute plus 99 g of solvent.

iv. Mole fraction The mole fraction of a substance is the ratio of the number of moles of the particular substance to the total number of moles of all substances present in the solution. For example, the mole fraction (N_1) of a solute in two component systems, is expressed as:

$$N_1 = \frac{n_1}{n_1 + n_2}$$

where n_1 and n_2 are the number of moles of the solvent and solute respectively.

Preparation of Reagent Solutions

The following types of solutions are prepared in the laboratory:

1. Saturated solution
2. Per cent solution
3. Molar solution
4. Molal solution
5. Normal solution

Saturated solution A saturated solution is one which holds as much solute as it can, e.g. preparation of saturated solution of sodium chloride.

About 100 ml of water is taken in a beaker. Sodium chloride is added in small portions to this with constant stirring. This addition is continued till some crystals are left undissolved. Now the solution is saturated with sodium chloride, filled in a bottle and one more spatula of the salt is added.

Per cent solution A per cent solution is one which contains a known weight of the substance in a specified volume of its solvent. If the percent solution contains solid substance then it is denoted as weight/volume percent solution (w/v) and if it contains liquid then it is denoted as volume/volume percent solution (v/v), e.g. preparation of 5% (w/v) sodium chloride solution.

Five grams of NaCl is dissolved in 100 ml of its solution. To prepare it, about 70 ml of water is taken in a beaker; exactly 5 g of sodium chloride is added to it. It is completely dissolved. It is then transferred to a 100-ml volumetric flask or in a measuring cylinder with stopper. The volume in the volumetric flask is made up to 100 with distilled water. The volumetric flask is stoppered, inverted and swirled several times for uniform mixing. It is then transferred to a reagent bottle and labelled properly.

Molar solution A molar solution is one in which a litre solution contains 1 gm molecular weight of the substance. It is denoted as "1M" solution. When the weight of the solute dissolved in one litre of its solution is equal to its molecular weight it represents 1M solution.

Molecular weight of a substance is obtained by adding the atomic weights of the elements in the proportion contained in the compound, e.g. NaCl.

Atomic weight of Na = 23

Atomic weight of Cl = 35.5

Molecular weight of NaCl = 58.5

If the molecule of a compound is hydrated, the weight of water is also summed up to the molecular weight of the compound, e.g. $COOH–COOH \cdot 2H_2O$, i.e., $C_2H_2O_4 \cdot 2H_2O$.

$$\text{Molecular weight} = (12 \times 2) + (1 \times 2) + (16 \times 4) + 2\,(1 \times 2 + 16)$$
$$= 24 + 2 + 64 + 36$$
$$= 126$$

Molecular weight of oxalic acid = 126

When the weight in grams taken is equal to its molecular weight, it is called gram-molecular weight. 1 molar solution is also called as 1 gram-mole or 1 mole.

Molal solution A molal solution is one which is prepared by dissolving one gram-molecular weight of a substance in one thousand grams of its solvent.

Example 58.5 g sodium chloride is dissolved in 1000 g of water to get 1 molal solution of sodium chloride solution. This type of solution is seldom used.

Normal solutions Normal solutions are used extensively in the laboratory. The word "Normal" is denoted by the letter 'N'. So 1 Normal solution is denoted as 1 N, 2 Normal as 2 N and N/100 as 0.01 N solutions and so on.

A Normal solution is one which contains one gram equivalent weight of solute dissolved in one litre of its solution.

Example If the equivalent weight of a substance is 20 then 20 g is the gram equivalent weight. When 20 g of that is dissolved in a little of water and then the volume made up to 1 litre, it is 1 N solution. Likewise, 40 g of that substance in 1 l, 10 g of it in 1 l and 2 g of it in 1 l respectively are 2 N, 0.5 N and 0.1 N solutions.

To prepare a normal solution, equivalent weight is calculated as follows:

Equivalent weight of an acid Acid is one which has ionizable or replaceable hydrogen ions (H). If an acid has one, two or three replaceable H^+, it is called a monobasic acid, dibasic acid, and a tribasic acid respectively. A monobasic acid forms one type of salt, dibasic two types and tribasic acid forms three types of salts on treating with a base, e.g. HCl forms only NaCl. So it is monobasic. H_2SO_4 forms $NaHSO_4$ and Na_2SO_4 and so it is dibasic.

$$\text{Equivalent weight of an acid} = \frac{\text{Molecular weight}}{\text{No. of replaceable } H^+}$$

i.e., for monobasic acids,

$$\text{Molecular weight} = \text{Equivalent weight}$$

$$\text{Equivalent weight of HCl} = \frac{36.5}{1} = 36.5$$

whereas,

$$\text{Eq. wt. of } H_2SO_4 = \frac{\text{Molecular weight}}{2}$$

$$= \frac{98}{2} = 49$$

i.e.,

$$1M\ H_2SO_4 = 2N$$

Likewise

$$\text{Equivalent weight of } H_3PO_4 = \frac{\text{Molecular weight}}{3}$$

i.e.,

$$1M\ H_3PO_4 = 3\ NH_3PO_4$$

Now, to find the weight to be taken for a known volume V litres of an acid of strength 'N' of equivalent weight 'E':

Wt. for V litres of N Normal $= E \times N \times V$ g.

Equivalent weight of a base is the molecular weight divided by the number of replaceable hydroxyl ions.

Example NaOH has one replaceable OH ion and hence,

$$\text{Equivalent weight of NaOH} = \frac{\text{Molecular weight}}{1}$$

$$= \frac{40}{1}$$

Calcium hydroxide $Ca(OH)_2$ has 2 replaceable hydroxyl ions and hence,

$$\text{Equivalent weight of } Ca(OH)_2 = \frac{\text{Molecular weight}}{2}$$

$$= \frac{74}{2}$$

$$= 37$$

The number of electrons, which they give or take during a reaction, determines the equivalent weights of salts such as $AgNO_3$ and $KMnO_4$.

Example $AgNO_3$ gives 1 electron. Therefore, equivalent weight of $AgNO_3$ is molecular weight divided by 1.

Potassium permanganate takes 5 electrons, therefore the equivalent weight is molecular weight divided by 5. Equivalent weights of some common chemicals are given in the Table 2.4.

Table 2.4 Equivalent weights of some common chemicals

Compound	Mol. wt.	Eq. wt.	% w/w
Acetic acid, glacial	60.050	60.050	100
Hydrochloric acid	36.461	36.461	37
Sulphuric acid	98.078	49.039	96
Oxalic acid	126.067	63.033	
Sodium hydroxide	40.000	40.000	
Sodium carbonate	106.000	53.000	
Potassium permanganate	158.038	31.607	
Nitric acid	63.030	63.030	70

UNITS USED IN LABORATORY CALCULATIONS

A number of laboratory calculations are involved in the preparation of reagents or computation of test results. Few examples are discussed here.

Ratios and Dilutions

A ratio is a relationship in number or degree between two things. Dilutions, which are ratios, express the relationship between a part of a solution and the total solution. Dilutions are used in haematology, serology and in biochemistry. For example, if the instruction calls for preparing 10 ml of diluted serum (1 : 10), one has to mix 1 ml of serum with 9 ml of saline.

Note In the expression of 1 : 10, the second numeral is the final volume. Thus 1 : 1 means no dilution.

In another example, the instruction calls for mixing 0.5 ml of blood with 9.5 ml of diluent for cell count. The dilution or the relationship between the aliquot (0.5 ml) and the total volume (0.5 + 9.5 = 10 ml) is expressed in the ratio of 10/0.5 = 20 times or 1 : 20.

Weight Relationship of Hydrated and Anhydrous Salts

For preparation of a reagent solution using a chemical, which is in hydrated form instead of anhydrous form, or vice versa, the following simple formula may be followed.

$$\frac{\text{Molecular weight of hydrated form}}{\text{Molecular weight of anhydrous form}} = \frac{\text{Weight of hydrated form}}{\text{Weight of anhydrous form}}$$

CONVERSION OF SOLUTION STRENGTHS

The solution strengths can be converted from one form to another as shown below:

$$\text{mEq/litre} = \frac{(\text{mg/100 ml}) \times 10}{\text{Eq. wt.}}$$

$$\text{mmol/litre} = \frac{(\text{mg/100 ml}) \times 10}{\text{Mol. wt.}}$$

$$\text{mmol/litre} \times \text{Valance} = \text{mEq/litre}$$

$$\text{Molarity} = \frac{\%\ \text{concentration} \times 10}{\text{Mol. wt.}}$$

$$\text{Normality} = \frac{\%\ \text{concentration} \times 10}{\text{Eq. wt.}}$$

Photometric Calculations

Concentration of the unknown can be calculated by the following formula if the absorbance readings of the standard and the unknown are determined from the photometer.

$$\text{Conc. of unknown} = \frac{\text{Absorbance of unknown} \times \text{Concentration of standard}}{\text{Absorbance of standard}}$$

This formula is applicable only when the test specimen and the standard are identically treated. The unit for the concentration of the unknown will be the same as the unit for the standard.

If the test solution and the standard solution are not identically treated, the above formula will change and will be written as follows:

$$\text{Conc. of unknown (mg/dl)} = \frac{A_u}{A_s} \times Q_s \times \frac{100}{\text{Volume of specimen}}$$

where,

A_u = absorbance of unknown,
A_s = absorbance of standard and
Q_s = quantity of standard (mg).

Titrimetric Analysis

$$\text{Conc. of unknown} = \frac{\text{Titration value of unkonwn (ml)}}{\text{Titration value of standard (ml)}} \times \text{Conc. of standard}$$

The unit of the concentration of the unknown will be the same as the standard.

REVIEW YOUR LEARNING

1. What are the basic units of weight, volume, and length?
2. Define the expression of concentration of the following solutions:
   - i. one molar
   - ii. weight/volume per cent
   - iii. normal
   - iv. osmolar
   - v. molal
   - vi. parts per million
   - vii. per cent by weight
3. Define mole fraction.
4. Explain the preparation of the following solutions.
   - i. Saturated solution
   - ii. Per cent solution
   - iii. Molar solution
   - iv. Molal solution
   - v. Normal solution

MICROSCOPY

Microscope is an indispensable tool for the observation of cells and their components. Z. Janssen and his nephew H. Janssen developed the first generally useful light microscope. The development of the conventional microscope at the end of the 16th century lead to a great step forward in science, particularly in biology and medicine. In the beginning though, the microscope was mainly a toy in rich homes, later many important discoveries followed. Antoni van Leeuwenhoek, in 1670s, was able to observe a variety of different types of cells, including sperms, red blood cells, and bacteria, using a microscope that magnified objects up to about 300 times their actual size. The cell achieved its current recognition as the fundamental unit of all living organisms because of observations made with the light microscope.

HISTORICAL DEVELOPMENT OF MICROSCOPE

The history of microscopes can be traced back to the 8th century B.C., when a rock crystal was used as a lens. In the 5th century B.C., Euclid investigated the properties of curved reflecting surface, which were used later on in the microscope. The use of convex lenses for magnification of small objects was introduced in the 13th century A.D. Leonardo da Vinci recommended the use of lenses for the study of small objects. Janssen and Hans designed the simplest compound microscope. Janssen's compound microscope contains two tubes, which contain lenses that slide together for focusing (Figure 3.1).

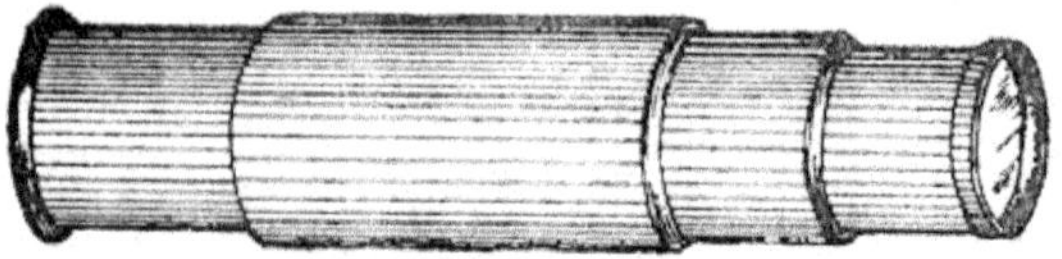

Figure 3.1 Janssen's compound microscope

Janssen's microscope magnified the objects to nine times. Robert Hook made a compound microscope, which was composed of an objective lens, a field glass, and an eye lens. He used lamp as the source of illumination and to intensify the light he used a bull's eye condenser in the microscope (Figure 3.2).

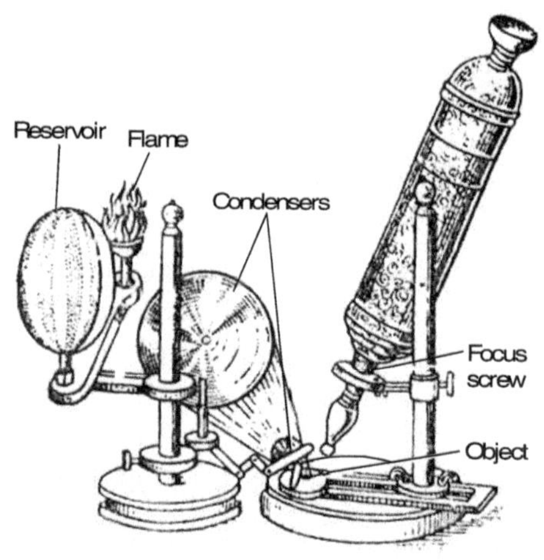

Figure 3.2 Hook's compound microscope

Hook's microscope had the magnification power of 10 to 42 times. A. V. Leeuwenhoek in 1673, constructed a simple microscope by mounting a lens between two flat pieces of metal and adding a pivoted joint for holding the specimen (Figure 3.3). This microscope had a magnification power of up to 300 times.

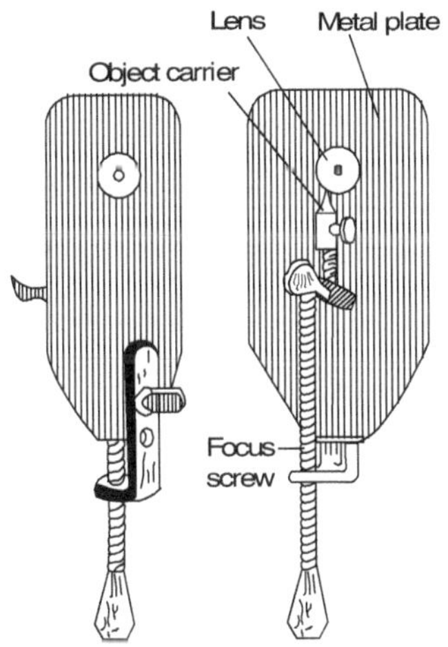

Figure 3.3 Leeuwenhoek's microscope

Some of the antique and vintage microscopes on display at the Billings Microscope Collection at the Medical Museum of the Armed Forces Institute of Pathology, Washington, DC are also discussed here. M. Moreau was a French microscope designer who is known for his Monkey Microscope, which was designed in 1850. In this type of microscope there is a simple compound lens with mirror (Figure 3.4).

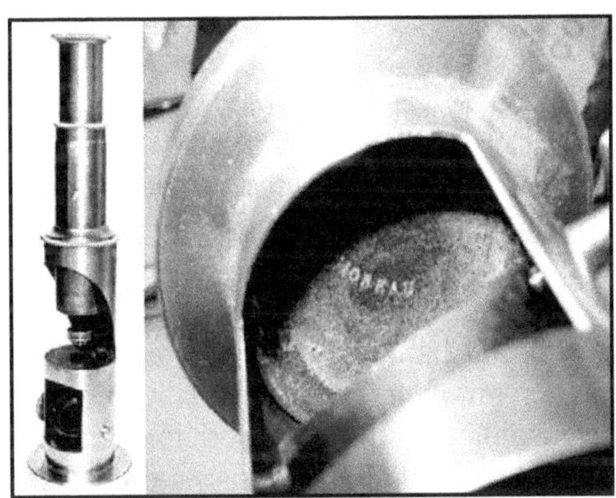

Figure 3.4 Moreau microscope

Older style E. Leitz Wetzler microscope has a triple nosepiece with three objectives and two eyepieces. Also included is a very rare original mechanical stage fitted on the stage platform. The substage is a condenser with iris diaphragm, included is an extra condenser fitting with several different options for light settings (Figure 3.5).

Figure 3.5 Leitz Wetzler microscope

The brass Bausch and Lomb microscope was manufactured in 1897 in Rochester, New York. A fully extended microscope is 13½″ high, 3¾″ wide and 5¼″ long. There is one brass 10× objective with 0.25 NA. It includes two stage clips and ocular. The hinged folding handle is made of ivory and is 1″ long. On top of the handle is the folding brass lens holder approximately ¼″ in diameter that houses the optic. Both the handle and lens holder fold onto the 2″-double-hinged brass stem. A brass nut adjusts the spiked specimen holder by moving it to and from the object (Figure 3.6).

Figure 3.6 Bausch and Lomb microscope

Wilson designed a screw-barrel type of compound microscope. This microscope had threads at the observing end to which many lenses of different magnifications were screwed. The other end had a condenser (Figure 3.7). The compound microscope of 20th century is the microscope of much improved and modified kind.

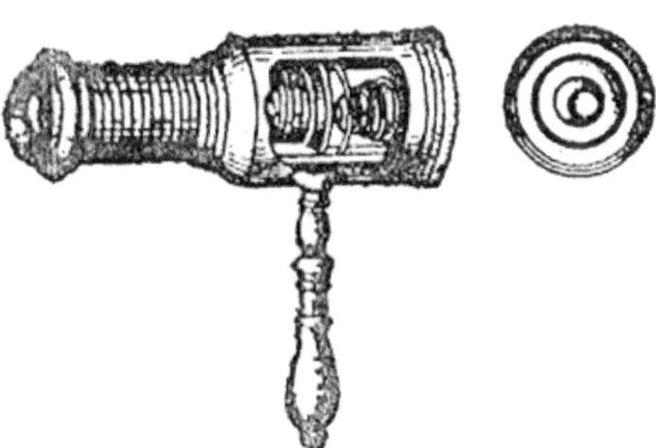

Figure 3.7 Wilson's screw-barrel type microscope

COMPONENTS OF A MICROSCOPE

The major components of a microscope are light source, mirror and lens, condenser system, diaphragm, objective lens and eyepieces. The components of monocular and binocular microscopes are shown in the Figures 3.8 a and b.

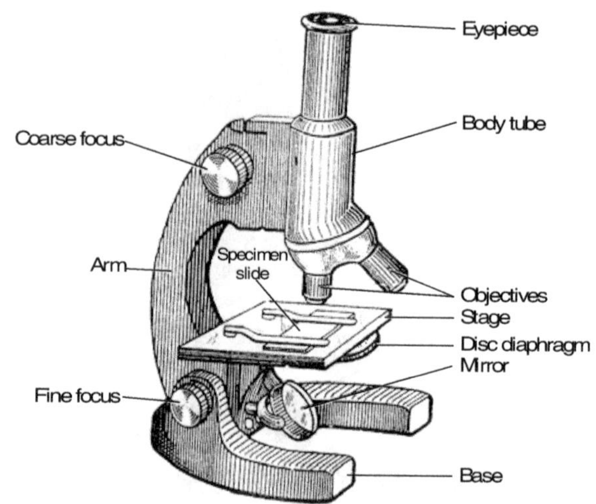

Figure 3.8 (a) Components of a monocular microscope

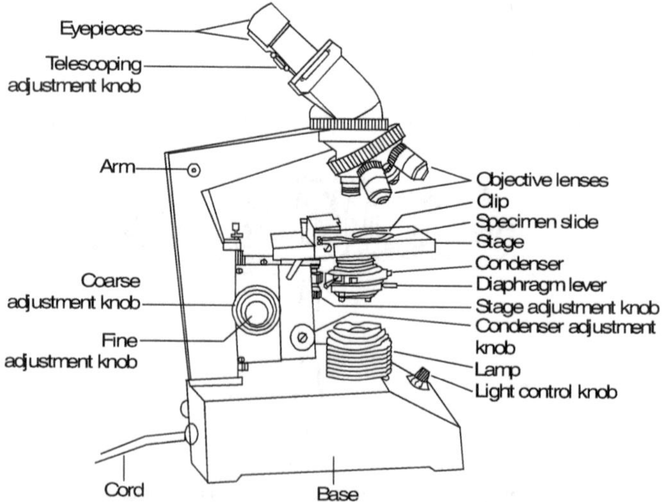

Figure 3.8 (b) Components of a binocular microscope

Light Source

Microscopes use light as a source of illumination. Illumination of the object could be by using either transmitted light or incident light. Incident light is mainly used in fluorescence microscope. Transmitted light is used in all types of light microscope. Transmitted light passes through the object and the image is captured by the lens system. Light source may be sunlight or artificial light. Advantages of using sunlight as light source are—(a) it is easily available (b) bright light is available during day (c) complete spectrum is represented by sunlight. The disadvantages are—(a) intensity is variable by itself (b) non-availability during night (c) lot of heat energy is emitted.

Light is a spectrum of electromagnetic waves. When a parallel beam of light is passed through a glass, light is returned (deflection) which may occur in two manners:

1. Light angle is unaltered when plane mirror is used and the phenomenon is called **reflection**.
2. Using non-uniform mirror surfaces may alter the angle of the light, which leads to the phenomenon called **diffraction**. Diffraction is a process of breaking up of a beam of light into a series of black and white bands or colour bands of the spectrum depending on the property of the deflecting beam.

In the transmitted light, the angle is generally altered. This would lead to two phenomena namely

i. Divergence or dispersion—intensity of light is less.
ii. Convergence or focus—intensity of light is very high.

Convergence and divergence depend on the construction of lens. These two properties are also called refraction. **Refraction** of light is the bending of light due to change in the angle of transmitted light. The rays are now called oblique rays. Light bends due to the difference in **refractive index** (RI) of the medium and glass or lens. Higher the RI, larger will be the angle of oblique rays transmitted. When the angle of oblique ray is larger, the working distance is more. The resolution power of objective is directly related to its light-collecting function. When the half angle of the cone of light is larger, the light collecting function is increased, **numerical aperture** (NA) of objective lens is increased, and ultimately the resolution is greater.

Numerical aperture NA is a measure of light collection function. Light is focused on the specimen by the condenser lens and then collected by the objective lens of the microscope. NA is determined by the angle of the cone of light entering the objective lens (α) and by the refractive index of the medium (usually air or oil) between the lens and the specimen (Figure 3.9).

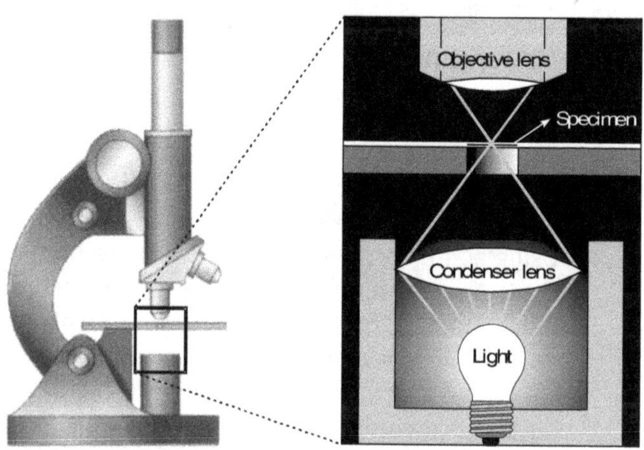

Figure 3.9 Numerical aperture

$$NA = \eta \sin \alpha$$

where,

η = maximum RI of various materials between specimen and the objective lens and

α = the half angle of oblique rays that enter the front lens of the object.

The maximum NA of the objective lens for use in air (dry objective) is 0.95. It is important that an objective should accept rays from the object, at as wide an angle as possible since a greater proportion of the ray diffracted within the specimen is then available for image formation. In general, increase in the primary magnification or power of an objective is accompanied by an increase in numerical aperture.

The NA of the condenser system can influence the effective NA of the objective. But the pre-condition is that the NA of the condenser should not be less than that of the objective lens.

$$\text{Effective NA(objective)} = \frac{NA_{objective} + NA_{condenser}}{2}$$

Increase in NA is directly proportional to enhancing image quality, i.e.,

$NA_{condenser} \geq NA_{objective}$ to achieve maximum effective resolution.

Resolution is the primary task of any microscope. It is nearly a linear function of the numerical aperture of the objective lens. **Resolution** is defined as the formation of sharply defined and distinct images of closely adjacent structural details on the specimen. **Resolving power** depends upon the NA of the objective lens and, wavelength (λ) of the light.

Artificial light source Incandescent lamp, carbon arc, xenon arc, zirconium lamp, mercury vapour lamp, etc. are used as artificial light sources.

1. **Incandescent lamp** It is the most commonly used electric lamp, which consists of tungsten filament. This lamp has medium intensity of illumination.

   Advantages

   i. Due to medium intensity, less amount of heat is produced.
   ii. Produces continuous spectrum of light.
   iii. Light intensity can be adjusted or varied depending on the requirement.
   iv. Very small lamps (6 V or 12 V) can be placed inside the microscope. The lamp contains neon.

Disadvantage

It cannot be used where high intensity of light is required.

2. **Carbon arc** Light is established between vertical and horizontal carbon electrodes. It can produce very high intensity of light.

   Disadvantages

   i. It is hazardous even in low intensity.
   ii. Not easy to handle.
   iii. Dangerously high voltage exists in electrode.
   iv. Moisture or drops of water may bring about the problem of electric shock.

3. **Xenon arc** The lamp has tungsten filament as in incandescent lamp and xenon gas. The lamp also has quartz envelope. This lamp can produce a visual spectrum of day light quality.

   Disadvantages

   i. It is very expensive.
   ii. Embitterment of quartz due to over use.
   iii. It is likely to explode during usage.

4. **Zirconium lamp** It is a secondary light source. The filament is made up of zirconium, and inert gas is filled at low pressure with glass covering. It produces bright light with continuous output. The continuous spectrum of this lamp is particularly an attractive feature. It is mainly used for magnification purposes in the microscope.

5. **Mercury vapour lamp** The electrodes are sealed in a quartz envelope and the area within the envelope is filled with mercury vapour. When current passes through it, a line spectrum is produced with discrete wavelength characteristics. The light rays coming out of this lamp have more light in the UV range than in the other range. It is mainly used in fluorescence microscope.

Mirror and Lens

The mirror reflects the light when sunlight is used as the light source. Light source other than sunlight requires both the mirror and lens. As the mirror increases the light intensity, the lens brightens the light. The lens collects the light rays coming directly from the light source and also reflects the rays for convergence.

Two types of mirrors are used

1. Plano mirror—reflects the parallel beam of light. There is no focusing of the light.
2. Concave mirror—the reflected light is converged by the mirror. As a result, a cone of light is formed. At the apex of cone the intensity of the light is maximum.

When a condenser is used, plano mirror should be used. Use of concave mirror along with condenser system makes the condenser system inefficient, i.e., when condenser system is not used, concave mirror is preferable.

Condenser System

Condenser system directly facilitates the resolving power of the microscope. Resolution is achieved by optical means. During resolutions, aberrations that occur are minimized or ruled out by a process called optical correction. The condenser system is located between the mirror and stage hence it is also called substage condenser. It can also be located between the focal lens and the stage.

The function of the condenser system is to collect, converge and regulate the light rays coming from the light source on the specimen. The lens system of a condenser is an assemblage of three types of lenses.

1. Biconcave lens
2. Planoconvex lens
3. Apochromatic lens

The NA of the condenser system is 1.4.

Filters are used to allow a particular spectrum of light, i.e., to cut down the specific wavelength of light and enable the complete spectrum of light. Blue and green filters are commonly used. A ring-like structure is present just below the condenser system to place the filters. Ground glass or opaque glass can also be used instead of filter, which maintains the uniformity in the light intensity.

Diaphragm

It helps to obtain bright light from the light source. Two types of diaphragms are used.

They are

1. **Field diaphragm** It is placed in front of the light source.
2. **Aperture diaphragm** It is placed just beneath the substage condenser. This ensures the transmittance of only high intensity of light. By selectively adjusting the diaphragm, one can enhance the NA of the condenser lens system.

Specimen Stage

The specimen stage is of square or round type. Square type of stage may be immovable or movable; immovable stage possesses clips to hold the slides; movable stage can be moved with the help of the gliders. The stage can be moved forward, backward and sidewards. Round stage, is seen in polarizing microscope, which can be moved through 360°.

Objective Lens System

The objective lens system is located at the lower end of the tube facing the object. The functions of the objective lens system are resolution and magnification. Resolution can be enhanced by increase in NA. Magnification could be achieved by increase in NA and increase in the number of lenses.

The images formed are of two types:

1. *Virtual image* The image observed at a distance of 250 mm below the eyepiece lens is called virtual image. The image formed is resolved by the objective lens. It is not yet magnified by the eyepiece lens.
2. *Real image* The image that can be observed by the eye lens on eyepiece is called real image. The real image represents total magnification of the microscope M_t.

Aberrations in images are of three types—Aplanatic, spherical, and chromatic aberrations.

1. *Aplanatic aberration* In this type, the image is curved on one plane due to the converging property of the lens system (Figure 3.10).

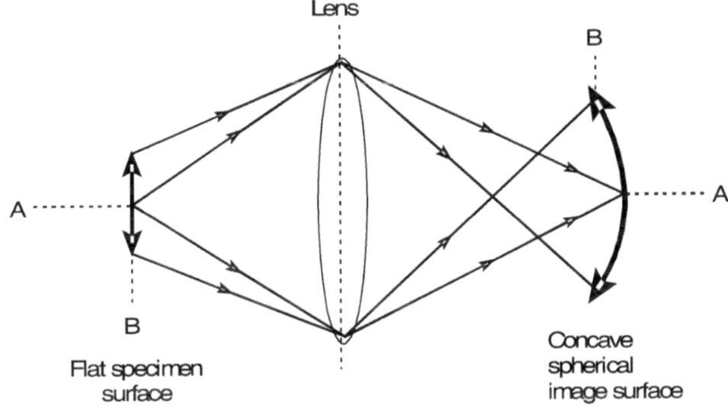

Figure 3.10 Aplanatic aberration

2. *Spherical aberration* The image will have spherical appearance. The best focus is only at the centre. In the other region, distorted, blurred or hazy image is produced. This is due to the light-converging property of the lens system. Due to the difference in the RI of lens in different points, the rays refracted will not be focused on the same point (Figure 3.11). Coverslip can also add to the problem. Higher the thickness of the coverslip, greater is the problem of spherical aberration. Spherical aberration is due to differential refraction of light rays by different lens systems.

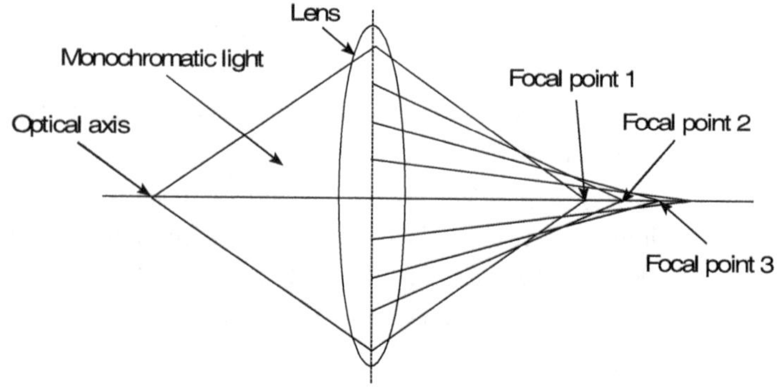

Figure 3.11 Spherical aberration

The method of optical correction, which reduces or completely eliminates aplanatic or spherical aberration is by selection of a particular wavelength of light. This approach is called achromatic type. Here two types of lenses called crown lens and flint lens are used. The curvature of flint lens exactly matches with the curvature of the convex crowned lens. Flint lens has double refractive capacity than the crowned lens and is helpful in compensation of differential refraction of light rays. With the result, a clear flat image, which is free from spherical and aplanatic aberrations, is produced. Achromatic lens also reduces chromatic aberration to some extent.

3. *Chromatic aberration* It is the defect due to the dispersion of light. As a result, the light is broken up into different component colours because the lens system has different RI as compared to RI of each colour or each wavelength of light. The usual colours are red, blue, and green. Consequently colour fringes surround the image (Figure 3.12).

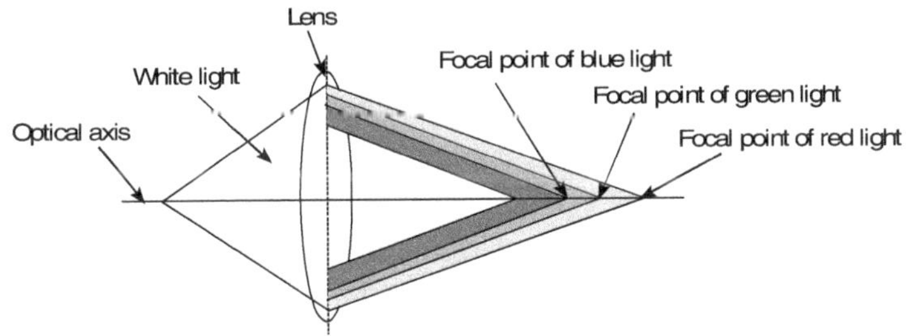

Figure 3.12 Chromatic aberration

The optical correction for chromatic aberration is called apochromatic type of correction—two meniscus lenses surround the crowned lens. The margins of this lens are highly reduced. Therefore, differential refractive property is highly reduced along with the reduction of RI. With the result chromatic aberration is reduced. The image is also free from aplanatic and spherical aberration.

Three types of objective lenses that are used in microscopes are listed in Table 3.1.

Table 3.1 Types of objective lenses used in microscopes

Lens type	Corrected for	Still suffer
Aplanatic	Flat field to some extent	Spherical and chromatic
Achromatic	Field curvature, no spherical aberration, chromatic aberration slightly reduced	Chromatic aberration
Semi-apochromatic	Aplanatic, no spherical aberration, corrected for two colours—blue and green	Red colour persists
Apochromatic	Aplanatic, no spherical and chromatic aberration (red, blue, green)	

Oil immersion objective lens Using a medium of high RI between the objective lens and the object can increase the effective NA of an objective. Most of the immersion objectives are made for cedar wood oil or a similar medium but water immersion lenses are particularly valuable to the biologists for whom they are commonly made with an aperture of 1.0 and a magnification of 50×. Such objectives combine good resolution with a fairly long working distance and relative insensitivity to variations in the thickness of the cover glass. 63× and 100× objective lenses are called oil immersion objective lenses. Usually these lenses are provided with a coloured ring towards their lower side. This coloured ring indicates that the lens should be used with oil. Without oil, there is no sharp focus.

Immersion media Distilled water was the earliest immersion fluid. The most commonly used medium is highly refined cedar wood oil (RI 1.5) or a synthetic equivalent, which has optical properties very similar to those of cover glass material. Its viscous nature enables it to be retained in the desired position between the lens and cover glass but cedar wood oil becomes rather gummy by oxidation when left in contact with the air and this is accompanied by an increase in the RI. It should be removed from lenses and specimens immediately after use by rubbing gently with a clean lens tissue or soft cloth. The immersion oil prevents the dispersion of light rays. Hence objective lens can receive maximum amount of light rays, which ultimately result in high resolution. There are various alternative immersion fluids, which are non-drying, but many of these have higher colour dispersion and are liable to upset the colour correction of the objective.

Eyepiece

The eyepiece is located at the upper end of the microscope tube. The eyepiece is used for examining the primary image formed by the objective, and sometimes also for projection purposes and for facilitating measurement of the image of some feature of the object. All the objective lens systems and eyepiece lenses are positive lenses; they enhance the magnification. Magnification is usually 5–10 times. Eyepiece lens is called second lens system. The power of magnification is engraved in both eyepiece lens system and objective lens system. Therefore, for a given microscope the two important parameters are the resolving power and the magnifying capacity.

Resolution = NA of objective + NA of condenser

$$M_t = M_o \times M_e$$

where,

M_t = total magnifying capacity of the microscope,
M_o = magnifying capacity of the objective lens and
M_e = magnifying capacity of the eyepiece lens.

The eyepiece consists of two lenses both of which are of planoconvex type. A single eyepiece lens used to examine the primary image will produce a bundle of emergent rays with an angle greater than that which can be accepted by the observer's eye, thus rendering simultaneous observation of the whole of the primary image impossible. A second lens is therefore introduced.

Types of eyepiece lens

1. **Huygenian eyepiece** Most modern eyepiece lenses are of the Huygenian type in which the field lens is arranged a little in front of the primary image to reduce the obliquity of the extreme rays. This results in a slight reduction in the size of the primary image, which is now formed behind the eye lens.

   Huygenian eyepiece consists of two simple planoconvex lenses called field lens (lower one) and eye lens. The function of field lens is to capture and bend light rays received from the objective lens, towards the axis of the microscope, whereas eye lens further magnifies the image.

   In Huygenian eyepiece, some spherical and chromatic aberrations are introduced by the lenses, and the apparent field of view is small with some field curvature but their most serious limitation is their small eye clearance (eye clearance is the distance between the eye lens and the pupil of the eye), which is particularly inconvenient for an observer wearing spectacles. Nevertheless, these eyepieces satisfy most requirements and are widely used with achromatic objectives of low and medium power.

2. **Ramsden or Kellner eyepiece** These eyepieces are called positive type, which simply means that the primary image lies beneath the field lens rather than between the field and eye lenses. Such lenses are useful for measurement and comparison purpose since the size of the primary image is not affected by the field lens. Accessories like reticule and ocular micrometer, can be used only when the lens type is Ramsden. They have large eye clearance.

Magnification of eyepiece Eyepieces range in power from 4× to 20× or more. With magnification less than 4× (i.e., 2.5×) the eyepiece becomes rather long and the field of view is usually limited by the diameter of the field lens. Eyepieces of magnification greater than 15× have small eye clearance and are difficult to use for long periods without eye strain and may magnify the primary image more than is justified by the resolution of the objective. For most work, eyepieces of magnification 6× to 12× are sufficient.

In a binocular microscope, a prism is placed to divert the light rays and this forms a split image, for which higher intensity of light is required.

Photomicrographic System

Photomicrography is concerned with the photography of an enlarged image of small objects (microphotography is the process by which miniature photographs of comparatively large objects are taken, as in document copying). For photomicrography, high intensity light is required and low speed films are preferable. A camera may be applied to the eyepiece of the microscope preferably with the camera lens removed (Figure 3.13) and the image may be focused upon the ground glass screen.

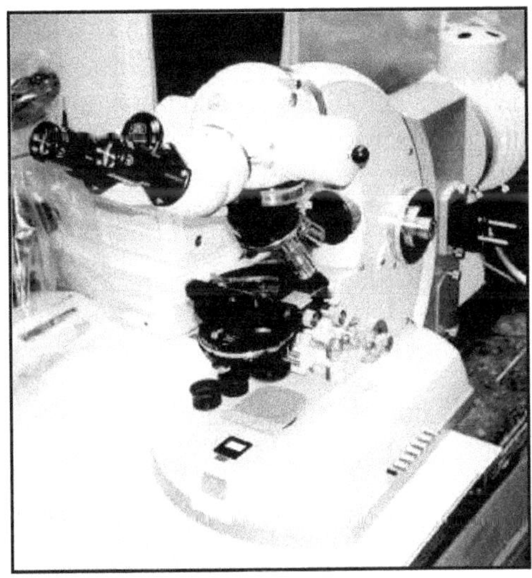

Figure 3.13 Photomicroscope

Achromatic objectives are adequate for monochrome objects, but apochromatic objectives are necessary for the monochrome photography of most colour objects. Apochromatic objectives must be used for colour photomicrography to eliminate coloured fringes due to chromatic aberration.

Use of filters during photography Transparent optical filters permit elimination or reduction of some parts of the spectrum. For most applications of microscopy, a filter of some sort is essential.

1. Blue filters—recommended for colour photographs.
2. Green filters—recommended for black and white photographs.
3. Interference filters—give selective transmission of narrow band of wavelength anywhere in the visible spectrum. They are thus suitable for the production of substantially monochromatic light from a continuous spectrum. In other words an interference filter gives a higher light transmission for the same colour range or alternatively a more selective colour for the same amount of light transmission.

4. Heat filters—whenever a high power light source is employed, a heat absorbing filter is necessary to protect the lenses and filters used in the illuminating light and in the microscope.

TYPES OF MICROSCOPES

SIMPLE MICROSCOPE

A simple microscope consists of only one lens or magnifying glass held in a frame, usually adjustable and often provided with a stand for conveniently holding the object to be viewed and a mirror for reflecting the light (Figure 3.14).

Figure 3.14 Simple microscope

BRIGHT-FIELD MICROSCOPE

This is the simplest type of light microscope. It is also called compound microscope. The light microscope remains a basic tool of cell biologists, with technical improvements allowing the visualization of ever-increasing details of cell structure.

Principle

When light rays are passed through a material, light absorption takes place, i.e., the material will have the tendency to absorb the light depending on its physiochemical properties. This light absorption is independent of wavelength (λ). The phenomenon is called differential absorption of light. Those components of the cell, which have absorbed more amount of light naturally, transmit less light; consequently, they appear dark and denser. And if light absorption is less, the material appears bright or less intense. The stained or coloured objects are resolved well by bright-field microscope. The coloured object may be a whole organism or a section.

The components of a bright-field microscope are the following:

1. Light source
2. Mirror/lens
3. Condenser system
4. Iris diaphragm
5. Stage
6. Objective
7. Focus adjustments
8. Eyepieces

A compound microscope differs from a simple microscope in having two sets of lenses, one known as objective and the other as the eyepiece (ocular). The image seen by the eye is the product of the magnification of this system. Accurate focusing is attained by a special screw appliance known as a fine adjustment. The light path in the ordinary light compound microscope is shown in the Figure 3.15. The specimen on a microscope slide is placed just outside the principal focus of the objective lens, which has a shorter focus. This lens produces a real image, which is formed inside the principal focus of the ocular lens. The eye looking through the eyepiece sees a magnified virtual image of the real image. The eyepiece lens is thus used as a magnifying glass to view the real image. Compound microscopes are necessary for viewing and examining very small organisms like bacteria.

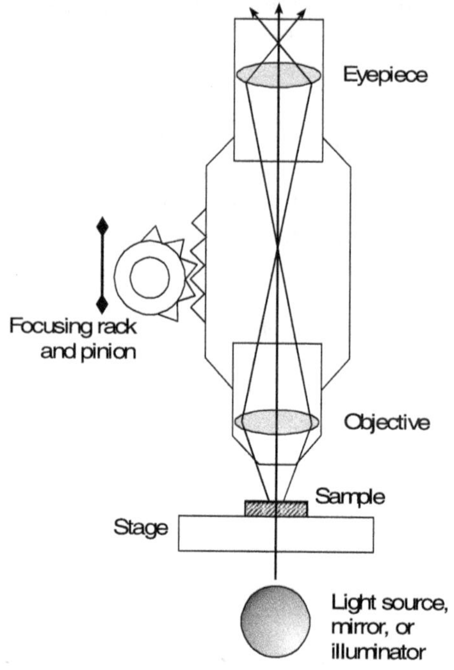

Figure 3.15 Optical path in light microscope

The limitation of the compound microscope is that the objects, which give poor intensity contrast in the transmitted light rays, cannot be resolved properly. Staining techniques enhance the light absorbing capacity of the object. However, smaller objects like cellular organelles cannot be resolved satisfactorily even after the staining.

PHASE CONTRAST MICROSCOPE

The development of the phase contrast microscope has provided an ingenious technique to make highly transparent objects more visible without killing them. The phase contrast microscope was invented by the Dutch physicist Frederick Zernike in the early 1930s, an accomplishment for which he was awarded the Nobel Prize in physics.

Principle

The phase contrast microscope converts the different degrees of phase retardation of light rays passing through a translucent material into amplitude differences to create an image with increased contrast. The phase contrast technique is capable of providing a high contrast image of bacterial cells based on changes in refractive index (light refraction) and not on differential absorption as in bright-field microscopy. Differences in light absorption are often negligible between living cells and their surrounding nutrient medium, as well as between the various intracellular components and plasma membranes, rendering these entities barely visible when observed by bright-field illumination. Phase contrast microscopy takes advantage of minute refractive index differences within cellular components and between unstained cells and their surrounding aqueous medium to produce contrast in these and similar transparent specimens. The light wave transmitted by the material can exist in two forms: (i) Unretarded, (ii) Retarded, when compared to incident light. Retardation may result in transmission of many different rays with different angles. Retardation is also called as phase change. The light absorbing ability of the specimen or object and the resulting phase change of transmitted light wave together determine the degree of intensity of contrast of the image.

Components of Phase Contrast Microscope

The major components of phase contrast microscope are the annular disc and phase plate (Figure 3.16).

Annular disc (aperture stop) It is a disc with a transparent circular ring (annulus). It is placed just below the condenser. The function is to transmit the hollow beam of light or to produce an annulus of light (a ring of light). It avoids hitting of bright light rays on the object.

Phase plate or phase ring It is a black ring. It is placed in the rear focal plane of the objective lens. It will block the unretarded rays entering into the objective lens. (It can retard or advance the phase of light waves traversing it).

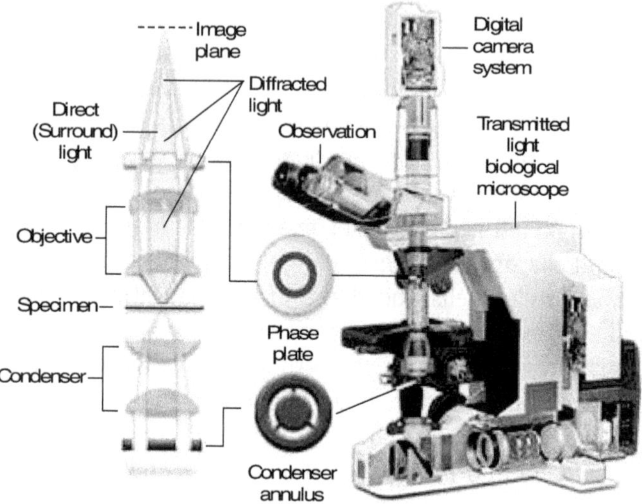

Figure 3.16 Phase contrast microscope configuration

The diameter of the annular disc and phase plate should be (or nearly) identical so that phase plate can be superimposed on annular disc. With the help of telescopic eyepiece, whose eye lens can be moved up and down, the image of the annular disc and phase plate can be seen and both can be superimposed. When the phase plate is superimposed on annular disc, a hollow cone of light is produced. Phase plate permits the entry of only retarded light rays. Unretarded light rays are prevented. So the primary image formed by the objective lens has only retarded light rays. The light rays coming from the centre of the phase plate will give a very crisp image when compared to that of the periphery.

In the bright-field microscope, both unretarded and retarded light rays are received and as unretarded rays have high intensity, the objects with high intensity contrast can be viewed clearly; whereas in the phase contrast microscope, the retarded light rays are allowed to form images and hence poor intensity contrast object can be viewed.

Phase contrast produces an image with highly contrasting bright and dark areas against a neutral grey background. As a result, internal structure of cells is often better visualized by phase contrast microscopy than with bright field optics. When the deviated and undeviated light is recombined by the objective, they differ in phase by one-half wavelength. Hence, the background will appear dark; the actual intensity depends on the relative intensities of the deviated and undeviated light. The object itself introduces a difference because its RI differs from that of the medium. This phase difference, when superimposed on the difference created by the phase plate, results in a loss of destructive interference. Hence, the object appears brighter than the background. The only requirement that a sample must meet for phase contrast microscopy is that it is non-absorbent.

Operating Instructions

1. Using telescopic eyepiece, check whether annular disc and phase plate can be seen and whether they are superimposed or not.
2. Superimpose the two discs by moving the annular disc. Phase plate is fixed whereas annular disc can be moved at 360°.
3. Remove the telescopic eyepiece and replace it with the normal eyepiece. If the objective lens is engraved with "Ph" it indicates that it is provided with phase plate.

Applications

1. Objects with poor intensity contrast can be resolved well.
2. Image of the unstained object can be obtained.
3. It is widely used in microbiology and tissue culture research to detect bacteria, cellular organelles and other small entities in living specimens.
4. It allows the visualization of living unstained cells and their organelles.
5. For the detection of bacterial components such as endospores and inclusion bodies containing poly β-hydroxy butyrate, polymetaphosphate, sulphur and other substances.

DIFFERENTIAL INTERFERENCE CONTRAST (DIC) MICROSCOPE

The interference microscope (Figure 3.17) is based on the principle similar to that of the phase microscope, but has the advantage of giving quantitative data. Wave theory has explained many physical properties of light. Light is a bundle of electromagnetic radiation and does not always travel in a straight line, but can bend around objects kept in its path. This was evident when a parallel beam of light is passed through a sharply defined image, and did not produce correspondingly a sharp image. In other words, it produced a blurred shadow edge. Blurring is due to fringes of light rays, which in turn is due to the bending of light rays. This modification of a beam of light rays as it passes an aperture or an obstacle is called diffraction.

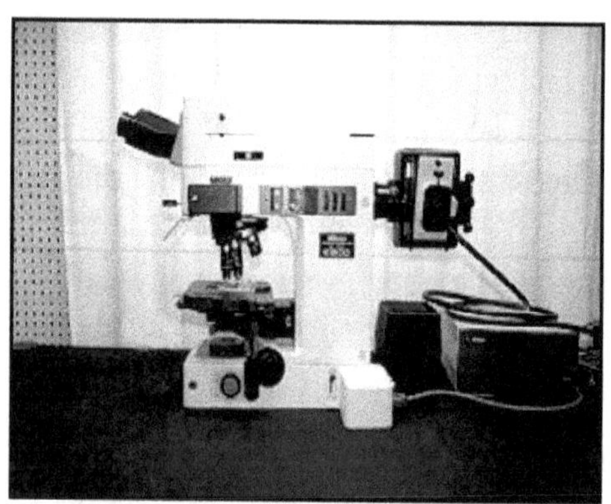

Figure 3.17 Differential interference microscope

When a beam of light is allowed to pass through a small slit, it is observed that maximum light passes straight and the region appeared bright or they formed a brighter image. In other words, when the light rays were in phase, the region is bright. But a variation in the intensity of light could be seen below this point. This is because the light rays are out of phase. Greater the degree of the light rays that are out of phase, greater will be the darkness. When the light rays pass through the obstacle or a small slit there is a modification in the light path. This modification or bending of light in its path results in light rays that are out of phase. Light rays interfere with each other destructively and the region appears darker.

To explain the mechanism of interference, a monochromatic light is allowed to pass through two slits and these slits are placed at equal distance from the light source. When a monochromatic light is allowed to pass through two narrow slits, a pattern of light fringes appear around the dark area. When the light passes straight due to the presence of an obstacle in the light path, it is expected that the region would be dark. But it is observed that the region appears brighter around the darker area.

Light rays are in phase on certain points on light path. When the light rays are in phase they interfere with each other constructively or in other words they supplement each other, thus producing a region of brightness. When the light rays are out of phase, they interfere with each other destructively or they neutralize each other and the field appears dark. The production of brightness by light rays due to their subsequent phases is known as 'interference'.

Principle

In the interference microscope the light emitted by a single source is split into two beams: One is sent through the object; the other bypasses the object. The two beams are then recombined and interfere with one another, as in the phase microscope. In comparison with the direct beam, the beam that has crossed the object is retarded, which means that it has undergone a phase change. This retardation (Γ) is determined by the thickness of the object (t) and the difference between the refractive indices of the object (n_o) and of the surrounding medium (n_m).

$$n_o - n_m = \Gamma / t$$

Interference microscopy permits the simultaneous determination of the thickness of the object, the concentration of the dry matter and the water content by successive measurements of optical phase difference in two media of known refractive indices.

Structural Components

Most of the components in interference microscope are same as that of bright-field microscope. The components specific to interference microscope are

- i. Polarizer
- ii. Birefringence plate
- iii. Half-wave plate
- iv. Analyser

Working Principle

1. A beam of light from an ordinary light source will have vibrations in many different directions. When allowed to pass through a polarizer, the light is polarized. It means that the wave transmitted in light will vibrate in a definite direction or single pattern.
2. In the interference microscope this polarized light is divided or split into two beams or waves by birefringence plate 1.
3. These waves are further rotated along each other by a half-wave plate. Such waves in a rotated state enter into the specimen or object.
4. Specimens having very large difference in retardation of waves across the field of view would generate interference fringes between the two rays.
5. The two rays coming out in between the interference fringes are combined by the birefringence plate 2 and the analyser produces an interference image of the object.

Advantages over Phase Contrast Microscope

1. Interference microscope has the advantage of giving quantitative data.
2. It permits detection of small, continuous changes in refractive index.
3. The variation of phase can be transformed into such vivid colour changes that a living cell may resemble a stained preparation.

Applications

1. To observe cells in culture. Live cells can be observed and the morphology and surface or boundary feature of the culture cells can be examined using interference microscope.
2. Attachment and spreading behaviour of cells in live condition can be easily seen and detected with interference microscopy.
3. The interference microscope gives the thickness of the object.
4. The concentration of dry matter and water content can be calculated.
5. The lipid, nucleic acid and protein contents of the cells can be determined.

VIDEO-ENHANCED DIFFERENTIAL INTERFERENCE-CONTRAST MICROSCOPY

The power of the light microscope has been considerably expanded by the use of video cameras and computers for image analysis and processing. Such electronic image-processing systems can substantially enhance the contrast of images obtained with the light microscope, allowing the visualization of small objects that otherwise could not be detected. For example, video-enhanced differential interference-contrast microscopy (Figure 3.18) has allowed visualization of the movement of organelles along microtubules, which are cytoskeletal protein filaments with a diameter of only 0.025 μm.

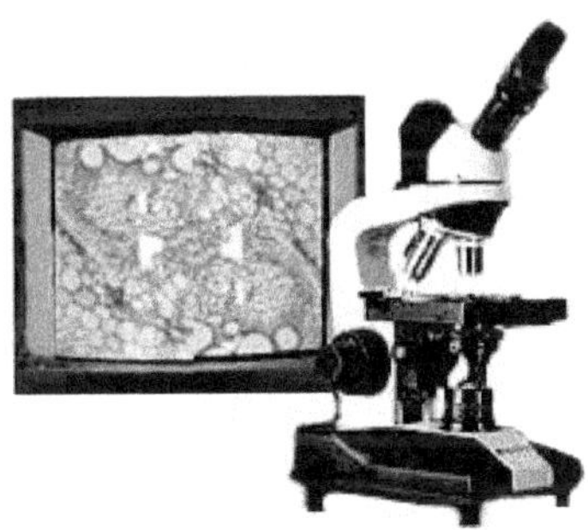

Figure 3.18 Video-enhanced differential interference-contrast microscope

Fundamentals of Video Imaging

Optical images produced in the microscope can be captured using either traditional film techniques, digitally with electronic detectors such as a charge-coupled device (CCD), or with a tube-type video camera. When a dynamic event must be recorded in real time, a video camera is often the most suitable resource for the task.

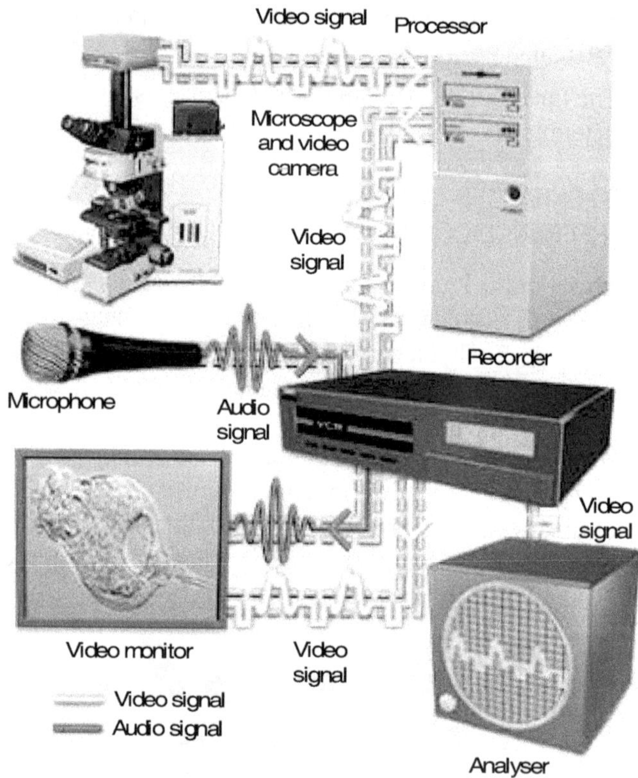

Figure 3.19 Schematic diagram of video microscopy

The primary function of a video is to produce an electrical signal by scanning an optical image of a dynamic scene, then transmitting the information to a receiver housed in a remote location for viewing or recording of the original event. Important features and components of video microscopy are shown in the Figure 3.19. An optical event is first captured in the microscope and translated into an electrical video signal by an electronic camera attached to the trinocular tube. The signal may be processed or recorded, then sent to an analyser or converted back into a two-dimensional image on a video monitor. Addition of an audio signal to the video information (by recording sound on a VCR or similar device) allows concurrent annotation of comments.

Among the important features of video to be considered are the facts that the optical image is generally a dynamic (as opposed to static) two-dimensional array, which must be transmitted over some distance utilizing a single channel or cable. Either before or after transmission, the video signal may be processed or stored prior to being displayed on a monitor that must immediately follow the original scene in real time. The widespread appeal of broadcast and cable television (and other video applications) has led to the development of an ingenious series of electro-optical and electronic devices, which have been of great benefit to video microscopy.

In video microscopy, the signal is usually transmitted over very short distances through coaxial cables rather than broadcast through a wireless system. This configuration is referred to as closed-circuit TV (CCTV), which shares much in common with wireless or broadcast TV, with some deviation in standards and convention.

Video Signal Generation

The conversion of a two-dimensional image captured through a microscope into a recoverable train of electrical impulses is accomplished by sequentially scanning narrow strips of the optical image. In principle, the microscope image can be scanned by any electro-optical detector that is capable of rapidly converting light intensity into an electrical voltage or current. Many of the characteristics of video signals can be understood by display of the scanned electrical signal on a suitable oscilloscope. Figure 3.20 illustrates the relationship between individual scan lines from an optical image and the resulting video signal. The specimen is a live rotifer imaged using differential interference contrast (DIC) microscopy. Sequential scanning of the rotifer image occurs from left to right (for example: A to A′ then B to B′, etc.) and incrementally downward starting at the upper left-hand corner of the image and finishing at the lower right. As each horizontal trace is scanned, it produces an electrical signal relating the image brightness to a corresponding value measured in volts (the video signal).

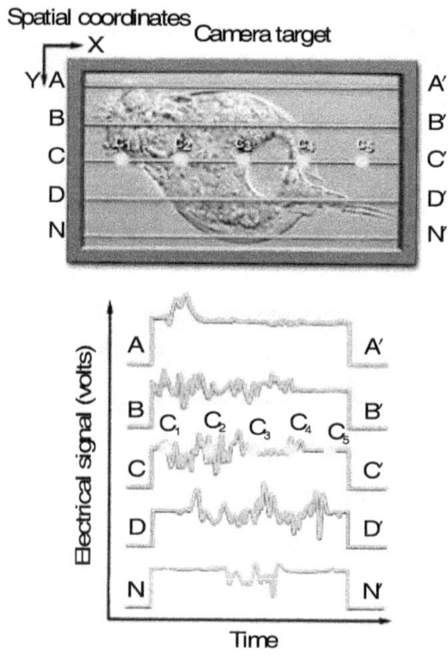

Figure 3.20 Scan line and video signal relationship

The graph presented in the lower portion of Figure 3.20 shows a plot of the electrical signal generated by scanning specific areas of the image versus time. The location of a specific area on the image, such as c_1 or c_2, corresponds to its point in time in the video signal (C_1 or C_2). Although only five scan lines are illustrated in the figure, the entire image is actually scanned. Because the amplitude of the electrical signal is proportional to the brightness of the image being scanned, the current or voltage at a particular point in the signal corresponds to light intensity of the image at that point (voltage c_1 corresponds to intensity C_1). The narrow horizontal strip in the image represented by the C horizontal scan line is scanned at constant speed so that the distances $C–C_1$, $C–C_2$ and $C–C_3$ in the scanned image are proportional to the time intervals $c–c_1$, $c–c_2$, and $c–c_3$ in the electrical signal. The result is signals from a single horizontal scan line which is time-dependent, is the electrical representation of the corresponding narrow strip in the optical image.

In order to construct a complete two-dimensional image, video scan lines must cover the entire image area, and the resulting electrical signal must contain the voltage amplitude distribution of all points on all of the scan lines. The electrical signal output from scan lines is formatted serially (in sequence) as a single linear signal rather than as an array of parallel signals. This means that each individual scan line (A–A′, B–B′, etc.) is added to the sequence to produce a single linear stream of electrical pulses that can be conveyed along a single coaxial cable or transmitted (broadcast) by electromagnetic waves through the atmosphere on a single channel.

As mentioned above, a two-dimensional optical image is transformed into a series of electrical pulses to produce a video signal. Standard convention dictates that the image is scanned as "seen on the video monitor" from the upper left to the right, and then repeated on the next line down, similar to reading English text. Starting with the upper left-hand corner (as illustrated in Figures 3.20 and 3.21), the image is first scanned horizontally along line A–A′ from left to right. Upon reaching point A′, the scanning spot is made to "fly back" to the beginning of the next scan line positioned at point B. The flyback (or retrace) operation occurs much faster than the scan speed, and the video signal is blanked during the flyback to prevent trace A′–B (Figure 3.21 dashed lines) from contributing to the signal. After completing the loop, the scan continues, at the standard horizontal scanning velocity, along trace B–B′ until it again flies back from B′ to C. This cycle is continued until all areas of the image have been sequentially scanned.

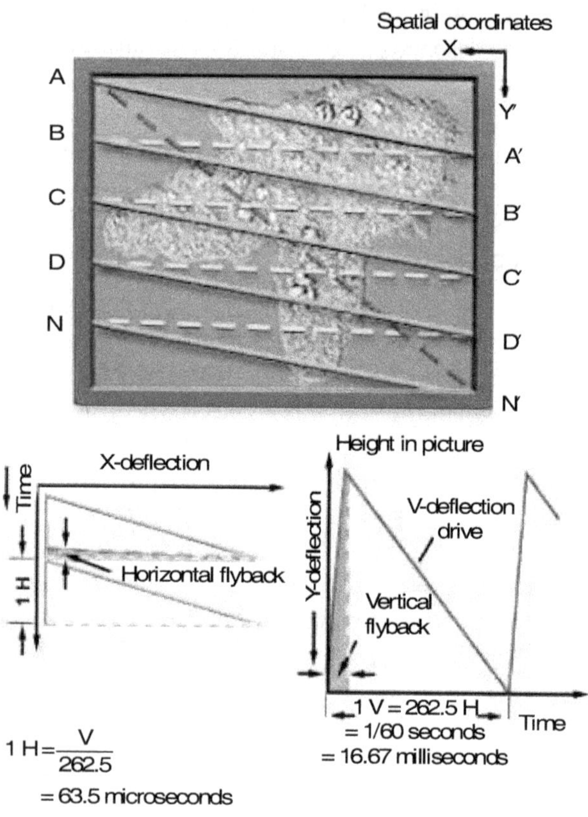

Figure 3.21 Horizontal and vertical scan lines

The horizontal scan (H) lines are not actually horizontal, but instead slant downward slightly to the right by a vertical distance that is equal to the width of an individual scan line. In this sense, the scanning spot is also undergoing a vertical scan (V), moving downward at a slower but constant speed. At the end of the scan sequence when the last horizontal scan (N–N′) has been completed, the scanning spot flies back from N′ at the lower right-hand corner to point A in the upper left. After the vertical flyback, the process starts over again to ultimately yield the video raster.

Horizontal and vertical movements of the image scanning beams in a video camera and monitor are generated by either magnetic or electrostatic deflectors, which provide the constant-velocity scans needed to drive the horizontal scan lines, vertical deflection, and rapid flybacks. Scanning frequencies for the H scan are much higher, producing faster rates and repeats that occur many times during a single vertical (V) scan.

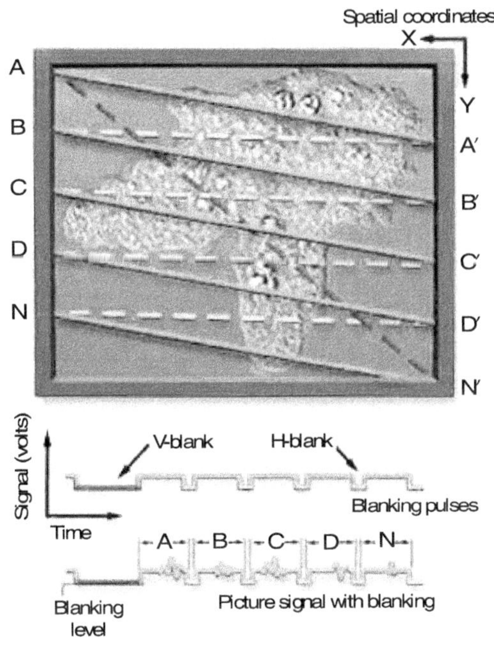

Figure 3.22 Video blanking pulses

During flybacks, the signal voltage of retraces is lowered to a blanking level, essentially turning off the scanning electron beam, to avoid adding extra lines that would degrade the image. These blanking pulses actually start a very short time before the H and V flybacks to ensure that blanking is complete, as illustrated in Figure 3.22.

In solid-state cameras such as charge-coupled devices (CCDs) and active pixel sensor CMOS imagers, display devices utilize arrays of fixed picture elements (termed pixels) to capture and display images. Although the detector and display are scanned in a different manner with solid-state devices, the video signal is identical to that produced by tube-type cameras.

FLUORESCENCE MICROSCOPE

A fluorescence microscope is basically a conventional light microscope with added features and components that extend its capabilities. A conventional microscope uses light to illuminate the sample and produce a magnified image of the sample, whereas, a fluorescence microscope uses a much higher intensity light to illuminate the sample. This light excites fluorescence species in the sample, which then emit light of a longer wavelength. A fluorescent microscope also produces a magnified image of the sample, but the image is based on the second light source, i.e., the light emanating from the fluorescent species, rather than from the light originally used to illuminate, and excite, the sample.

British scientist **Sir George G. Stokes** first described fluorescence in 1852 and was responsible for coining the term when he observed that the mineral fluorspar emitted red light when it was illuminated by ultraviolet excitation. Stokes noted that fluorescence emission always occurred at a longer wavelength than that of the excitation light. This shift towards longer wavelength is known as **Stokes' shift**. Early investigations in the 19th century showed that many specimens (including minerals, crystals, resins, crude drugs, butter, chlorophyll, vitamins, and inorganic compounds) fluoresce when irradiated with ultraviolet light. However, it was not until the 1930s that the use of fluorochromes was initiated in biological investigations to stain tissue components, bacteria, and other pathogens. Several of these stains were highly specific and stimulated the development of the fluorescence microscope.

The technique of fluorescence microscopy has become an essential tool in biology and the biomedical sciences, as well as in materials science due to attributes that are not readily available in other contrast modes with traditional optical microscopy. The application of an array of fluorochromes has made it possible to identify cells and sub-microscopic cellular components with a high degree of specificity amid non-fluorescing material.

The phenomenon where a molecule after absorbing radiation emits radiation at a longer wavelength is known as fluorescence. Light of shorter wavelength excite specific molecules in the specimen and hence electrons are excited to a higher energy level. When they relax to a lower level, they emit light with longer wavelength. At a particular wavelength, the excitation would be at maximum and emission might also be great at a particular wavelength. Excitation maxima and emission maxima are two characteristics that vary with the type of substance.

The substance which exhibits the property of fluorescence is called **fluorophore** and the phenomenon is **auto-fluorescence** or **primary fluorescence**. Some substances, for example chlorophyll, show auto-fluorescence, and readily illuminate and produce red light. Another type of fluorescence is the **secondary fluorescence** or **induced fluorescence**. Most of the biological materials do not possess the property of auto-fluorescence and hence certain chemicals, dyes or stains are used to induce fluorescence. Chemicals like glutaraldehyde, formalin, etc. when mixed with certain substances induce fluorescence. Fluorescent dyes are directly taken up by the cells. They are incorporated and concentrated in specific subcellular compartments. The commonly used dyes are fluorescein isothiocyanate (FITC) with excitation maxima 495 nm and emission 525 nm and Rhodamine isothiocyanate with excitation maxima 550 nm and emission 580 nm. If FITC is used for preparing the substance to exhibit, fluorescence, green colour is exhibited while if rhodamine is used, red to pink colour is obtained. Mixtures of these two dyes can also be used. A number of fluorochromes such as auramine O, acridine yellow, and acriflavine are used as fluorescence probes for nucleic acids, especially DNA.

There are two basic types of illumination used in fluorescence microscopes. i) transmitted light illumination is used in diascopic fluorescence or trans-fluorescence microscope and ii) epi-illumination is used in epi-fluorescence microscope. The optical pathway of trans-fluorescence microscope and epi-fluorescence microscope is shown in the Figures 3.23 a, b.

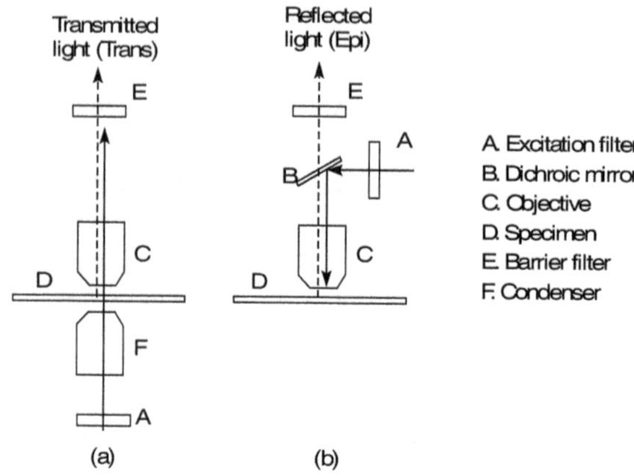

Figure 3.23 (a) Optical pathway in trans-fluorescence microscope
(b) Optical pathway in epi-fluorescence microscope

In trans-fluorescence microscope, either mercury-vapour arc lamp or quartz-halogen-tungsten filament lamp is used as the source of illumination. The specimen is excited by light passing through the condenser lens, and the fluorescent emission is captured by the objective lens. Transmitted light illumination fluorescence microscope uses either bright-field condenser or dark-field condenser. When a bright-field condenser is used to illuminate the object with excitation light, separation of excitation light from fluorescence light becomes difficult, as both directly enter the objective. Hence, dark-field condenser is used to illuminate the object. The dark-field condenser creates a cone of excitation light, so that oblique rays of light illuminate the object. The fluorochromes in the specimen get excited and emit light rays of longer wavelength. These emitted long-wavelength rays enter the objective. The barrier filter placed between the objective and ocular blocks the light of short wavelength (below 400 nm) and hence these short-wavelength rays do not reach eyepiece.

In an epi-illumination fluorescence microscope, mercury vapour lamp is used as the light source. Two electrodes are sealed in quartz glass chamber filled with mercury vapour. A very intense spectral light is produced with short wavelength. These lamps can be used for medium to high intensity illumination. High intensity is desirable, as this would enable us to isolate the specific wavelength, which is required. Light passes through an excitation filter to select light of the wavelength (e.g. blue) that excites the fluorescent dye. A dichroic mirror then deflects the excitation light down to the specimen. The light passes through the objective before striking on the specimen. The fluorescent light emitted by the specimen (e.g. green) again passes through objective then to the dichroic mirror and a second filter (barrier filter) to select light of long wavelength emitted by the dye. This type of excitation-emission configuration, in which both the excitation and emission light travel through the objective, is called epifluorescence. The key to the optics

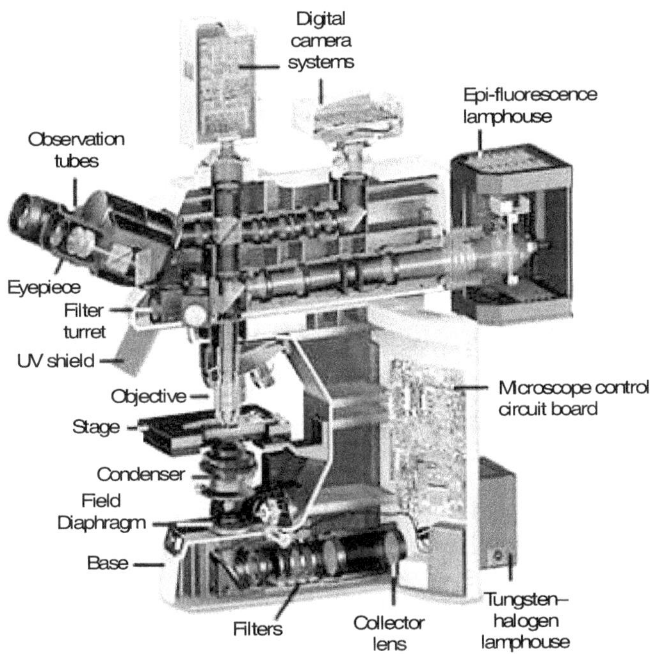

Figure 3.24 Cutaway diagram of an epi-fluorescence microscope equipped for both transmitted and reflected fluorescence microscopy

in an epifluorescence microscope is the separation of the illumination (excitation) light from the fluorescence emission emanating from the sample. A **dichroic mirror** is used to separate the excitation and emission light paths. Within the objective, the excitation emission share the same optics. The **excitation** light reflects off the surface of the dichroic mirror into the objective. The fluorescence **emission** passes through the dichroic to the eyepiece or detection system.

Illustrated in Figure 3.24 is a cutaway diagram of a modern epi-fluorescence microscope equipped for both transmitted and reflected fluorescence microscopy.

Immunofluorescence is another technique that involves the use of antibodies to which a fluorescent marker has been attached.

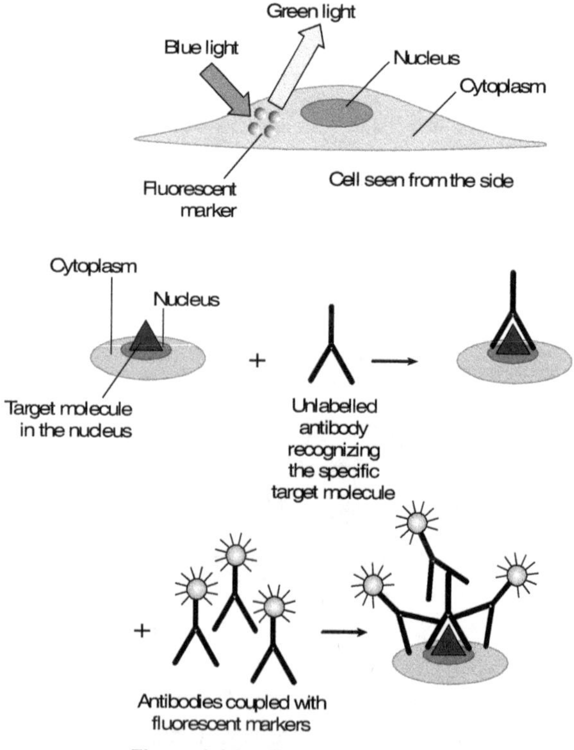

Figure 3.25 Tagging of proteins

Antibodies are molecules that recognize and bind selectively to specific target molecules in the cell. The fluorescent signal can be amplified by using an unlabelled primary antibody and detecting it with labelled secondary antibodies. It is possible to modify cells so that they create their own fluorescing molecules. The protein molecules are tagged with a fluorescing marker (Figure 3.25). When a specific protein is modified in this way, the location of that protein can be studied. It is also possible to watch the movements of the proteins and its interactions with other cellular components inside the cell.

Applications

Studying the dynamics in the cell is essential for understanding cell function. Fluorescence microscopy is one of the most used approaches in studying the location and movement of molecules and sub cellular components in the cell. Fluorescence microscope can be used to examine materials with auto fluorescence, e.g. trachea, chitin, egg shells. Most of the cellular components do not fluoresce themselves. Fluorescent markers are therefore introduced. The structure and molecular components of the cells can be examined after introduction of dyes.

The recent development of biosensors based on genetically encoded variants of green fluorescent protein (GFP), coupled with advances in digital, multi-mode, epi-fluorescence microscopy, has introduced new powerful tools for observing protein dynamics and protein–protein interactions at high spatial and temporal resolution within living cells.

POLARIZING MICROSCOPE

This method is based on the behaviour of certain components of cells and tissues when they are observed under polarized light. Polarized light is produced when the light waves lie in one plane. If the material is 'isotropic', polarized light is propagated through it with the same velocity, independent of the impinging direction. These materials exhibit limited degree of phase effects or phase change (e.g. cubic crystals, glasses, amorphous materials). This results in one major and few minor retarded waves. Such substances or materials are characterized by having the same index of refraction in all the directions.

In an anisotropic material the velocity of propagation of polarized light varies. Such materials are called **birefringent** because it presents two different indices of refraction corresponding to the respective different velocities of transmission. When light is passed through such materials, it resolves into two plane polarized waves that vibrate in mutually perpendicular direction and therefore one wavefront is retarded with respect to the other. This kind of retardation depends on the nature of the material, the specific orientation of the material in the light path and the thickness of material. Thus the polarization microscope is used to resolve anisotropic material.

Components of Polarizing Microscope

The polarizing microscope is like a light microscope except that the condenser and ocular are equipped with polarizing optics, (i.e., double refracting crystal of calcite or quartz or a sheet of polarized plastic film) each of which transmits only plane-polarized light. The polarizing microscope is shown in the Figure 3.26. The main components of polarizing microscope are (i) Polarizer (ii) Stage (iii) Accessory slots and (iv) Analyser.

1. **Polarizer** Polarizer is a prism or filter with a selective absorption film coated with some orientation dye material which is designed to transmit light in one given direction and to send it into the object. It is located just below the condenser. It is also called as first polarizer.
2. **Stage** Stage is of a rotating type, usually circular with a hole in the centre. The object can be rotated on the rotating stage at 360°.

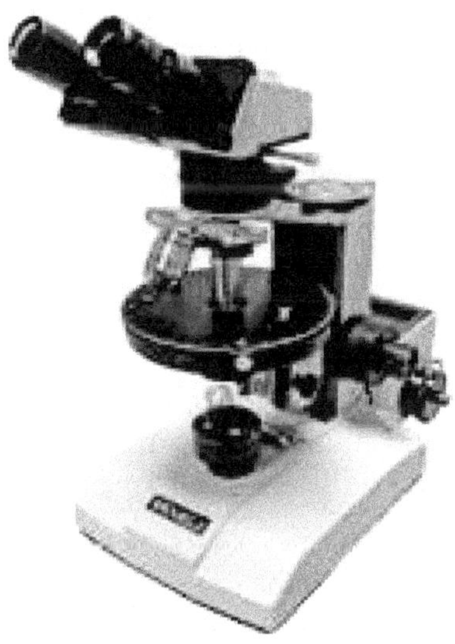

Figure 3.26 Polarizing microscope

3. Accessory slots It is a half plate and is used to cut off few light waves that are vibrating in various directions. It is used to get better contrast of the image of the object. Its location is just above the objective lens.
4. Analyser It is a type of polarizing filter that will transmit the polarized light in a single direction. Its position is above the accessory slots. Maximum light transmission is obtained when analyser is set parallel to the polarizer.

Working Principle

A polarizer allows the light to vibrate in one single direction. When the polarized light passes through the object (anisotropic material), light vibrates in different directions. Polarized light, which has undergone some modifications after passing through anisotropic material, is allowed to pass through the analyser, which depolarizes it by changing the angle of refraction. Interfering waves are either cancelled or reinforced. When interfering waves get cancelled the object is dark, and when interfering waves get reinforced, the object appears bright. The polarizer and analyser remain adjusted in such a fashion that when the analyser is rotated to 360°, the visual field alternates between bright and dark at every 180° turn. The two positions of maximum light transmission are obtained when the analyser is set parallel to the polarizer.

Applications

1. The birefringence of some biological materials such as muscle fibres, hair, and plant fibres is large enough to permit them to be conveniently studied by the orthoscopic technique and the retardation may be determined by quartz- or gypsum-sensitive tin plates.
2. Very useful to observe the contents of anisotropic materials in fresh preparations.
3. Polarized light technique may be used in biological research not only to identify particular features, but also to conduct analysis as in determining the lipid content of lipoprotein.
4. In certain cells accumulation of salts in the form of crystals is seen. To observe the nature of such crystals this microscope is used.
5. Polarization microscope is used to observe the sections of insect cuticle and certain egg shells of insects with heavy mineral deposition.
6. The greatest value of polarized light microscopy remains in the field of examination of more or less transparent crystalline materials by transmitted light. It is thus applicable for the identification of rocks, refractories and ceramics.
7. Metals and minerals may also be examined by reflected polarized light.

DARK-FIELD MICROSCOPE

A dark-field microscope is very similar to bright-field microscope except that the light from the light source does not enter directly into the objective lens. A dark-field microscope permits the light rays which are heavily retarded to form the image. Thus objects which retard light strongly are only viewed.

Principle

In dark-field microscopy, light passing through the condenser lens is restricted to far off-axis rays (Figure 3.27). To view a specimen in a dark field, an opaque disc is placed underneath the condenser lens, so that only light that is scattered by objects on the slide can reach the eye. If an objective is located in the condenser focal plane, diffracted rays from the objective may fall into the objective lens and become visible in the ocular lens. Hence, the specimen is displayed as a bright object on a dark background.

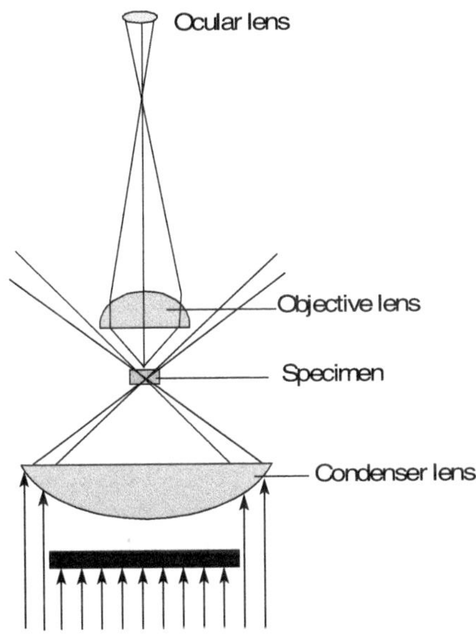

Figure 3.27 Optical pathway in dark-field microscope

Components

1. It utilizes a condenser system of relatively high NA, i.e., higher than that of the objective lens.

2. A device called a central opaque stop would exclude all direct light when NA is equal to (or) less than that of the objective lens. The dark-field stop allows a hollow cone of light instead of a solid cone to shine on the specimen. The only light that reaches the objective lens is the light reflected from the specimen. Consequently, the field of view is dark while objects that reflect light are bright.

Applications

Dark-field microscopy is used in several biological and medical investigations. These include

1. Initial examination of suspensions of cells such as yeast, bacteria, small cell and tissue fractions including cheek epithelial cells, chloroplasts, mitochondria, and even blood samples.
2. Initial survey and observation of pond water samples, hay or soil infusions.
3. Examination of lightly stained prepared slides.
4. Determination of motility of cells in culture.

CONFOCAL MICROSCOPY

Confocal microscope was invented by Marvin Minsky (Figure 3.28 a, b). Confocal microscopy combines fluorescence microscopy with electronic image analysis to obtain three-dimensional images. A small point of light, usually supplied by a laser, is focused on the specimen at a particular depth. The emitted fluorescent light is then collected using a detector, such as a video camera. Before the emitted light reaches the detector, however, it must pass through a pinhole aperture (called a confocal aperture) placed at precisely the point where light emitted from the chosen depth of the specimen comes to a focus. Consequently, only light emitted from the plane of focus is able to reach the detector. Scanning across the specimen generates a two-dimensional image of the plane of focus, a much sharper image than that obtained with standard fluorescence microscopy. Moreover, a series of images obtained at different depths can be used to reconstruct a three-dimensional image of the sample.

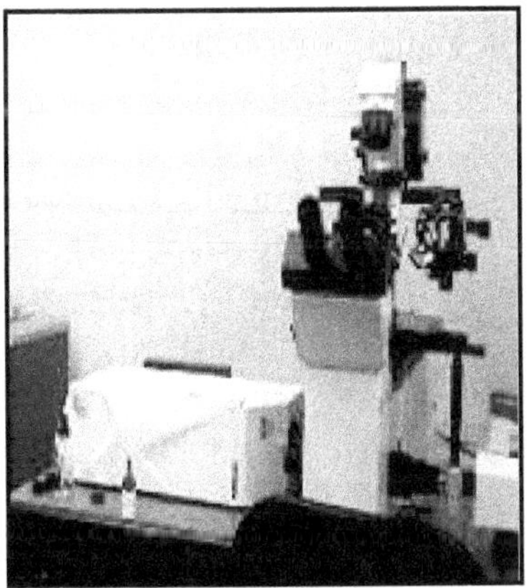

Figure 3.28 (a) Inverted microscope, with a confocal microscope attachment

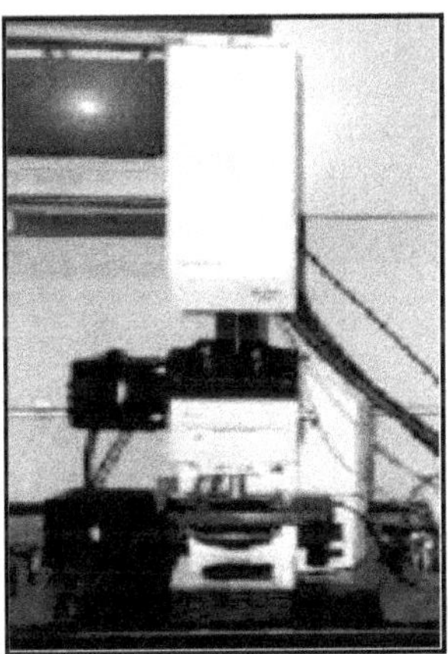

Figure 3.28 (b) Confocal microscope attachment mounted on top of the upright microscope

Working Principle

The imaging principle in a confocal laser scanning microscope (CLSM) is very different from conventional light microscopes. In conventional microscopes, the whole specimen is continuously illuminated by a condenser. Via the objective lens and the ocular an image is visible to our eyes. The key idea of the working principle of the CLSM is to restrict the illumination area by using an illumination point source. An image of the entire specimen surface is produced by a scanning mechanism. In addition to the small aperture used for illumination, a second small aperture is used in front of the detector. The intermediate image of the specimen in the back-focal plane of the objective lens is simultaneously located in the focal plane of the collector lens which gives rise to the name "confocal". This leads automatically to a further important characteristic of the CLSM. Light rays with an origin off the optical axis and off the focal plane are focused in a different plane, and hence, are excluded from image formation by the pinhole aperture (Figure 3.29). Therefore, the CLSM is capable of focusing in the third dimension (the depth of the sample) and by scanning different focal planes, one may produce a three-dimensional image of a three-dimensional specimen.

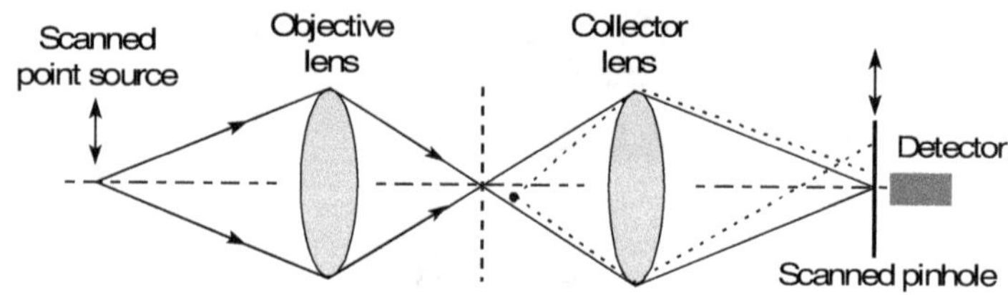

Figure 3.29 Optical path in confocal microscope

TWO-PHOTON EXCITATION MICROSCOPY

Two-photon excitation microscopy is an alternative to confocal microscopy that can be applied to living cells. The specimen is illuminated with a wavelength of light such that the excitation of the fluorescent dye requires the simultaneous absorption of two photons (Figure 3.30). The probability of two photons simultaneously exciting the fluorescent dye is only significant at the point in the specimen upon which the input laser beam is focused, so fluorescence is only emitted from the plane of focus of the input light. This highly localized excitation automatically provides three-dimensional resolution, without the need for passing the emitted light through a pinhole aperture, as in confocal microscopy. Moreover, the localization of excitation minimizes damage to the specimen, allowing three-dimensional imaging of living cells.

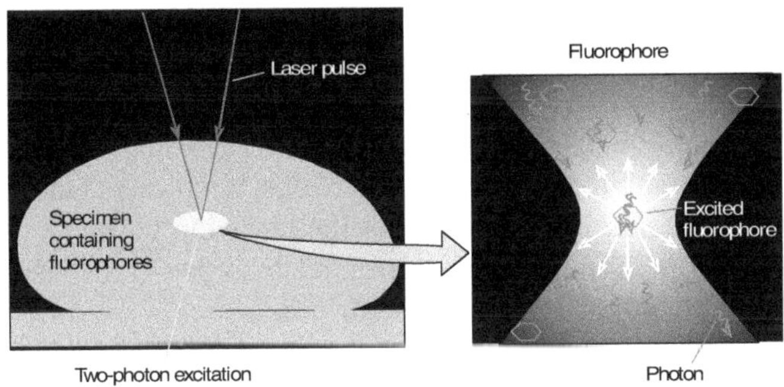

Figure 3.30 Two-photon excitation and excited fluorophore in two-photon excitation microscopy

ELECTRON MICROSCOPY

Electron microscope has become indispensable in microbiological research. The high resolving power of the instrument enables the visualization of the fine structure of bacteria, virus, cells and tissues, which is not possible with the light microscope. Electron microscopy was developed in the 1930s and first applied to biological specimens by Albert Claude, Keith Porter, and George Palade in the 1940s and 1950s. The electron microscope can achieve a much greater resolution than that obtained with the light microscope. The illumination source in electron microscope is a white hot tungsten filament which emits electrons. The wavelength of electrons is shorter than that of light. The wavelength of electrons in an electron microscope can be as short as 0.004 nm, about 100,000 times shorter than the wavelength of visible light. Theoretically, this wavelength could yield a resolution of 0.002 nm, but such a resolution cannot be obtained in practice, because resolution is determined not only by wavelength, but also by the numerical aperture of the microscope lens. Numerical aperture is a limiting factor for electron microscopy because inherent properties of electromagnetic lenses limit their aperture angles to about 0.5 degrees, corresponding to numerical apertures of only about 0.01. Thus, under optimal conditions, the resolving power of the electron microscope is approximately 0.2 nm. Moreover, the resolution that can be obtained with biological specimens is further limited by their lack of inherent contrast. Consequently, for biological samples the practical limit of resolution of the electron microscope is 1 to 2 nm. Although this resolution is much less than that predicted simply from the wavelength of electrons, it represents more than a hundred-fold improvement over the resolving power of the light microscope. Two types of electron microscopy—transmission and scanning are widely used to study cells.

TRANSMISSION ELECTRON MICROSCOPE (TEM)

In principle, viewing specimens under a transmission electron microscope is similar to the observation of stained cells with the bright-field light microscope. Specimens are fixed and stained with salts of heavy metals, which provide contrast by scattering electrons. A beam of electrons is then passed through the specimen and focused to form an image on a fluorescent screen. Electrons that encounter a heavy metal ion as they pass through the sample are deflected and do not contribute to the final image, so stained areas of the specimen appear dark.

Specimens to be examined by transmission electron microscopy can be prepared by either **positive** or **negative staining**. In **positive staining**, tissue specimens are cut into thin sections and stained with heavy metal salts (such as osmium tetroxide, uranyl acetate, and lead citrate) that react with lipids, proteins, and nucleic acids. These heavy metal ions bind to a variety of cell structures, which consequently appear dark in the final image. Alternative positive-staining procedures can also be used to identify specific macromolecules within cells. For example, antibodies labelled with electron-dense heavy metals (such as gold particles) are frequently used to determine the subcellular location of specific proteins in the electron microscope. This method is similar to the use of antibodies labelled with fluorescent dyes in fluorescence microscopy.

Negative staining is useful for the visualization of intact biological structures, such as bacteria, isolated subcellular organelles, and macromolecules. In this method, the biological specimen is deposited on a supporting film, and a heavy metal stain is allowed to dry around its surface. The unstained specimen is then surrounded by a film of electron-dense stain, producing an image in which the specimen appears light against a stained dark background.

Metal shadowing is another technique used to visualize the surface of isolated subcellular structures or macromolecules in the transmission electron microscope. The specimen is coated with a thin layer of evaporated metal, such as platinum. The metal is sprayed onto the specimen from an angle so that surfaces of the specimen that face the source of evaporated metal molecules are coated more heavily than others. This differential coating creates a shadow effect, giving the specimen a three-dimensional appearance in electron micrographs.

The preparation of samples by **freeze fracture**, in combination with metal shadowing, has been particularly important in studies of membrane structure. Specimens are frozen in liquid nitrogen (at $-196^{\circ}C$) and then fractured with a knife blade. This process frequently splits the lipid bilayer, revealing the interior faces of a cell membrane. The specimen is then shadowed with platinum, and the biological material is dissolved with acid, producing a metal replica of the surface of the sample. Examination of such replicas in the electron microscope reveals many surface bumps, corresponding to proteins that span the lipid bilayer. A variation of freeze fracture called **freeze etching** allows visualization of the external surfaces of cell membranes in addition to their interior faces.

Optical System and Component Parts

The transmission electron microscope and its optical system are shown in the Figures 3.31 and 3.32. The electron microscope

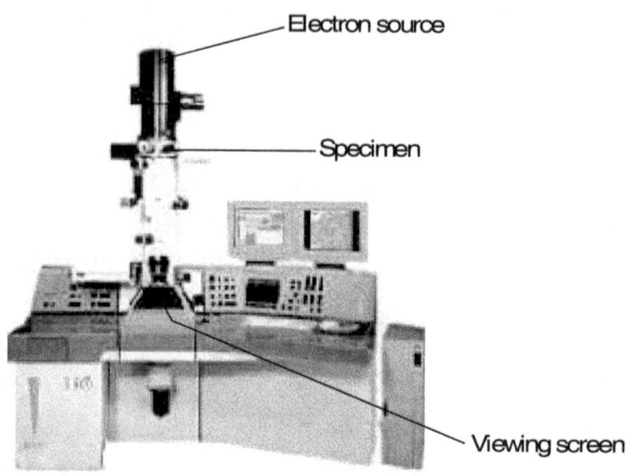

Figure 3.31 Transmission electron microscope

consists of a column of stack at the top of which is mounted the source of illumination, the electron gun (Figure 3.33) which emits electron from a hot tungsten wire filament. Beneath this filament a cathode shield is placed. A high voltage, which can be varied from 50 to 100 kV, is applied to the anode. The life of tungsten wire filament is limited and it usually requires to be replaced after 15 hours of viewing. A pencil beam of electrons moving at high velocity is projected through the hole in the anode and onwards down the stack. The high accelerating voltage used must be stabilized to ensure uniform velocity of electrons. The stack is completely evacuated and a vacuum is maintained. Focusing and magnification are achieved by electromagnetic lenses.

There is a condenser lens system, which bends the rays of electrons so that a parallel beam is directed onto the object placed below it. The electrons are scattered to a degree that is proportional to the thickness and density of the various parts of the specimen. An objective lens gathers the scattered electrons through a very small aperture and brings them to a focus where a real primary image is formed and is magnified about a hundred times. Two projector lenses, which have the function of the eyepiece of the light microscope, magnify a part of the primary image further 300

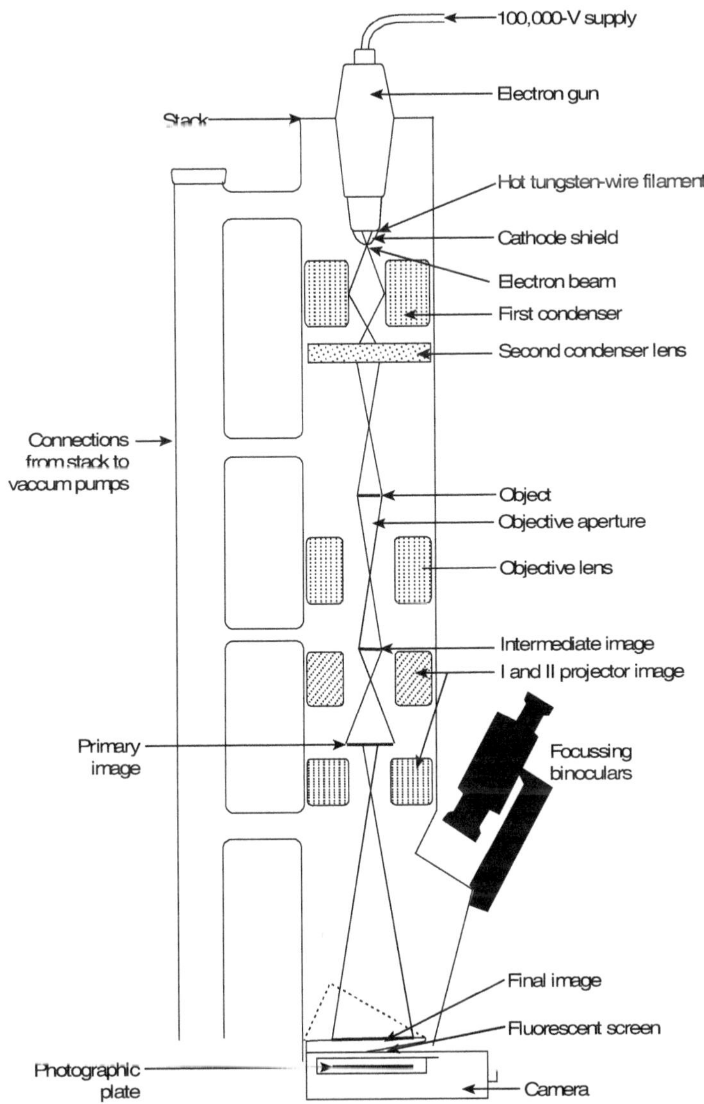

Figure 3.32 Optical system and components of an electron microscope

to 500 times. The focal length of the electromagnetic lenses can be changed by varying the current flowing through the lens and thus a continuously variable magnification is obtained. The final image is observed on a fluorescent screen situated at the lower end of the stack and is viewed through a glass window. The screen can be withdrawn by a lever to allow the electrons to impinge on a photographic plate or film held in a camera. Owing to the high resolving power of the electron microscope, it is possible to take negatives at very high magnification.

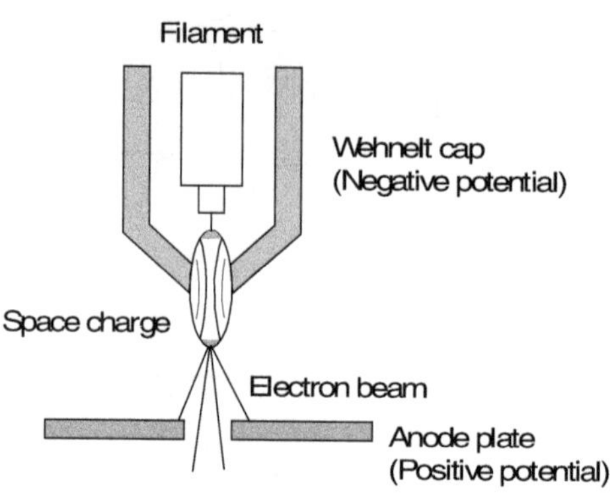

Figure 3.33 Electron gun

The specimen to be examined is placed in a special holder and then introduced into the stack between the condenser and objective lenses through an airlock. Finally when a vacuum has been restored it is lowered into position by a lever and its examination can be made. A 10× binocular microscope is used to facilitate accurate focusing before taking photographs. The controls of the electron microscope are accommodated in panels on its desk. They include dials for variation of the magnification, a mechanical stage to move the specimen, switches and meters to control and check the voltage used, vacuum gauges and controls for electron optimal alignment.

SCANNING ELECTRON MICROSCOPE (SEM)

The second type of electron microscopy, scanning electron microscopy, is used to provide a three-dimensional image of cells. In scanning electron microscopy the electron beam does not pass through the specimen. Instead, the surface of the cell is coated with a heavy metal, and a beam of electrons is used to scan across the specimen. Electrons that are scattered or emitted from the sample surface are collected to generate a three-dimensional image as the electron beam moves across the cell. Because the resolution of scanning electron microscopy is only about 10 nm, its use is generally restricted to studying whole cells rather than subcellular organelles or macromolecules.

Optical System and Component Parts

The scanning electron microscope and its optical system are shown in the Figures 3.34 and 3.35. The virtual source at the top represents the electron gun, producing a stream of monochromatic electrons. The stream is condensed by the first condenser lens (usually controlled by the coarse probe current knob). It works in conjunction with the condenser aperture to eliminate the high-angle electrons from the beam. The beam is then constricted by the condenser aperture (usually not user-selectable), eliminating some high-angle electrons. The second condenser lens forms the electrons into a thin, tight, coherent beam and is usually controlled by the fine probe current knob. A user-selectable objective aperture further eliminates high-angle electrons from the beam. A set of coils then scans or sweeps the beam in a grid fashion (like a television), dwelling on points for a period of time determined by the scan speed (usually in the microsecond range). The final lens, the objective, focuses the scanning beam onto the part of the specimen

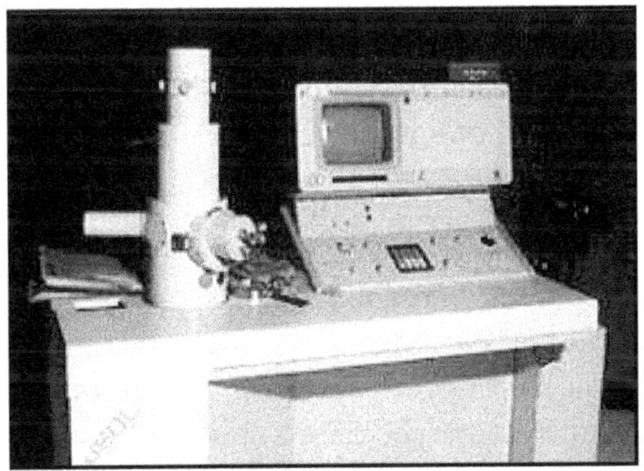

Figure 3.34 Scanning electron microscope

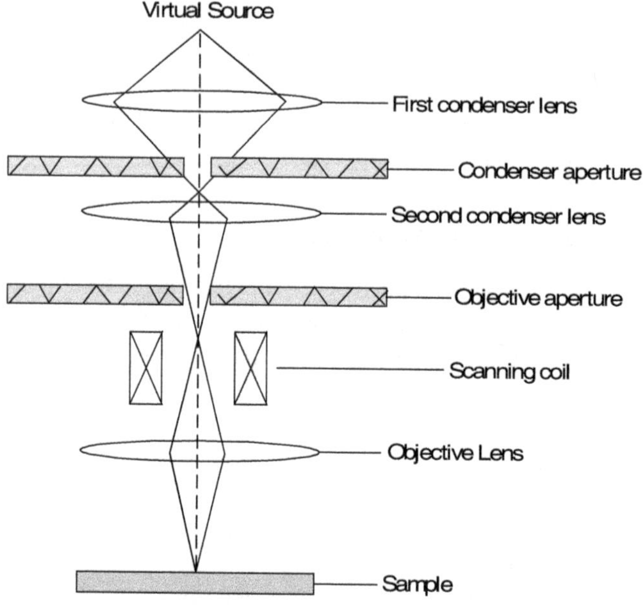

Figure 3.35 Optical system in scanning electron microscope

desired. When the beam strikes the sample (and dwells for a few microseconds) interactions occur inside the sample and are detected with various instruments. Before the beam moves to its next dwell point, these instruments count the number of interactions and display a pixel. A pixel is one of the many tiny dots that make up the representation of a picture in a computer's memory. This process is repeated until the grid scan is finished and then repeated. The entire pattern can be scanned 30 times per second.

SCANNING PROBE MICROSCOPY

The name **scanning probe microscopy** (SPM) encompasses several microscopic techniques like scanning tunnelling microscope (STM), atomic force microscope (AFM) and electrochemical scanning tunnelling microscope (EC-STM). These techniques are based on a common working feature: A mechanical probe sensor is scanned across an interface. During the scan, the probe sensor samples a specific signal, which is interpreted in terms of structure, electronic or force interaction information from the interface.

The first scanning probe microscope was invented in 1982 by Binnig and Rohrer with the scanning tunnelling microscopy (STM) technique for which they received the Nobel Prize in 1986 sharing it with E. Ruska for his achievements in electron optics and the invention of the electron microscope. The STM spreads fast into scientific labs over the world, because of its simple design, low costs, easiness to handle and the possibility to provide atomically resolved images of the surface of every electronically conducting sample. The application of the STM is restricted to conducting or semi-conducting samples. In 1986, Binnig *et al.* extended the field of application to non-conducting, e.g. biological, samples by introducing the atomic force microscope (AFM). The STM and AFM may be applied to samples in very different environments. These microscopes work under vacuum conditions as well as in air and, partly with specific modifications, in liquids. STM is used in electrochemical surface science where the STM is run within an electrochemical cell filled with electrolyte and it probes the interface between the working electrode and the electrolyte. This, type of STM is generally called as electrochemical scanning tunneling microscope (EC-STM).

1. SCANNING TUNNELLING MICROSCOPE (STM)

The STM allows scientists to visualize regions of high electron density and hence infer the position of individual atoms. The STM has higher resolution than the atomic force microscope (AFM). Both the STM and the AFM fall under the class of scanning probe microscopy instruments. It is used to obtain images of conductive surfaces at an atomic scale 2×10^{-10} m or 0.2 nm. It can also be used to alter the observed material by manipulating individual atoms, triggering chemical reactions, and creating ions by removing individual electrons from atoms and then reverting them to atoms by replacing the electrons.

The STM is a non-optical microscope which employs principles of quantum mechanics. An atomically sharp probe (the tip) is moved over the surface of the material under study, and a voltage, is applied between probe and the surface. Depending on the voltage, electrons will tunnel (this is a quantum-mechanical effect) or jump from the tip to the surface (or vice-versa depending on the polarity), resulting in a weak electric current. The size of this current is exponentially dependent on the distance between probe and the surface. Obviously, for a current to occur, the substrate being scanned must be conductive. Insulators cannot be scanned through the STM.

A servo loop keeps the tunnelling current constant by adjusting the distance between the tip and the surface (constant current mode). This adjustment is done by placing a voltage on the electrodes of a piezoelectric element. By scanning the tip over the surface and measuring the height (which is directly related to the voltage applied to the piezo element), one can thus reconstruct the surface structure of the material under study (Figure 3.36). High-quality STMs can reach sufficient resolution to show single atom. The STM will get the surface structure of a material within 2 nm range.

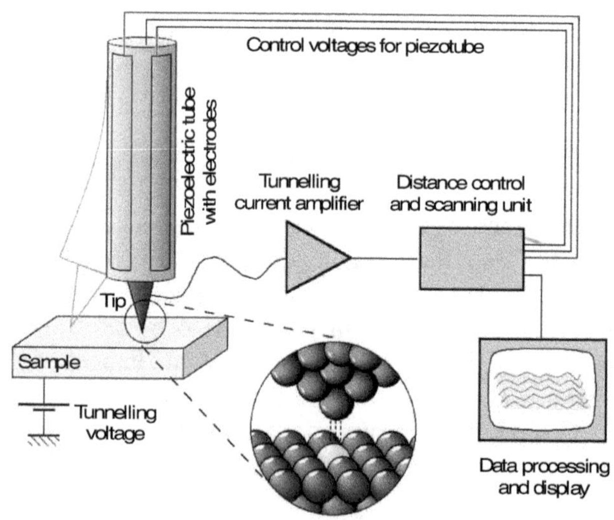

Figure 3.36 Schematic diagram of scanning tunnelling microscope

Use of the STM

The scanning tunnelling microscope is one of the most important tools for surface physics and surface chemistry, where it shows the structure of the topmost layer of atoms or molecules, e.g. defects and surface domain formation, morphology of thin films grown by various deposition techniques or modifications of surfaces by chemical processes.

For high resolution of metals and semiconductors, the STM is usually operated in ultra-high vacuum to avoid contamination or oxidation of the surface. Samples that are less sensitive to the atmosphere, such as graphite, gold, etc. can be imaged with high resolution under air.

Furthermore, the STM can be used to study charge transport mechanisms in molecules or other extremely small structures such as carbon nanotubes.

STM is also a tool for modification of surfaces through various methods such as indenting the tip or modification of the substrate by the electrons emitted from the tip. At low temperatures it is even possible to move single atoms with high accuracy by carefully "pushing" or "dragging" them with the tip of an STM. Since STM can be used as both a tool and an observation instrument on the nanometer scale, it has been vital for the emergence of the nanosciences.

2. ATOMIC FORCE MICROSCOPE

In the **atomic force microscope (AFM)** the tip as used in scanning tunnelling microscopy (STM) is replaced by a flexible **cantilever**. The cantilever is equipped with a sharp tip at one of its ends and the cantilever is scanned across the interface. Due to the topography of the probed sample and/or due to attractive or repulsive forces between the sample surface and the tip, the cantilever is bent up and down during scanning. The bending motion of the cantilever is detected by a laser beam reflected from the cantilever to a position-sensitive light detector (Figure 3.37). The AFM is working in three

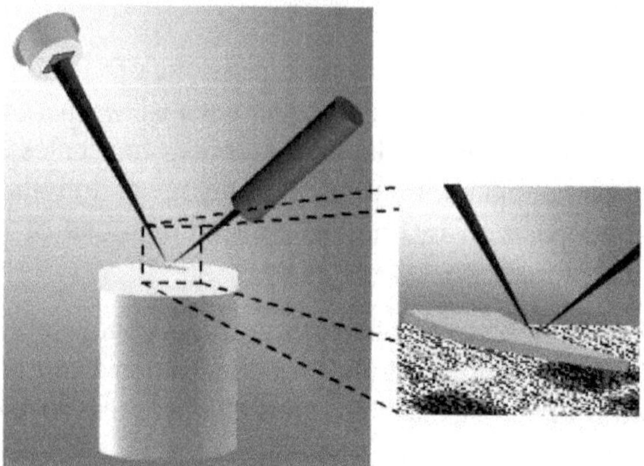

Figure 3.37 Cantilever in atomic force microscope

different modes: In its repulsive "contact" mode, the instrument lightly touches the sample with the tip at the end of the cantilever and the detected laser deflection measures hardsphere repulsion forces between the tip and the surface. In "non-contact" mode, the tip does not touch the sample and the AFM derives topographic images from measurements of attractive forces. In the "tapping" mode, the cantilever oscillates near its resonant frequency using a piezoelectric crystal. As the oscillating cantilever begins to intermittently contact the surface, the cantilever oscillation is necessarily reduced due to energy loss caused by the tip–surface interactions. The reduction in oscillation amplitude is used to identify and measure surface features.

3. ELECTROCHEMICAL SCANNING TUNNELLING MICROSCOPE

The general working principle of an **electrochemical scanning tunnelling microscope (EC-STM)** (Figure 3.38) is similar to that of a scanning tunnelling microscope, with the difference that the EC-STM is designed to run in an electrochemical cell. Figure 3.39 shows a model drawing of an STM-tip dipped into an electrolyte-filled electrochemical cell during image recording. The probed sample is the working electrode (WE). The reference electrode (RE) and the counter electrode (CE) ensure the control over

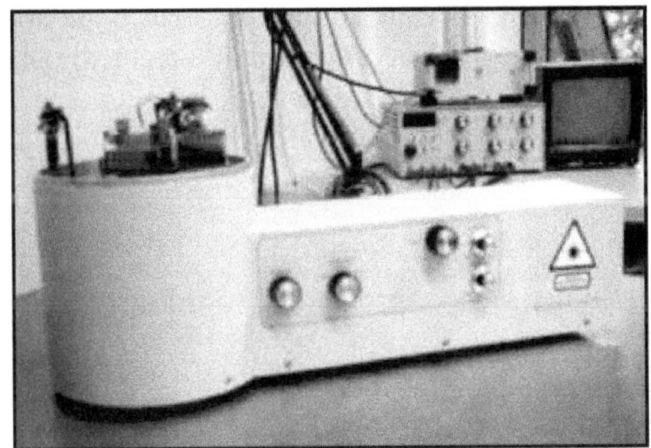

Figure 3.38 Electrochemical scanning tunnelling microscope

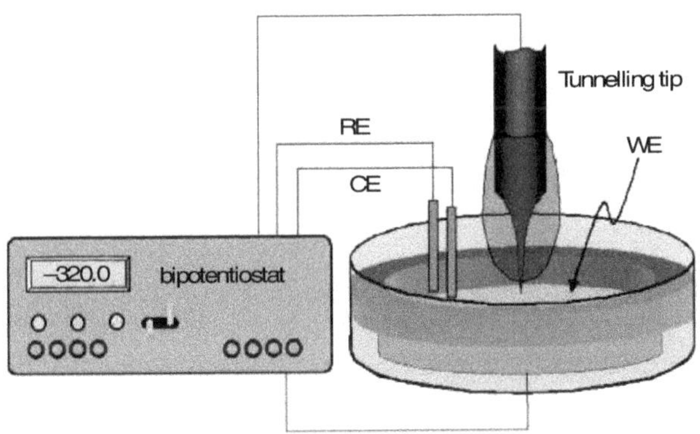

Figure 3.39 A model drawing of an STM tip dipped into an electrolyte-filled electrochemical cell during image recording

electrochemical processes within the cell and at the working electrode surface. For the tunnelling process a bias is applied between the WE and the tip. Hence, the tip itself serves as a further electrode within the cell. Typical electrochemical currents are of the order of μA, much too large to detect an underlying tunnelling current down to nA or even pA. Therefore, any influence of electrochemical currents at the tip must be avoided. This is provided by a special tip coating which leaves merely the foremost part of the tip in contact with the liquid. The 4-electrode arrangement in the electrochemical cell is controlled by a bipotentiostat. EC-STM is used in the field of electrochemistry.

REVIEW YOUR LEARNING

1. Define/explain the following:
   - i. Numerical aperture
   - ii. Resolving power
   - iii. Stokes' effect
   - iv. Two-photon microscopy
   - v. Electron gun
   - vi. Metal shadowing
   - vii. Chromatic aberration
2. Explain the principle, instrumentation and applications of the following:
   - i. Light microscope
   - ii. Phase contrast microscope
   - iii. DIC microscope
   - iv. Video microscope
   - v. Fluorescence microscope
   - vi. Polarizing microscope
   - vii. Dark-field microscope
   - viii. Confocal microscope
   - ix. TEM
   - x. SEM
   - xi. Scanning tunnelling microscope
   - xii. Scanning probe microscopy (SPM)
   - xiii. Atomic force microscope (AFM)
3. Write a brief note on the following:
   - i. Photomicrography
   - ii. Image analysis
   - iii. Condenser system
   - iv. Diaphragm
   - v. Types of image aberrations

BALANCE

A balance is an important instrument in the laboratory, which is used to measure the weight of a substance. Different types of balances with different sensitivities are used to weigh substances. The physical balance is less sensitive than the analytical balance.

ANALYTICAL BALANCE

Analytical balances are more accurate, with a sensitivity of 0.1 mg or lower. Most analytical balances have a maximum weight limit; beyond the tolerance point the balance should not be used for weighing substances. The balance is enclosed in a glass case to avoid air draft. The balance should be placed on a firm table, preferably made of concrete, to minimize disturbance during weighing. There are two basic types of analytical balances, the double-pan type and the single-pan type.

DOUBLE-PAN ANALYTICAL BALANCE

In double-pan analytical balance, two pans are suspended from a cross-beam; material to be weighed is put on the left pan and counterweights are put on the right pan. The parts of the analytical balance are shown in the Figure 4.1.

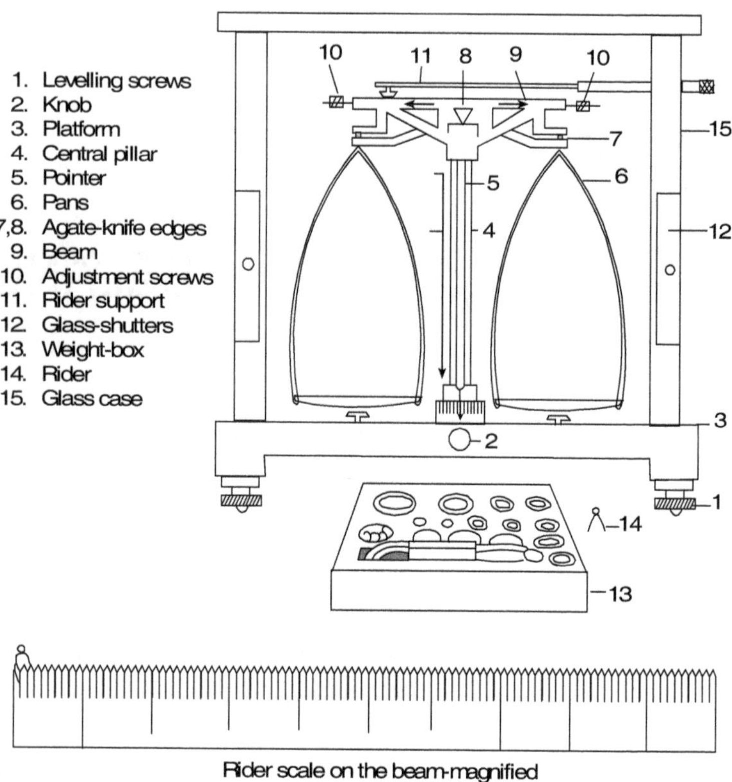

Figure 4.1 Double-pan analytical balance

With each double-pan analytical balance, there is a set of weights. The weights are well protected from dust and moisture. The weights should be handled carefully with the help of forceps and are placed on the right pan. The weight box is provided with a loop of wire of the shape λ. This is meant for use on the rider scale. Counterweights of less than 100 mg are manipulated by the rider which is placed on the cross beam. The use of rider makes possible to weigh objects correct to fourth decimal place of gram. On the rider scale each mg is divided into 10, each portion corresponds to 0.1 mg. For example, to weigh a substance of 5.2121 g, the weights used are 5 g, 200 mg, 10 mg and the last two digits 2.1 mg, so the rider is placed on the first tooth after digit '2' on the rider scale.

Thus, in weighing 5.2121 g the following weights are used:

Gram weight	5 g
mg weight	200 mg
	10 mg

Rider scale: 1st division after '2 mg'.

Analytical balances are expensive and must be handled carefully. During the procedure of weighing the following precautions are taken:

1. Select a balance that suits the requirement.
2. Never put the substance directly on the pan. Use a watch glass or weighing paper; beakers and other containers are also used, provided they are not too heavy.
3. Keep the balance in an area which is least disturbed.
4. Weigh all substances at room temperature.
5. Load and unload the balance only when the pan is arrested.
6. If the standard weights are to be placed manually, always use forceps to pick up the weights.
7. Always balance the empty pans before using the balance. In the double-pan balance, a screw is attached to each end of the cross-beam, which is screwed out (increases weight on that side). Initial adjustment of the unloaded balance to a reading of zero is necessary.

SINGLE-PAN ANALYTICAL BALANCE

In recent years, single-pan balances have been introduced which are replacing the older double-pan balances. These have internal counterweights, which are added or removed by turning a knob on the outside of the balance case. More improved balances have digital and electronic operation of the weighing process (Figure 4.2 a, b).

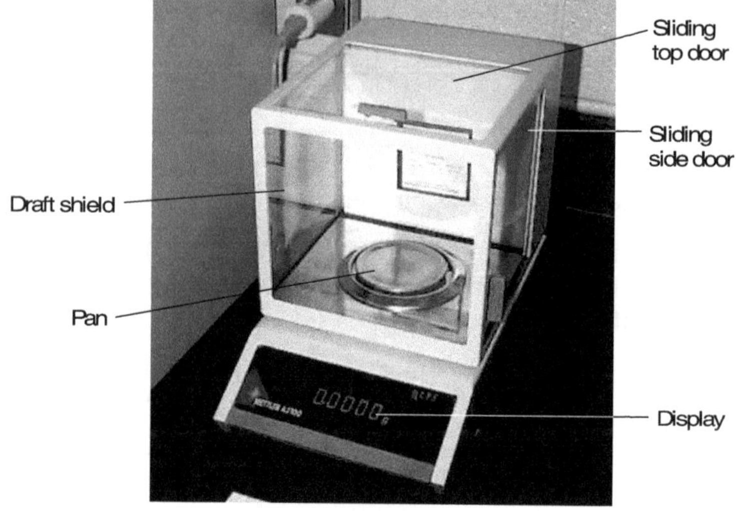

Figure 4.2 (a) Mono-pan electronic digital balance

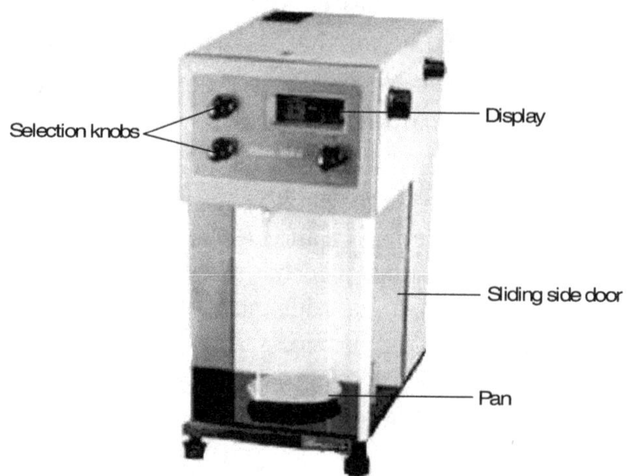

Figure 4.2 (b) A high-accuracy mono-pan electronic digital balance

The digital mass balances are very sensitive instruments used for weighing substances to the milligram (0.001 g) level. The direct reading balances employ the principle of the constant load balance. At the one end of the beam, pan and ring weights are there for weighing the objects and in the other end there is a balance weight (load). When a sample is placed it breaks the balance (equilibrium). Therefore the corresponding amount of weights must be removed or added to restore the original balance equilibrium. The operating knob is located in front on the base of the balance and this aims at not only reducing vibration and increasing the efficiency of the balance but also enables the users to work on the balance for longer hours with more comfort.

How to Use the Balance

The balance weighing pan is surrounded by a glass draft shield. The doors on the right, left, and top of the shield slide open to allow access to the weighing pan. When weighing objects, the draft shield doors are closed to prevent changes in air pressure (and measured mass). The front of the balance has a bar and a display window. To turn most balances on, the bar is pressed once and after a few seconds for the digital display reads "0.000 g". However, some balances require us to press "On/Off" and then press the small circle to the right of the display for the digital display to read 0.000 g.

To weigh an object, slide open one of the draft doors and place the object in the centre of the pan. Always use tongs, clamps, or a tissue to handle solid objects or liquid containers. Do not use your fingers, as oil and water from them initially adds a few milligrams to the mass that then partially evaporates, resulting in an error. Close the door and wait a few seconds for the digital display to read a constant mass. Lighter objects (< 75 g or 200 g, depending on the scale) are weighed to the milligram level (0.001 g). Heavier objects (> 75 g or 200 g, depending on the scale) are weighed to the centigram level (0.01 g). All numbers displayed should be recorded as they indicate the sensitivity of the scale (they are all significant figures). A balance in the "centigram mode" will not automatically return to the milligram mode. To reset the balance, press the bar until the display reads "0.000 g". The balance will now measure in the milligram range until a heavy object is again placed on the pan.

Always use a container or weighing paper when weighing a chemical; do not place any chemical directly onto the balance pan.

Many substances are corrosive and will ruin the sensitive pan and balance mechanisms in just a few minutes. Waxed weighing paper, plastic weighing boats, small beakers, watch glasses, small vials, etc. are all convenient containers for weighing chemicals. Additional care must be taken when weighing liquids. If possible, flasks containing liquids must be sealed with stoppers to prevent spilling or evaporation during weighing.

Methods for Weighing

The two common methods used to weigh a chemical are "weighing by difference" and "taring the balance".

Weighing by difference The mass of the chemical is calculated by subtracting the weight of an empty container from the total weight of the container and chemical. Place an empty container on the pan, close the draft shield doors and wait a few seconds for the display to read a constant mass. Record the mass of the empty container to three decimal places; do not round off. Remove the container from the pan, keep the chemical into the container, and record the mass of both container and chemical. The mass of the chemical is the difference of the two recorded masses. Remember to handle the container with tongs or tissue; moisture from fingers can cause an error in the apparent mass.

Taring the balance The balance is set to ignore the mass of the container so the mass of the added chemical is measured directly. Place the empty container on the pan and close the draft shield. Wait a few seconds for the display to register a constant mass. Press the bar so the display reads "0.000 g". The balance is now set to "ignore" the mass of the container (a process called "taring" the balance). Now if a chemical is added to the container, the balance displays only the mass of that chemical. When the container and chemical is removed from the pan, a negative weight will be displayed. (This negative weight is the mass of the original empty container, which the balance was instructed to ignore.) To erase this weight from memory, press the bar again. The display should read "0.000g".

Problems

If the displayed mass does not remain constant, make sure all doors of the draft shield are closed. If the mass still does not stabilize, the reason may be one of the following:

- A hot object causes convection currents inside the draft shield resulting in mass fluctuations. Objects must be cooled to room temperature before weighing.
- A solid may be adding or losing weakly bound waters of hydration or may not be completely dry.
- A liquid may be evaporating; cap the container to prevent loss.

Maintenance

Use containers when weighing chemicals and always weigh objects at room temperature. Keep the draft shields closed. Do not disturb the instrument or change the levels. Use the sable brush attached to each balance to clean the pan and surrounding area after weighing chemicals. Use paper towels to clean any liquid spills.

Avoid contamination when weighing chemicals. Always grasp a chemical container with the label toward the palm of the hand to prevent chemical drips (from the lip of the container) obscuring the label and safety information. Pour out a small amount of the chemical into a clean beaker first and use a clean spatula to transfer the chemical from beaker to weighing container. If any chemical in the beaker is not used, it should be placed in the collection receptacle for solids in the hood. (The chemical will be used when purity is not a necessity.) Never pour a chemical back into a reagent bottle. This practice ensures the purity and integrity of the chemical in the original container.

PHYSICAL BALANCE

The physical balance is used for relatively crude weight measurements with accuracy up to 10 mg to 100 mg. They are faster and easier to weigh on and are cheaper than analytical balances.

TRIPLE-BEAM SINGLE-PAN PHYSICAL BALANCE

The triple-beam single-pan balance (Figure 4.3) is more common in clinical laboratories. It has the advantage of not requiring a large set of weights like the older type of double-pan balance.

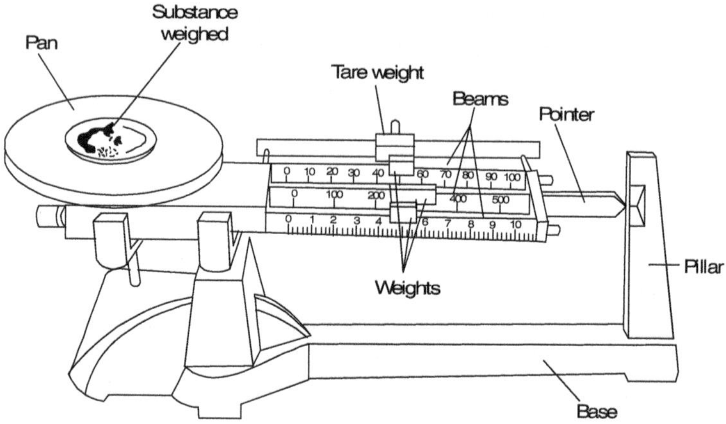

Figure 4.3 Triple-beam single-pan balance

The base of the balance holds the pan, the beam and the pillar. The pillar is located on the side of the balance which has a '0' mark for taring and balancing. To weigh a substance, place the container with the substance to be weighed on the pan. The arms or beams in front hold weights of different ranges. Place the sliding weight on the notches on the beam which avoids the use of several standard weights. The pivot is the hinged part of the balance, which counterbalances the weight of the substance on the pan with the weights on the beams. The pointer swings on the pillar of the balance and when rested to '0', it indicates that the weight is balanced. The poising nut adjusts the balance before weighing the substance; at the end of the adjustment, all the slide weights are at '0' and the pointer is at '0'. A tare beam is supplied to some of the improved triple-beam balances; this is used to "zero" the container to hold the substance during weighing. The three beams of a triple-beam balance are illustrated in Figure 4.4.

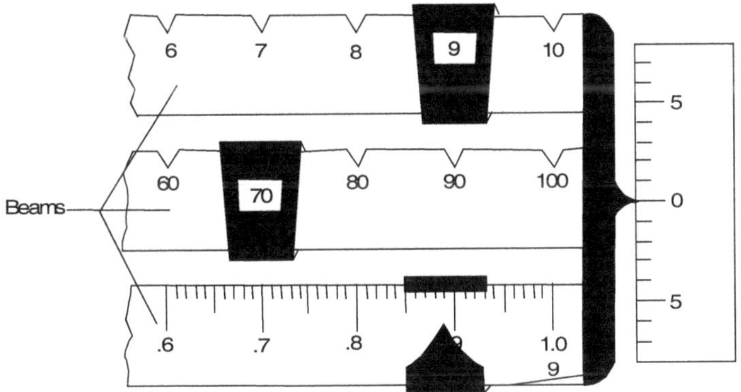

Figure 4.4 Three beams of a triple beam balance. The reading is 79.89

DOUBLE-PAN PHYSICAL BALANCE

The double-pan physical balance consists of two pans (Figure 4.5). The substance to be weighed is placed on a pan, which is counterbalanced by known weights on the other side. The following procedure is adopted to weigh the substances in the double-pan physical balance.

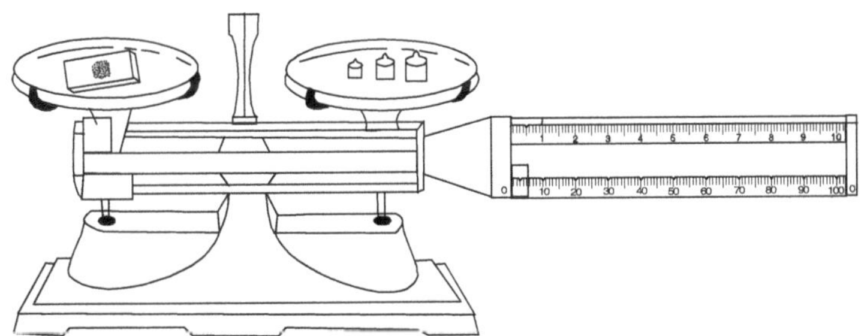

Figure 4.5 Double-pan physical balance

1. *Zero setting*

   a) Move the poising nut to the middle of its screw and push all the weights to the '0' position, the extreme left notch.

b) Check that the pointer is swinging freely. If it is touching the side of the pillar, move the pointer little forward. The pointer should move equally to both sides of the '0' mark in the centre. If not, move the poising nut for the 'zeroing' of the balance.

2. *Determining the container weight*

   a) Model with tare weight   Put the container on the pan and move the tare weight until the pointer swings equally to both sides of '0'. The balance is again poised and the weight of the container is nullified.

   b) Model without tare weight   Find out the weight of the container by moving the weights on the arms at different ranges until the pointer shows equal swing on both sides of '0'. For example, if after the weight setting for the container, the middle beam weight (range 0 to 500 g) is at '0', tare beam weight (range 0 to 100 g) at 10, and the front beam weight (range 0 to 10 g) at 5.1, the weight of the container is 15.1 g.

3. *Calculation of final weight*

   Make note of the weight of the container (15.1 g, in the above example) and add this weight to the required weight of the substance (e.g. 384.2), which comes to 384.2 + 15.1 = 399.3 g.

4. *Setting the weight*

   a) Set the middle beam weight (range 0 to 500 g) to the 300 g position.

   b) Then set the tare beam weight (range 0 to 100 g) to 90 g position.

   c) Finally, set the front beam weight (range 0 to 10 g) to 9.3 g position.

Note   Always check the scale of the graduation on each beam. If there are 10 divisions between 0 to 1 g on the front beam (0 to 10 g range), each division is equivalent to 0.1 g to 100 mg. If there are 5 divisions, it will be equivalent to 0.2 g or 200 mg.

This type of physical balance is commonly used in the clinical laboratories.

REVIEW YOUR LEARNING

1. Explain the components, operation procedure and use of
   i. Analytical balance
   ii. Physical balance
2. What does "taring the balance" mean?
3. What is the meaning of "weighing by difference"?
4. What should you check if the display does not give a constant mass value?

CENTRIFUGE

One of the most common equipments used to separate materials into subfractions in a biochemistry lab is the centrifuge. A centrifuge is a device that spins liquid samples at high speeds and thus creates a strong centripetal force causing the denser materials to travel towards the bottom of the centrifuge tube more rapidly than they would under the force of normal gravity. In other words, centrifuge is a device for separating particles from a solution based on their size, shape, density, viscosity of the medium and rotor speed. In biology, particles usually refer to the cells, subcellular organelles, viruses, biomolecules such as proteins and nucleic acids, etc.

BASIC PRINCIPLES OF SEDIMENTATION

The rate of sedimentation is dependent upon the applied centrifugal force, directed radially outwards.

Relative Centrifugal Force (RCF)

When an object moves in a circle at a steady angular velocity, it experiences a force, F, directed outwards. F is determined by the square of the angular velocity of the rotor (ω, in radians per second) and the radial distance of the particle from the axis of rotation (r, in cm).

$$\text{Centrifugal force} = (\text{angular velocity})^2 \times \text{radius}$$

$$F = \omega^2 r \qquad (1)$$

The common way of expressing rotor speed is in terms of revolutions per minute (rev min^{-1} or rpm). Since one revolution of the rotor is equal to 2π radians, its angular velocity can be expressed as,

$$\omega = \frac{2\pi \text{ rev min}^{-1}}{60} \tag{2}$$

Substituting the value of ω in equation 1,

$$F = \frac{4\pi^2 (\text{rev min}^{-1})^2 r}{3600}$$

However, F is expressed as a multiple of the earth's gravitational field (g = 981 cm s^{-2}).

The ratio of the weight of the particle in the centrifugal field to the weight of the same particle when acted by gravity alone is known as the relative centrifugal field (RCF) which is commonly referred to as the 'number times g'. Hence,

$$\text{RCF} = \frac{4\pi^2 (\text{rev min}^{-1})^2 r}{3600 \times 981}$$

$$\text{RCF} = (1.118 \times 10^{-5})\,(\text{rev min}^{-1})^2 r$$

From the above relationship, it is clear that RCF depends upon the rpm and the radius of rotation, r. If r is a constant for a given rotor, then variations in rpm alone determine the variations in RCF.

Sedimentation Rate

In the centrifugation process, the particles will sediment progressively with time towards the bottom of the sample tube. To simplify mathematical terminology we assume all biological materials to be spherical particles. The sedimentation rate of a given particle depends upon a number of factors such as density of the particle (ρ_p), radius of the particle (r_p), the density of the medium (ρ_m) and the viscosity (η) of the suspending medium. A mathematical expression relating to all these factors for sedimentations of a rigid spherical particle is given below:

$$v = \frac{2}{9} \times \frac{r_p^2\,(\rho_p - \rho_m)}{\eta} \times g$$

where,

v = rate of sedimentation,
g = gravitational field and
$\frac{2}{9}$ = shape factor constant for a spherical particle.

From the above equation, it is clear that the sedimentation rate of a given particle is proportional to its size, the difference in density between the particle and the medium and to the applied centrifugal field. Sedimentation rate becomes zero when the density of the particle and the medium are equal. It increases when the force field increases and decreases when the viscosity of the medium increases. However, since the equation involves the square of the particle radius, it is apparent that the size of the particle has the greatest influence upon its sedimentation rate. Some other factors that affect the sedimentation rate are i) characteristics of the centrifuge and ii) concentration of the suspension.

Svedberg Unit or Sedimentation Coefficient

The sedimentation rate or velocity (v) of a particle can be expressed in terms of its sedimentation rate per unit of centrifugal field. It is commonly referred to as its sedimentation coefficient, S. Since sedimentation rate studies are performed using a wide variety of solvent–solute systems or at different temperatures, the value is affected by temperature, solution viscosity and density. Therefore, these values are corrected to the sedimentation constant theoretically obtainable in water at 20°C. The equation for standard coefficient is

$$S_{20w} = \frac{S_{obs}(1 - \bar{v}\,\rho_{20w})}{(1 - \bar{v}\,\rho_T)} \times \frac{\eta_T}{\eta_{20}} \times \frac{\eta}{\eta_{20}}$$

where,

S_{20w} = the standard sedimentation coefficient,
S_{obs} = the experimentally measured sedimentation coefficient,

η_T/η_{20} = the relative viscosity of water at temperature T compared with that at 20°C,
η/η_{20} = the relative viscosity of the solvent to that of water,
ρ_{20w} = the density of water at 20°C,
ρ_T = the density of the solvent at temperature T(°C) and
$\bar{v}$ = the partial specific volume of the solute.

The basic unit of sedimentation coefficient is 1×10^{-13} sec. This is also termed as one Svedberg unit (S), in recognition of T. Svedberg's pioneering work. Generally, the larger the molecule or particle, the larger is its Svedberg unit and hence the faster is its sedimentation rate. Svedberg units for viruses are 40 to 1000S, for lysosomes 40000S and for mitochondria 20×10^3S to 70×10^3S.

TYPES OF CENTRIFUGES

Centrifuges may be classified into four major groups—small bench centrifuges, large capacity refrigerated centrifuges, high-speed centrifuges, and ultracentrifuges.

Small Bench Centrifuges

Hand centrifuge Hand centrifuge is manually operated consisting of two centrifuge tube holders (Figure 5.1). Hand centrifuges are used to sediment the larger particles for simple experiments.

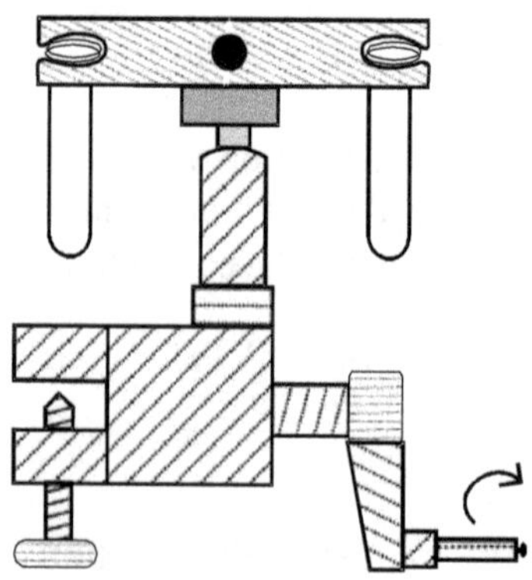

Figure 5.1 Hand centrifuge

Clinical centrifuges These are very simple and small and hence can be placed atop a desk (Figure 5.2). They are normally used to isolate red blood cells, yeast cells or bulky precipitates of chemical reactions. Their maximum speed is usually 3000 rev min^{-1} with maximum relative centrifugal force of 7000 g. They do not usually have any temperature regulatory system. They are useful for the separation of large volumes of crude preparation. The rotors are mounted vertically on a rigid shaft. Therefore, centrifuge tubes must be placed diametrically opposite to each other after balancing their weights accurately.

Figure 5.2 Clinical centrifuge

Large Capacity Refrigerated Centrifuges

These have a maximum speed of 6000 rev min^{-1} and produce a maximum relative centrifugal field approaching 6500 g. They have refrigerated rotor chambers and vary only in their maximum carrying capacity, all being capable of utilizing a variety of interchangeable swinging-bucket and fixed-angle rotors. Large total capacity centrifuges are also available which in addition to accommodating smaller tubes, are also capable of holding bottles (Figure 5.3). These instruments are most often used to collect substances that sediment rapidly, for example erythrocytes, coarse or bulky precipitates, yeast cells, nuclei and chloroplasts.

Figure 5.3 Large capacity refrigerated centrifuge

High-speed Refrigerated Centrifuges

These centrifuges are available with maximum rotor speeds in the region of 25000 rev min^{-1}, generating a relatively centrifugal field of about 60000 *g*. They generally have a range of interchangeable fixed-angle and swinging-bucket rotors. They are equipped with refrigeration equipment to remove the heat generated due to friction between the air and the spinning rotor. The temperature can easily be maintained in the range 0°C to 40°C (Figure 5.4). These instruments are most often used to collect microorganisms, cellular debris, larger cellular organelles and proteins precipitated by ammonium sulphate. They cannot generate sufficient centrifugal force to effectively sediment viruses or smaller organelles such as ribosomes.

Figure 5.4 High speed refrigerated centrifuge

Continuous flow centrifuges The continuous flow centrifuge is a relatively simple high-speed centrifuge. The rotor is long and tubular, through which particles suspended in medium flow continuously. As the medium enters the rotating rotors, particles are deposited against its wall and excess medium overflows through an outlet port. The major application of this type of centrifuge is in the harvesting of bacteria or yeast cells from large volumes of culture medium.

Ultracentrifuges

Ultracentrifuges are of two types—preparative and analytical ultracentrifuge.

Preparative ultracentrifuge Preparative ultracentrifuges are capable of spinning rotors to a maximum speed of 80,000 rev min^{-1} and can produce a relative centrifugal field of up to 60,000 *g*. The rotor chamber is refrigerated, sealed and evacuated to minimize any excessive rotor temperature being generated by frictional resistance between the air and the spinning rotor. The temperature monitoring system is more sophisticated than in simpler instruments, employing an infrared temperature sensor that can continuously monitor rotor temperature and control the refrigeration system. An over-speed control system is also incorporated into these instruments to prevent operation of the rotor above its maximum rated speed and there are electronic circuits to detect rotor imbalance. In order to minimize vibration caused by slight motor imbalance that may arise due to unequal loading of the centrifuge tubes, ultracentrifuges are fitted with a flexible drive shaft system. For safety reasons, rotor chambers of both high-speed and ultracentrifuges are always enclosed in heavy armour plating.

Airfuge An air-driven, tabletop preparative ultracentrifuge, called an airfuge, is available. This is capable of acceleration of a magnetically suspended 3.7-cm diameter rotor, accommodating 6×175 mm^3 tubes on a friction-free cushion of air in a non-evacuated chamber. The rotor speed is 100,000 rev min^{-1} (160,000 *g*). The airfuge has found applications in biochemical and clinical research where there is only small volume of samples requiring high centrifugal forces. Examples include macromolecules/ ligand binding-kinetic studies, steroid hormone receptor assays, separation of the major lipoprotein fraction from plasma, and deproteinization of physiological fluids for amino acid analysis.

Analytical ultracentrifuge Analytical ultracentrifuges are capable of operating at forces as great as 600,000 ± 100 *g* and with temperature control within approximately 0.1°C. Analytical ultracentrifuge basically consists of a motor, a centrifuge rotor which is present in a protective and armoured chamber and an optical system for recording the distribution of the sample in the ultracentrifuge cell. The rotor is kept in an evacuated and cooled chamber and is suspended on a wire coming from the drive shaft of the motor. The tip of the rotor contains a thermistor for measuring temperature. The thermistor makes electrical contact with the control circuit by means of a pool of mercury, which the rotor tip touches. The rotor chamber contains an upper condensing lens and a lower collimating lens. The lower lens allows the passage of the light so that the sample is illuminated by parallel light. The upper lens and the camera lens focus the light on the film (Figure 5.5). Several types of rotors are available. A rotor contains two cells, namely the analytical cell and the counterpoise cell.

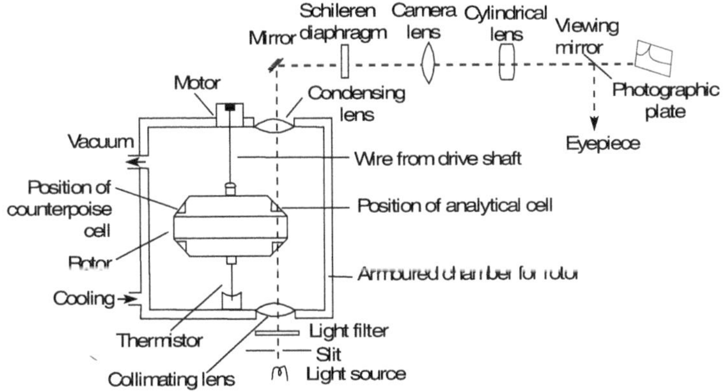

Figure 5.5 Diagram of analytical ultracentrifuge

Three types of optical systems are available. They are

1. Ultraviolet light absorption system
2. Schlieren optical system
3. Rayleigh interference system

In the ultraviolet light absorption system, light of a suitable wavelength is passed through the moving analytical cell containing the solution under analysis. The intensity of the transmitted light is recorded on a photographic plate.

In Schlieren optical system, when light passes through a solution having different density zones it is refracted at the boundary between these zones. The Schlieren optical system plots the refractive index gradient against the distance along the analytical cell, which is useful for sedimentation velocity measurements. However, it is not sufficiently sensitive to detect small concentration differences.

Rayleigh interference system employs a double-sector cell. One sector contains the solvent and the other the solution. The optical system measures the difference in refractive index between the reference solvent and the solution by the displacement of interference fringes caused by splits placed behind the two liquid columns. Each fringe traces a curve of the refractive index gradient against the distance in the cell. Since the position of the fringes is determined by solute concentration, it is possible to measure the concentration of the solute at any point along the cell.

Applications The analytical centrifuge is used for the following:

1. To determine relative molecular mass of macromolecules such as proteins and DNA.
2. To investigate the purity of DNA preparations, viruses and proteins.
3. To detect conformational changes in macromolecules such as DNA and proteins.

TYPES OF ROTORS

Rotors used for low-speed centrifugation are made up of brass or steel because they experience a much lower degree of stress. Rotors made of alloys of aluminium or titanium are used in high-speed centrifugation. The different types of rotors used for centrifugation process are given in the following sections.

Vertical Tube Rotors

The vertical tube rotor is a fixed zero-angle rotor. In this, tubes are aligned vertically in the body of the rotor at all times. The design and operation of the vertical tube rotor is shown in Figure 5.6. In this type of rotor, the pellet is deposited along the entire length of the outer wall of the centrifuge tube. The major disadvantage in this rotor is that the pellet tends to fall back into the solution at the end of centrifugation.

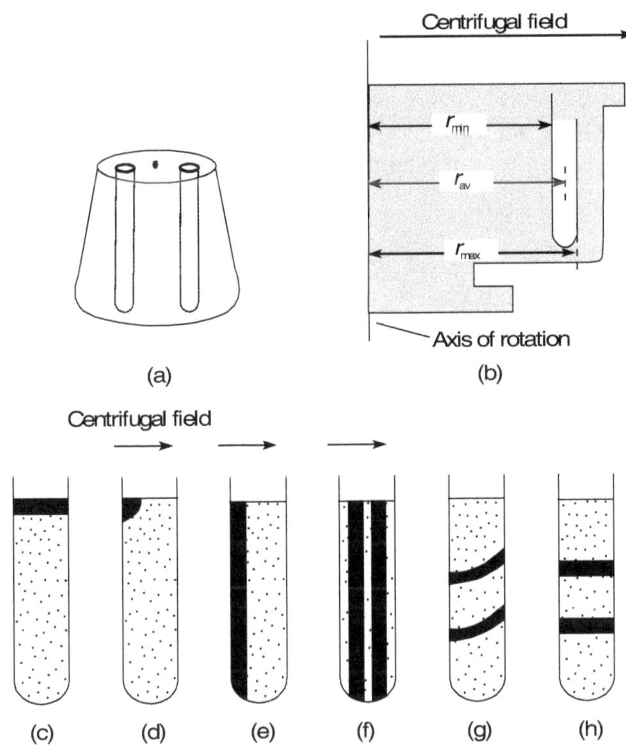

Figure 5.6 Design and operation of the vertical tube rotor. (a) Vertical rotor (b) Cross-sectional diagram of a vertical tube rotor. (c) The centrifuge tube is filled with gradient; the sample is layered on top and is then placed in the rotor. (d) As the rotor accelerates, the sample and gradient begin to reorient. (e) The sample and medium reorientation is complete. (f) Sedimentation and separation of particles occur during centrifugation. (g) Reorientation of separated particles and gradient occur during the rotor deceleration. (h) Rotor is at rest: bands of separated particles and gradient are fully reoriented.

Fixed-Angle Rotors

In fixed-angle rotors, the tubes are located in holes in the rotor body set at a fixed angle between 14° and 40° to the vertical. Under the influence of the centrifugal field, particles move radially outwards and have only a short distance to travel before colliding with the outer wall of the centrifuge tube. The design and operation of fixed-angle rotor is shown in Figure 5.7. They are used for the differential separation of particles whose sedimentation rates differ by a significant order of magnitude.

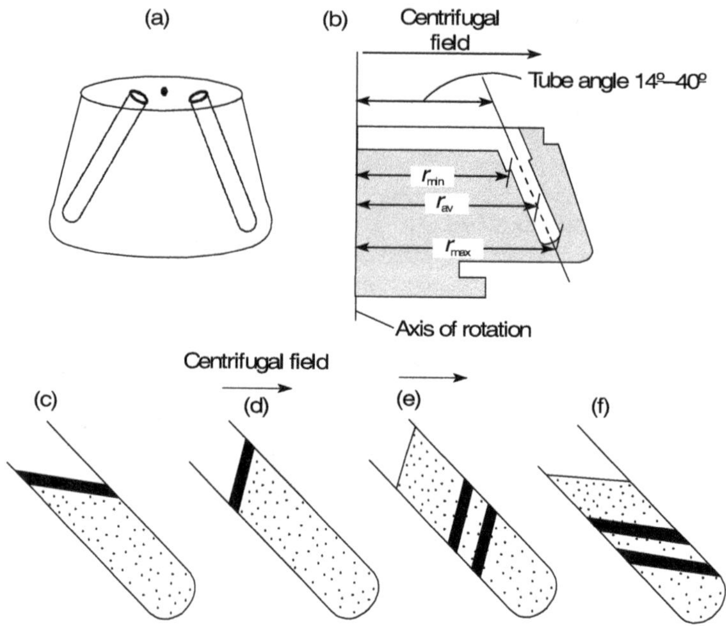

Figure 5.7 Design and operation of the fixed-angle rotor. (a) Fixed-angle rotor. (b) Cross-sectional diagram of a fixed-angle rotor. (c) The centrifuge tube, after being filled with gradient, is loaded with sample and then placed in the rotor. (d) During rotor acceleration, reorientation of the sample and gradient occur. (e) Sedimentation and separation of the particles occur during centrifugation. (f) Rotor is at rest; the gradient reorients and bands of separated particles appear.

Swinging-Bucket Rotors

This type of rotor has buckets. During acceleration of the rotor, they swing out from the vertical position to a horizontal position. They are then aligned perpendicular to the axis of rotation and parallel to the applied centrifugal field. The design and operation of swinging-bucket rotor is shown in Figure 5.8. An undesirable swirling effect that causes mixing of the tube contents are also produced during rotor acceleration and deceleration.

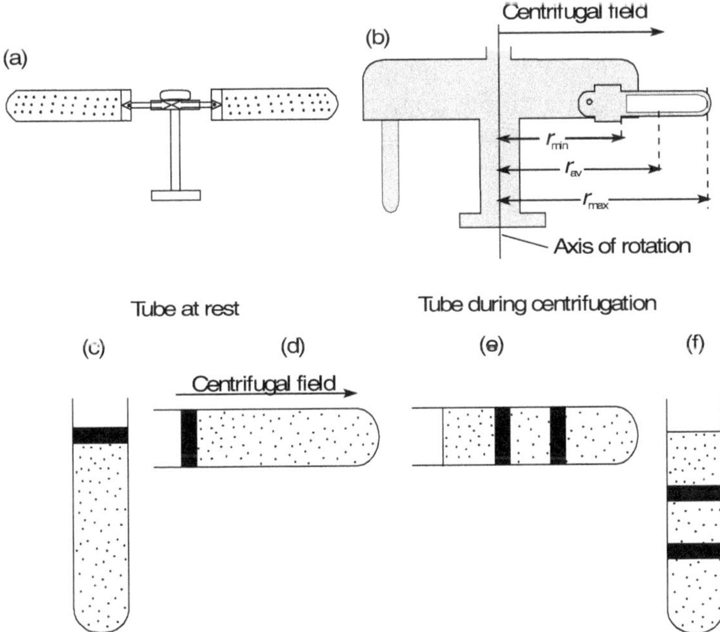

Figure 5.8 Design and operation of the swinging-bucket rotor. (a) Swinging-bucket rotor. (b) Cross-sectional diagram of a swinging-bucket rotor. (c) The centrifuge tube is initially loaded with gradient, the sample is then layered on top before the tube is placed in the bucket for attachment to the rotor. (d) During acceleration of the rotor, the rotor bucket reorients to lie perpendicular to the axis of rotation. (e) Sedimentation and separation of the particles occur during centrifugation. (f) At the end of centrifugation the rotor decelerates, the bucket coming to rest in its original vertical position.

Zonal Rotors

There are two types of zonal rotors, namely, the batch type and the continuous flow type. The batch type zonal rotor is extensively used. It is designed to minimize the wall effects that are encountered in swinging-bucket and fixed-angle rotors and to increase sample size. Low-speed batch rotors are designed to operate at near 5,000 rpm (5,000 *g*). The high-speed batch rotors are made of aluminium or titanium alloy and can operate at speeds up to 60,000 rpm (256,000 *g*) As shown in Figure 5.9, the density gradient is formed while the rotor is spinning; then the sample is layered and centrifuged until the isopycnic zonal layering of the particles is reached. At this moment, an injection of a denser sucrose solution pushes the layers toward the centre where they are collected in tubes of a fraction collector. Batch-type zonal rotors are used to remove contaminating proteins from a variety of preparations and for the separation and isolation of hormones, enzymes, ribosomal subunits, viruses and subcellular organelles from animal or plant tissue homogenate.

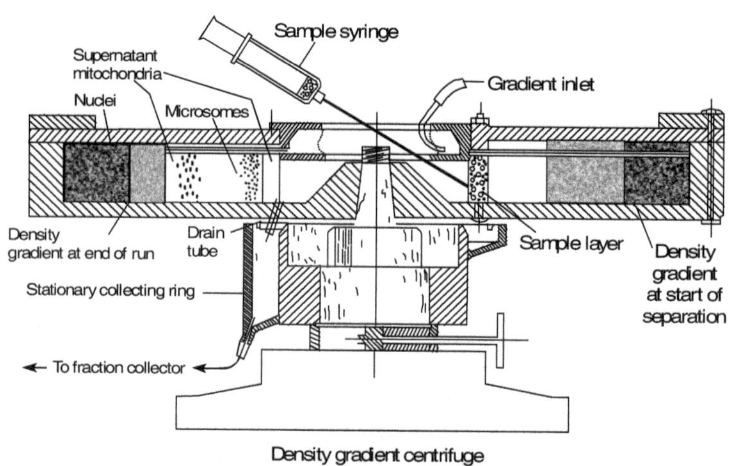

Figure 5.9 Schematic section through a zonal rotor

Continuous flow zonal rotors are designed for high-speed separation or relatively small quantities of solid matter from large volumes of suspension. The rotors are similar in shape to batch type zonal rotors but differ in their design because of the continuous fluid flow in the rotor. They are useful for the harvesting of cells and isolating viruses in large scale.

Elutriator Rotors

The elutriator rotor is a type of continuous flow rotor that contains recesses to hold a single conical-shaped separation chamber, the apex of which points away from the axis of rotation, and a by-pass chamber on the opposite side of the rotor that serves as a counterbalance and to provide the fluid outlet. Particles suspended in a uniform low-density medium are pumped into the rotor chamber and the rotor is spinned at a preselected speed (usually between 1000 and 3000 rpm). Since the separation chamber is conical in shape, larger particles accumulate towards the centrifugal end of the chamber where the liquid flow velocity is high, while the smaller particles accumulate towards the centripetal end of the chamber where the liquid flow velocity is low (Figure 5.10). Either by a stepwise decrease in rotor speed or by astepwise increase in liquid flow rate through the separation

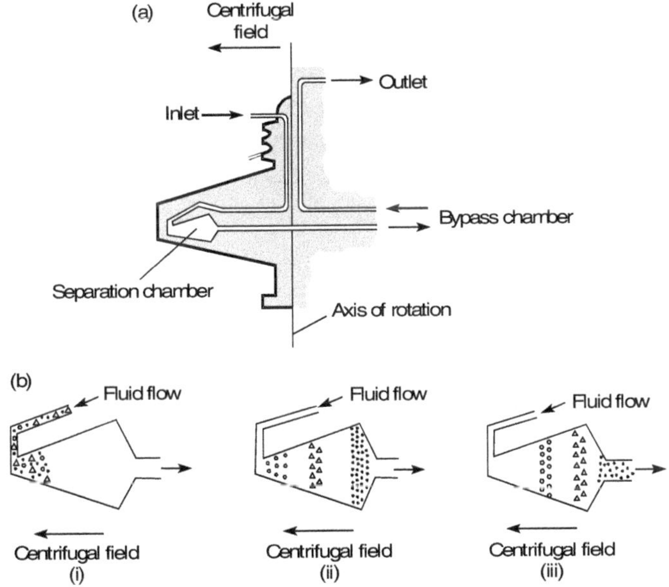

Figure 5.10 (a) Cross-section through an elutriator rotor and (b) the separation of particles in the separation chamber of an elutriator rotor by centrifugal elutriation

chamber, collection of the separated uniformly sized particles can be made centripetally in order of successively increasing diameter by elutriation from the chamber.

With the technique of centrifugal elutriation, the elutriator rotor has been used successfully to separate various cell types from mammalian testis and different types of monocytes and lymphocytes from human blood, to purify Kupffer and endothelial cells from sinusoidal liver cells and fat-storing cells from rat liver, and for the bulk separation of rat brain cells and the fractionation of yeast cell populations.

TYPES OF CENTRIFUGATION

Centrifugation is classified based on the (i) purpose of centrifugation (ii) speed at which centrifuge is operated and (iii) method of application of the samples. Based on the purpose, centrifugation is classified into two most common types—analytical and preparative centrifugation. Analytical centrifugation involves measuring the physical properties of the sedimenting particles such as sedimentation coefficient or molecular weight. The other form of centrifugation is called preparative and the objective is to isolate specific particles, which can be reused.

Analytical Centrifugation

Optical methods are used in analytical ultracentrifugation. Molecules are observed by optical system during centrifugation, to allow observation of macromolecules in solution as they move in the gravitational field. The samples are centrifuged in cells (tubes with quartz windows) having windows that lie parallel to the plane of rotation of the rotor head. As the rotor turns, the images of the cell (proteins) are projected by an optical system onto a film or a computer. The concentration of the solution at various points in the cell is determined by absorption of a light of appropriate wavelength (Beer's law is followed). This can be accomplished either by measuring the degree of blackening of a photographic film or by the pen deflection of the recorder of the scanning system and fed into a computer.

The analytical ultra centrifuge has found many applications in biology, especially in the field of protein chemistry and nucleic acid chemistry. This technique is used to determine the sedimentation coefficient, relative molecular mass and also to test the purity of macromolecules.

The relative molecular mass of macromolecules is determined either by sedimentation velocity method or sedimentation equilibrium method.

Sedimentation velocity method In sedimentation velocity method the sedimentation coefficient of the molecule is initially determined either by boundary sedimentation or zonal sedimentation. The equation used for calculating molecular weight by sedimentation equilibrium method is

$$M = \frac{RTS}{D(1-\bar{v}\rho)}$$

where,

M = relative molecular weight of the molecule,
D = diffusion coefficient of the molecule,
$\bar{v}$ = partial specific volume of the molecule,
ρ = density of the solvent at 20°C,
R = molar gas constant,
T = absolute temperature and
S = sedimentation coefficient.

The measured values of S and D are corrected to standard conditions of zero concentration of solute in water at 20°C. However the determination is complicated by difficulties encountered in the accurate determination of the diffusion coefficient of the particles and in correction in differences in viscosity and temperature. The determination of the relative molecular weight of a macromolecule using sedimentation velocity analysis is therefore less accurate and more time-consuming than determination by sedimentation equilibrium method.

Sedimentation equilibrium method Sedimentation equilibrium method is more versatile and accurate. This method can be used to determine relative molecular mass values ranging from a few hundred to several million. This versatility is due to the large range of centrifugal fields available for ultracentrifugation. In this method, the ultracentrifuge is operated until a balance is established between sedimentation and diffusion of material in the opposite direction. Molecular weight can be calculated using the equation

$$M = \frac{2RT \ln(C_2 / C_1)}{\omega^2 (1 - \bar{v}\rho)(x_2^2 - x_1^2)}$$

where,

C_1 and C_2 are the concentrations of solute at distances x_1 and x_2 respectively from the centre of rotation,
R = molar gas constant,
T = absolute temperature,
ω^2 = square of the angular velocity of the rotor,
$\bar{v}$ = partial specific volume of the molecule and
ρ = density of the solvent.

This technique does not require diffusion coefficient and hence the method is more convenient and widely used for the determination of molecular weight of proteins.

Preparative Centrifugation

There are many types of preparative centrifugation such as differential and density gradient centrifugation.

Differential centrifugation (cell fractionation) The process of separation of cell organelles is known as subcellular fractionation. To isolate a specific organelle, initially, the organs (liver, brain, or kidney) are homogenized in a suitable homogenizing medium at 4°C. The resulting suspension, containing many intact organelles, is known as a homogenate.

Fractionation of the contents of a homogenate is done by a classical biochemical technique called differential centrifugation. This

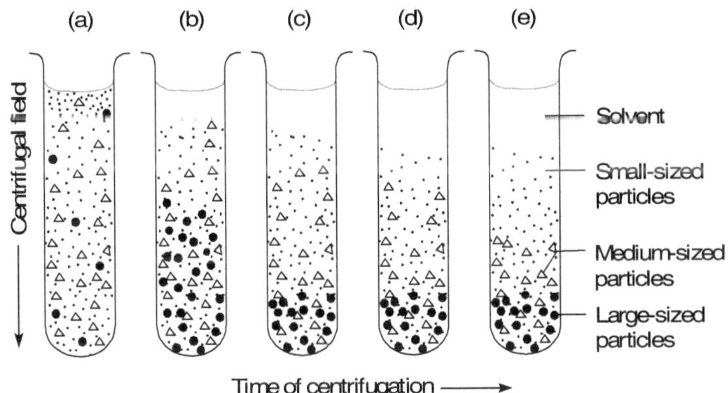

Figure 5.11 Differential sedimentation of a particulate suspension in a centrifugal field. (a) Particles are uniformly distributed throughout the centrifuge tube. (b) to (e) Sedimentation of particles during centrifugation is dependent upon their size, shape and density.

method is based upon the differences in the sedimentation rate of particles of different sizes and density (Figure 5.11). This method uses a series of four different centrifugation steps at successively greater speeds. Each step yields a pellet and a supernatant. The supernatant from each step is subjected to centrifugation in the next step. This procedure provides four pellets, namely, nuclear, mitochondrial, lysosomal and microsomal fractions (Figure 5.12). At the end of each step, the pellet is washed several times by resuspending in the homogenization medium followed by recentrifugation under the same conditions. This procedure minimizes contamination of other subcellular organelles and gives a fairly pure preparation of pellet fraction.

The purity of organelles obtained by differential centrifugation is measured by estimating some marker activity. A marker is one that is almost exclusively present in one particular organelle. A marker may be an enzyme molecule or a biochemical compound. Table 5.1 gives a list of various fractions, their functions and markers.

Table 5.1 Cellular fractions and their functions

Organelle	Function	Marker
Plasma membrane	Regulates entry and exit of compounds	5´ Nucleotidase
Nucleus	Site of DNA-directed RNA synthesis	DNA
Mitochondrion	Citric acid cycle, ammonia release for urea formation	Glutamate dehydrogenase
Lysosome	Site of many hydrolases	Acid phosphatase
Endoplasmic reticulum	Oxidation of many xenobiotics	Glucose 6-phosphatase
Cytosol	Enzymes of glycolysis, fatty acid synthesis	Lactate dehydrogenase

The microsomal fraction contains mostly a mixture of smooth endoplasmic reticulum and free ribosomes. The contents of the final supernatant correspond approximately to those of cytosol.

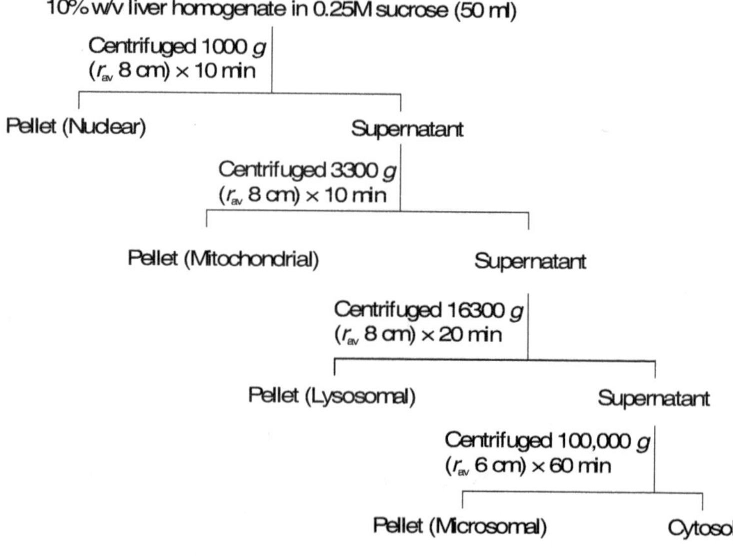

Figure 5.12 Cellular fractionation

Density gradient centrifugation Differential centrifugation uses a homogeneous medium for separation whereas density gradient centrifugation uses a medium that has gradients. The commonly used gradient materials and their applications are given in Table 5.2.

Table 5.2 Commonly used gradient materials and their applications

Gradient materials	Applications
Caesium chloride	Banding of DNA, nucleoproteins, viruses, plasmid isolation
Caesium sulphate	Banding of DNA and RNA, purification of proteoglycans
Sodium bromide	Fractionation of lipoproteins
Sodium iodide	Banding of DNA and RNA
Glycerol	Banding of membrane fragments, protein separation
Sucrose	Separation of subcellular particles, proteins, viruses and membranes
Ficoll	Separation of whole cells, subcellular particles and viruses
Dextran	Separation of whole cells, banding of microsomes
Bovine serum albumin	Separation of whole cells

In density gradient centrifugation, separation depends upon the buoyant densities of the particle. This method gives a much better separation than differential centrifugation. There are two types of density gradient centrifugation, namely, the rate-zonal technique and the isopycnic technique.

i. *Rate-zonal centrifugation technique* Rate-zonal centrifugation is a type of density gradient centrifugation in which particle separation is based upon

1. Differences in the size, shape and density of the particle,
2. The density and viscosity of the medium and
3. The applied centrifugal field.

The gradient used in rate-zonal centrifugation technique has the maximum density at the bottom but its density is lesser than the most dense sedimenting particle to be separated. The density gradient is shallow. It is produced by layering different samples of very narrow densities in decreasing order. Thus, a gradient is produced whose density continuously increases from the top towards the bottom of the sample tube. Centrifugation is then performed at a comparatively low speed for a short time. The sample particles travel through the steep gradient and form discrete zones depending upon their sedimenting rate (Figure 5.13). The centrifugation must be terminated before any of the zones reaches the bottom of the centrifuge tube. This method is extremely useful for the separation of proteins possessing nearly identical densities but differing slightly in their size. It is used for the separation of RNA–DNA hybrids, ribosomal subunits and subcellular organelles.

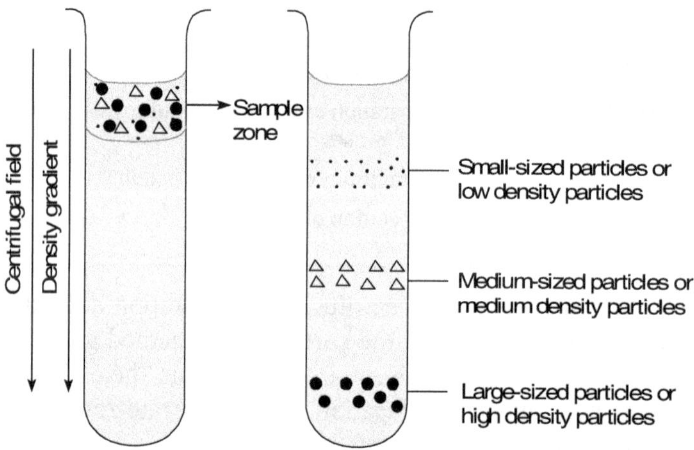

Figure 5.13 Rate zonal centrifugation

ii. *Isopycnic centrifugation* Isopycnic centrifugation depends only upon the buoyant density of the particle. It does not depend upon the shape or size of the particle and is independent of time. In isopycnic centrifugation, the maximum density of the gradient always exceeds the density of the densest particle. A continuous density gradient is always used. During centrifugation, sedimentation of the particle occurs until the buoyant density of the particle and the density of the gradient are equal. At this point of isodensity, no further sedimentation occurs, irrespective of how long centrifugation continues. This technique is used to separate particles of similar size but of differing density (Figure 5.14). Subcellular organelles such as golgi apparatus, mitochondria and peroxisomes can be effectively separated by this method.

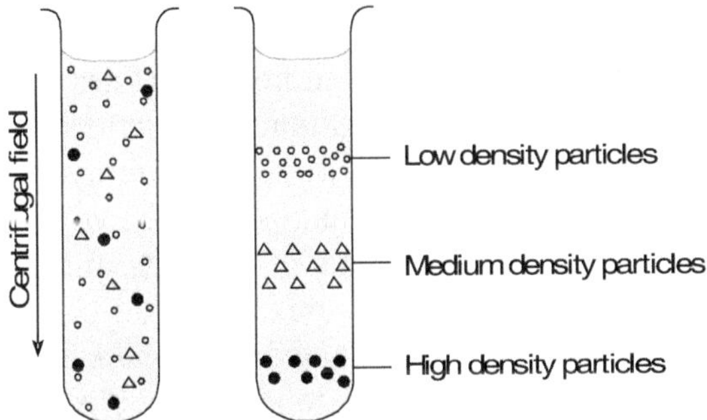

Figure 5.14 Isopycnic centrifugation using the equilibrium isodensity method. (a) Particles distributed homogeneously throughout the tube prior to centrifugation. (b) During centrifugation the gradient is allowed to establish itself, sample particles redistribute and band in a series of zones at their respective isopycnic positions.

Centrifugal elutriation In this technique the separation and purification of a large variety of cells from different tissues and species can be achieved by a gentle washing action, using an elutriator rotor. The technique is based upon differences in the equilibrium set-up in the separation chamber of the rotor, between the opposing centripetal liquid flow and applied centrifugal field being used to separate particles mainly on the basis of differences in their size. The technique does not employ a density gradient and has the advantage that any medium totally compatible with the particles can be used, for example buffered salt solutions or culture medium; because pelleting of the particles does not occur, fractionation of delicate cells or particles, between 5 and 50 μ diameter, can be achieved with minimum damage so that cells retain their viability. Separations can be achieved very quickly, giving high cell concentrations and a very good recovery yield.

A second system of classification of centrifugation is the rate or speed at which the centrifuge is operated. Ultracentrifugation is carried out at a speed faster than 20,000 rpm. Superspeed ultracentrifugation is at speeds between 10,000 and 20,000 rpm. Low-speed centrifugation is at a speed below 10,000 rpm.

A third method of defining centrifugation is by the way the samples are applied to the centrifuge tube. In moving boundary (or differential centrifugation), the entire tube is filled with sample and centrifuged. Through centrifugation, one obtains a separation of two particles but any particle in the mixture may end up in the supernatant or in the pellet or it may be distributed in both fractions, depending upon its size, shape, density, and conditions of centrifugation. The pellet is a mixture of all of the sedimented components, and it is contaminated with whatever unsedimented particles are in the bottom of the tube initially. The only component, which is purified, is the slowest sedimenting one, but its yield is often very low. The two fractions are recovered by decanting the supernatant solution from the pellet. The supernatant can be recentrifuged at higher speed to obtain further purification, with the formation of a new pellet and supernatant.

In rate-zonal centrifugation, the sample is applied in a thin zone at the top of the centrifuge tube on a density gradient (Figure 5.13). Under centrifugal force, the particles will begin sedimenting through the gradient in separate zones according to their size shape and density. The run must be terminated before any of the separated particles reach the bottom of the tube.

In isopycnic technique, the density gradient column encompasses the whole range of densities of the sample particles. The sample is uniformly mixed with the gradient material (Figure 5.14). Each particle will sediment only to the position in the centrifuge tube at which the gradient density is equal to its own density, and will remain there. The isopycnic technique, therefore, separates particles into zones solely on the basis of their density differences, independent of time.

SAFETY ASPECTS IN THE USE OF CENTRIFUGES

There are a number of safety precautions that must be adhered to when using any centrifuge and rotor.

1. Before running a centrifuge, check the classification details on the centrifuge to ensure that the rotor is safe to use in the centrifuge at hand.
2. Never use an alkali detergent on a rotor (most are highly alkaline—be sure to check before use).
3. Always clean and completely dry the rotor after every use. Any spilled materials, especially salts and corrosive solvents must be removed immediately with running water. Fixed angle rotors are stored upside down, to drain after thorough cleaning and rinsing. Swinging buckets have only the buckets cleaned and dried, and stored inverted and with the caps removed. Never immerse the rotor portion of a swinging bucket rotor. Inevitably the linkage pins will rust, as it is virtually impossible to remove all fluids from them.
4. Be especially careful not to scratch the surface of a rotor or bucket. Use plastic brushes only. Normal wire brushes will scratch the anodized surface of aluminum rotors, which will increase the likelihood of corrosion. The anodized layer is extremely thin and is the main defence against corrosion of an aluminum rotor.
5. Always use a proper centrifuge tube. Glass tubes are used in clinical centrifuges only. High Speed Corex tubes can be used up to 15,000 rpm (in SS34 rotor). All ultracentrifugation use employs nitrocellulose or polyallomer tubes. Nitrocellulose tubes will collapse in a strong centrifugal field if old.
6. Always fill the centrifuge tubes to the proper level. (Usually fill to within 1/2 inch of the top). The tubes are thin-walled and will collapse if improperly filled.
7. Always balance the rotor properly. Use a precision scale for most work. Always balance the tube with a medium that is identical to that being centrifuged, i.e., do not balance an alcohol solution with water, or a dense sucrose solution with

water. For swinging buckets, be sure the buckets are weighed with their caps in place, that the seals are intact and that the caps are secure. Be careful in the placement of tubes within a rotor to ensure proper balance. Check the manufacturer's guidelines for complex rotors that hold multiple tubes.

8. Ensure that the rotor is properly seated within the centrifuge. For swinging buckets, ensure that they are hanging properly. For preparative rotors, be sure the rotor cover is in place and properly screwed down, where appropriate. Never use a rotor without its lid.
9. Check that the centrifuge chamber is clean, defrosted and that all membranes or measuring devices are intact and functional (Beckman speed and temperature controls) and that the lid is securely closed.
10. Adjust acceleration rates, deceleration rates, temperature and rpm controls as appropriate. Set brake on or off as appropriate and check vacuum level where appropriate.
11. Start the centrifuge and set the timer. Do not attempt to open the centrifuge until the rotor has come to a complete stop.
12. Before opening the centrifuge, record the appropriate information in the centrifuge log.

If properly balanced and used, the rotor should accelerate smoothly and with a constant change in the pitch of the motor sound. In the case of any vibrations or unusual sounds the operator should immediately cease operation. Never leave the centrifuge until you are certain that it has reached its operating speed and is functioning properly. All rotors go through a minor vibration phase when they first start. There will be a minor flutter when the rotor reaches this vibration point; do not confuse this with a serious vibration caused by imbalance.

REVIEW YOUR LEARNING

1. Define relative centrifugal force.
2. Define Svedberg unit.

3. Discuss different types of centrifuges.
4. Describe different types of rotors used in centrifugation.
5. Describe differential centrifugation and its applications.
6. Explain the principle, methodology and applications of density gradient centrifugation.

pH AND pH METER

Sorenson in 1909 introduced the term pH to express the "power of hydrogen ion concentration". pH is defined as the negative logarithm of hydrogen ion concentration.

SORENSEN'S pH SCALE

Sorenson's pH scale covers a range of pH from 0–14. Thus, a solution of pH 7.0 is neutral (pure water). Sorenson's pH scale 0–14 is based on the ionic product of water.

ELECTROLYTIC DISSOCIATION OF WATER

Water can be considered to be a weak electrolyte.

It can dissociate as follows:

$$H_2O \rightleftharpoons H^+ + OH^-$$

The equilibrium constant for the dissociation of water is given by the equation

$$K_{eq} = \frac{(H^+)(OH^-)}{(H_2O)}$$

or

$$K_{eq}\,(H_2O) = (H^+)\,(OH^-)$$

At 25°C, K_{eq} is found to be 1.8×10^{-16}

In pure water, the molecular concentration of water (H_2O) is 1000/18, i.e., 55.5 moles/litre.

Therefore,

$$(H^+) \times (OH^-) = 55.5 \times 1.8 \times 10^{-16}$$

$$= 1.01 \times 10^{-14}$$

This product $(H^+) \times (OH^-)$ is called the ionic product of water, abbreviated as K_w.

Therefore, at 25°C,

$$K_w = 1.01 \times 10^{-14}$$

In pure water, equal number of hydrogen and hydroxyl ions exist, i.e., $(H^+) = (OH^-)$.

Therefore, $(H^+) \times (OH^-) = (H^+) \times (H^+) = (H^+)^2 = 1.01 \times 10^{-14}$

For ease of representation of (H^+) in numbers rather than as fractions, Sorenson developed the concept of pH. pH is defined as the logarithm of the reciprocal of hydrogen ion concentration.

$$pH = \log\frac{1}{(H^+)}$$

$$= -\log(H^+)$$

For pure water, since $(H^+) = 1.0 \times 10^{-7}$, pH will be 7.0.

ACIDS AND BASES

An acid is a proton donor and a base is a proton acceptor.

$$CH_3COOH \rightarrow CH_3COO^- + H^+$$

Acetic acid is hence a proton donor and an acid. Acetate (CH_3COO^-) is a proton acceptor and hence a base. The two together constitute a conjugate acid–base pair. In dilute aqueous solutions, an acid will dissociate to give a proton, which is taken up by a water molecule to form a hydronium ion, H_3O^+.

$$HA + H_2O \rightarrow H_3O^+ + A^-$$

Acids which have only a slight tendency to give up protons, are weak acids (e.g. acetic acid). Acids which give up their protons readily to water, are strong acids (e.g. HCl).

DISSOCIATION OF STRONG ELECTROLYTES

Strong acids, bases and their salts are called strong electrolytes. They dissociate almost completely in water.

$$HCl \rightarrow H^+ + Cl^-$$

$$NaOH \rightarrow Na^+ + OH^-$$

$$NaCl \rightarrow Na^+ + Cl^-$$

DISSOCIATION OF WEAK ELECTROLYTES

Weak acids, weak bases and their salts are called weak electrolytes. They dissociate only slightly in solution.

Taking acetic acid as an example

$$CH_3COOH \rightarrow CH_3COO^- + H^+$$

The equilibrium constant for this acid is 1.8×10^{-5}.

i.e.,
$$\frac{(H^+)(CH_3COO^-)}{(CH_3COOH)} = 1.8 \times 10^{-5}$$

The pH of a 1.0 molar solution of acetic acid can be calculated from the above equation to be 2.38.

HENDERSON–HASSELBALCH EQUATION

Henderson–Hasselbalch equation forms the basis for preparation of buffers. The quantitative relationship between pH, the buffering action of a mixture of weak acid with its conjugate base and the pK of the weak acid is given by Henderson–Hasselbalch equation.

The tendency of a weak acid (HA) to lose a proton (H^+) and form its conjugate base (A^-) is defined by the equilibrium constant K_a for the reversible reaction,

$$HA \rightleftharpoons H^+ + A^-$$

The equilibrium constant K_{ion} or $K_a = \frac{(H^+)(A^-)}{HA}$

$$K_a \times HA = (H^+)\,(A^-)$$

$$(H^+) = K_a \frac{HA}{A^-}$$

Taking logarithm on both sides,

$$\log(H^+) = \log K_a + \log \frac{HA}{A^-}$$

Multiplying both sides by (–1), we have

$$-\log(H^+) = -\log K_a - \log \frac{HA}{A^-}$$

But $-\log(H^+) = pH$; $-\log K_a$ is defined as pK_a and

$$-\log \frac{HA}{A^-} = \log \frac{(A^-)}{HA}$$

Substituting in the above equation,

$$pH = pK_a + \log \frac{(A^-)}{HA}$$

or

$$pH = pK_a \frac{(A^-)}{HA}$$

where,

pK_a is the dissociation constant of the weak acid HA and

(HA) and (A^-) are the molar concentrations of the weak acid and its conjugate base respectively.

BUFFERS

A buffer solution is one which resists a change in pH when an acid or base is added to it. Buffers are indispensable for performing many biological reactions *in vitro*. It is usually made up of a mixture of weak acid and its conjugate base (salt of the weak acid), e.g. acetic acid and sodium acetate. To such mixture, suppose, an alkali is added, it will react with the weak acid to form its salt.

$$NaOH + CH_3COOH \rightarrow CH_3COONa + H_2O$$

On the other hand, if an acid is added, this will be taken up by the base sodium acetate.

$$HCl + CH_3COONa \rightarrow CH_3COOH + NaCl$$

In either case, there is no increase in either H^+ or OH^-. The relative concentrations of the acetate and acetic acid are altered slightly.

Buffers play an important role *in vivo* also, because most of the biological reactions cannot operate efficiently and optimally at a wide range of pH. A constant pH is maintained in cells and body fluids by the buffering action of various biological substances such as phosphates, carbonates, bicarbonates, amino acids and proteins that are present in the cells and body fluids. Therefore, to maintain a constant pH for *in vitro* studies of biological reactions, a buffer is used. Buffers of desired pH and molarity are prepared based on an equation derived by Henderson and Hasselbalch.

Buffer Action and Buffer Capacity

The resistance of buffers to change in pH on addition of small amount of alkali or acid is known as **buffer action**. The magnitude of this action is called **buffer capacity** (β) which is measured by the amount of strong base or strong acid required to alter the pH by one unit

$$\beta = \frac{db}{d(pH)}$$

where,

db is the volume of strong base or acid and
d(pH) is the difference in pH.

The greatest buffering capacity is always at the pK_a values. Best buffering action could be achieved at pK_a values ± 1.

Properties of a Good Buffer System

Most of the biological reactions occur in the pH range 6 to 8. Certain factors have to be considered in selecting an appropriate buffer system for performing biological reactions. They are the following

1. The pK_a of the acid should be between 6 and 8, since the *in vivo* pH of biological system is 7.00 ± 0.4.
2. The buffer components should be highly soluble.
3. The buffer components should be easily available in their purest form.
4. The buffer components should be easily chemically stable at physiological pH.
5. As most of the enzyme reactions and biological components are estimated between 240 and 750 nm, the buffer components should show minimum or insignificant light absorption in the above region of the spectrum.
6. The buffer components should have minimal salt effects and should not interact with mineral cations that are often essential in most enzymatic reactions.
7. The buffer components should have minimal effects on dissociation due to concentration, temperature and ionic composition. (It is interesting to note that temperature markedly affects dissociation constants which shift with temperature).
8. The ionic components of the buffer should be excluded by biological membranes. Zwitterionic buffers like Morpholino Ethane Sulphonate (MES), N-(2-Hydroxyethyl piperazine-N´-2-Ethanesulphonate) (HEPES), Morpholino Propane Sulphonate (MOPS), etc. do not pass through the biological membrane system whereas acetate, carbonate and phosphate pass through the membranes. Thus, the zwitterionic buffers manage the pH of the suspending medium without interfering with internal pH of the organelles.

Phosphate Buffers

Phosphate buffers are widely used because of their very high buffering capacity. Sodium and potassium salts are highly soluble in water. Since these ions are strongly charged, high ionic strength is obtained without the need for excessive molarity. But it is not possible to prepare a phosphate buffer that has a low ionic strength and a high buffering capacity. By choosing appropriate mixtures of H_3PO_4/ $H_2PO_4^-$, $H_2PO_4^-/HPO_4^{2-}$ or HPO_4^{2-}/PO_4^{3-}, buffer solutions having a pH range from 2 to 12 can be prepared. Phosphate buffers have the following two major disadvantages.

i. They are somewhat toxic to mammalian cells *in vitro*.
ii. Ca^{2+} and Mg^{2+} ions are bound by phosphate ions.

Tris Buffers

A buffer that is widely used in the clinical laboratory and in biochemical studies is that prepared from tris (hydroxy methyl) aminomethane [$(OHCH_2)_3CNH_2$-Tris or THAM] and its conjugate acid [$(OHCH_2)_3CNH_3^+$]. It has a high solubility in physiological fluids. It is non-hygroscopic and does not absorb CO_2 appreciably. It does not appear to inhibit many enzyme systems. It has a pK close to physiological pH (pK_a for the conjugate acid = 8.08). These buffers are usually prepared by adding hydrochloric acid or glycine solution to a solution of Tris and by adjusting the pH to the desired value. Tris-Glycine buffer is commonly used in disc-gel electrophoresis. Tris also has the following drawbacks.

i. It reacts with some metal ions such as Cu^{2+}, Ni^{2+}, Ag^+ and Ca^{2+}.
ii. Its pH varies slightly with temperature.
iii. It reacts with certain electrodes.

pH METER

pH meter is a potentiometer which measures the voltage between two electrodes placed in a solution (Figure 6.1). pH meter is used for accurate pH measurements of solutions.

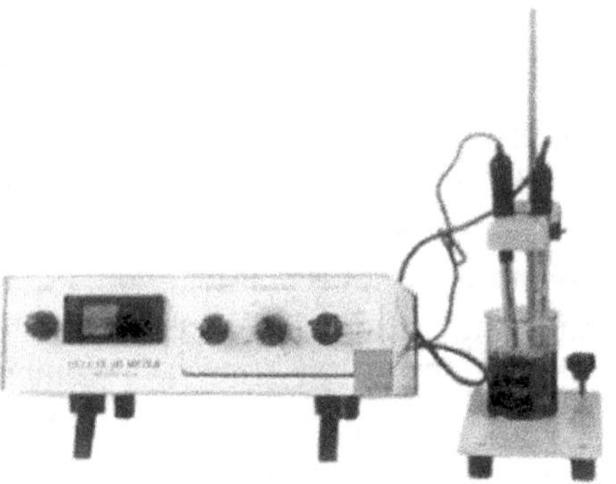

Figure 6.1 pH meter

Principle

An electrical potential is generated when a thin glass membrane separates two solutions of different H^+ ion concentrations. When a solution of unknown pH is used, the difference in potential between the glass electrode and the reference electrode is measured, amplified, and converted into a direct pH reading on the meter.

The two electrodes used in pH meter are a calomel electrode and a glass electrode. The calomel electrode is the external reference electrode whose electrical potential is always constant; whereas, the glass electrode is the standard test electrode whose electrical potential depends on the pH of the test solution.

The electromotive force (emf) of the complete cell (E) is given by

$$E = E_{ref} - E_{glass}$$

where,

E_{ref} is the potential of the reference electrode (calomel), which at normal temperature is + 0.250 V and

E_{glass} is the potential of the test electrode (glass) which depends on the pH of the test solution.

Then the pH of the solution can be determined by the following equation

$$pH = \frac{E_{glass} - E_{ref}}{0.0591} \text{ at } 25^{\circ}C$$

Substituting the value for E_{ref}

$$pH = \frac{E_{glass} - 0.250V}{0.0591} \text{ at } 25^{\circ}C$$

Calomel Electrode

The calomel electrode contains mercury, mercury chloride and saturated solution of potassium chloride (Figure 6.2). Each of these compounds exists in ionized state although the extent of ionization may vary widely. Their dissociation constants are:

a) $Hg \rightleftharpoons Hg^{+} + e^{-}$

$$K_a = \frac{[Hg^{+}][e^{-}]}{[Hg]}$$

b) $Hg_2Cl_2 \rightleftharpoons 2Hg^{+} + 2Cl^{-}$

$$K_b = \frac{[2Hg^{+}][2Cl^{-}]}{[Hg_2Cl_2]}$$

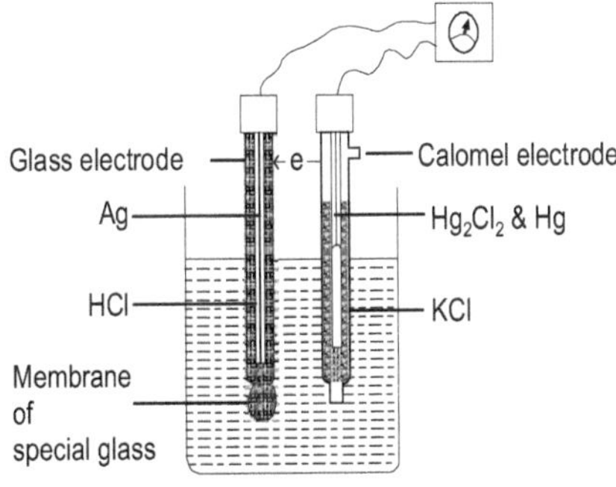

Figure 6.2 Glass and calomel electrode assembly for pH measurements

The calomel electrode is dipped in saturated solution of KCl. The electrical contact between the calomel electrode and the test solution is achieved by the KCl salt bridge through a fine capillary in the glass casing known as porous plug.

Glass Electrode

The glass electrode contains silver, silver chloride and 0.1M HCl solution (Figure 6.2). Their dissociation constants are as follows.

c) $Ag \rightleftharpoons Ag^{+} + e^{-}$

$$K_c = \frac{[Ag^{+}]\,[e^{-}]}{[Ag]}$$

d) $AgCl \rightleftharpoons Ag^{+} + Cl^{-}$

$$K_d = \frac{[Ag^{+}][Cl^{-}]}{[AgCl]}$$

This electrode is dipped in 0.1M HCl solution.

The calomel electrode is made of a thick glass that is impermeable to H^+ ions. Therefore, its potential is independent of pH. In contrast, in the glass electrode, the tip of the electrode is made of a special thin (0.05 to 0.1 mm) borosilicate glass bulb, which is permeable to H^+ ions only but not to other cations or anions.

All these equilibrium reactions from (a) to (d) are delicately balanced and when the electrodes are connected, the electrons will move from more positive electrode to the other.

If the electrodes are placed in a solution containing high concentrations of H^+ ions (low pH, acidity), the calomel electrode will not respond as it is not permeable to H^+. The H^+ ions pass through the glass membrane and neutralize the electrons of the electrodal reaction and hence electrons flow from calomel to glass. When the test solution has high concentrations of OH^- ions (high pH, alkalinity), the H^+ ions move out of the glass bulb rendering a momentary negative charge and hence the electrons flow from glass to calomel. Because of this passage of ions, an electrical potential develops across the glass electrode and calomel electrode, which results in the flow of current between the electrodes. The magnitude of this current depends on the concentration of H^+/OH^- ions in the test solution. In the pH meter, the current is fed into a calibrated dial in such a way that the dial reading directly gives the pH of the solution.

Combined Electrode

Combined electrode consist of a glass and a reference electrode in a single unit (Figure 6.3). The advantage of this type of electrode is that smaller volumes of solution can be measured. The disadvantage is its high cost and the fact that it must be discarded if one of the elements fails.

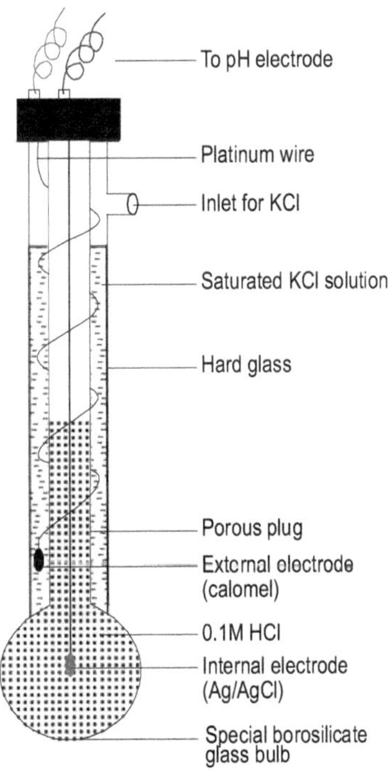

Figure 6.3 Combined pH electrode

Operation of a pH Meter

New or dry electrodes should be soaked in water or in a buffer of pH 6–7 overnight. The electrode can also be activated by soaking in 0.1M HCl for 12 to 24 hours. Electrodes should always be stored in distilled water or as per the manufacturer's recommendations. The pH meter is operated in the following way.

i. The 'temperature compensation knob' is fixed to adjust the temperature of the solution. This is essential since equilibrium constant of a reaction, i.e., in acid–base titration does vary with temperature.

ii. Remove the beaker containing H_2O, rinse the electrode with water and wipe gently with tissue paper. Dip the electrode gently into a standard buffer solution whose pH is accurately known. Make sure that the electrode does not touch the sides or bottom of the beaker. Adjust the pH, using the 'pH adjust knob'.

iii. Remove the standard buffer, rinse the electrodes with distilled water and wipe it gently with tissue paper.

iv. Dip the electrode into the solution whose pH should be found out. The dial directly reads the pH of the test solution.

v. After finding out the pH, wash and store the electrode in distilled water.

Factors that Affect pH Measurement

1. **The alkaline error** The ordinary glass electrode becomes somewhat sensitive to alkali metal ions and gives low readings at pH values greater than 9.
2. **The acid error** Values registered by the glass electrode tend to be somewhat high when the pH is less than 0.5.
3. **Dehydration** Dehydration of electrodes may cause erratic electrode performance.
4. **Error in the pH of the standard buffer** Inaccuracies in the preparation of the buffer used for calibration or any change in its composition during storage cause an error in subsequent pH measurements.

5. **Variation in junction potential** A fundamental source of uncertainty for which a correction cannot be applied is the junction potential variation resulting from differences in the composition of the standard and the unknown solution.
6. **Error in unbuffered neutral solution** Before determining the pH of any solution, the glass electrode should be thoroughly rinsed with water, since equilibrium between the bulk of the solution and the layer of solution at the surface of a membrane is achieved only slowly in poorly buffered neutral solutions. To avoid this error, both electrodes should be immersed in successive portions of the unknown solution for a long time until a constant pH reading is obtained.

Applications

The pH meter is used to measure pH of a given solution and to prepare buffer for biological research. Buffer solution resists a change in hydrogen ion concentration on addition of an acid or alkali. Buffer systems provide protection to cells against sudden changes in pH. They have a role in the maintenance of osmotic pressure between the cell and extracellular fluid. Blood contains a number of buffer systems. They maintain a constant blood pH of about 7.4. Enzymes exhibit a maximum catalytic action at some pH. Buffers are responsible for providing the desired pH. Phosphate buffers and Tris buffers are widely used in the clinical laboratory and in biochemical studies.

EXPERIMENTS USING A pH METER

Acid–Base Titration Curves and pK_a Values

I. Titration between a strong acid and a strong base

   1. Prepare 50 ml of 0.1M hydrochloric acid and take it in a 250 ml beaker.
   2. Prepare 100 ml of 0.1M sodium hydroxide.
   3. Switch on the pH meter, warm it for 10 minutes.

4. Calibrate the pH meter to the pH of the standard buffer.
5. Remove the beaker containing the standard buffer, rinse the electrodes with distilled water, and wipe it gently with a tissue paper.
6. Now introduce the beaker containing 50 ml of 0.1M HCl and note down the pH.
7. Add 2.0 ml of 0.1M NaOH, mix thoroughly with a glass rod or gently stir with a magnetic stirrer. Note down the pH.
8. Repeat the above steps (6–7) until the pH is ~10.
9. Plot the readings on a graph sheet (pH vs volume of 0.1M NaOH added) and connect all the points by a smooth line. This gives the titration curve for the acid (Figure 6.4).

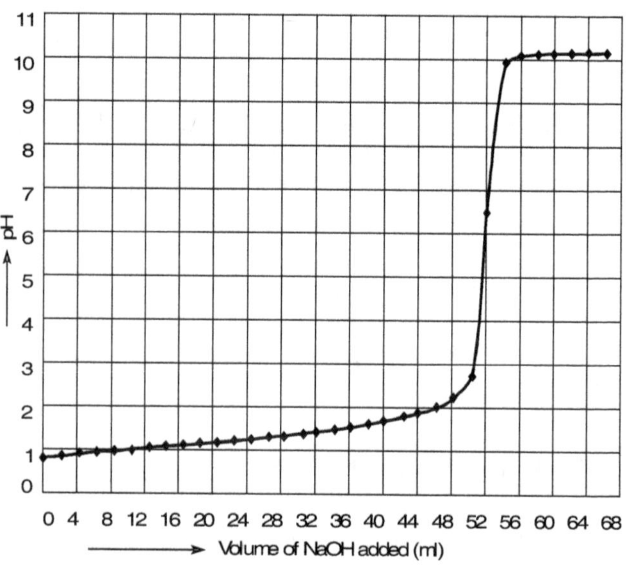

Figure 6.4 Titration between a strong acid and a strong base

II. Titration between a weak acid and a strong base

1. Prepare 50 ml of 0.1M acetic acid (a weak acid).
2. Prepare 100 ml of 0.1M sodium hydroxide.
3. Follow the steps as elaborated for the previous experiment.

4. Plot the readings on a graph sheet (pH vs volume of 0.1M NaOH added) and connect all the points by a smooth line. This gives the titration curve for a weak acid. The midpoint of inflexion gives the pK_a of the acid (Figure 6.5).

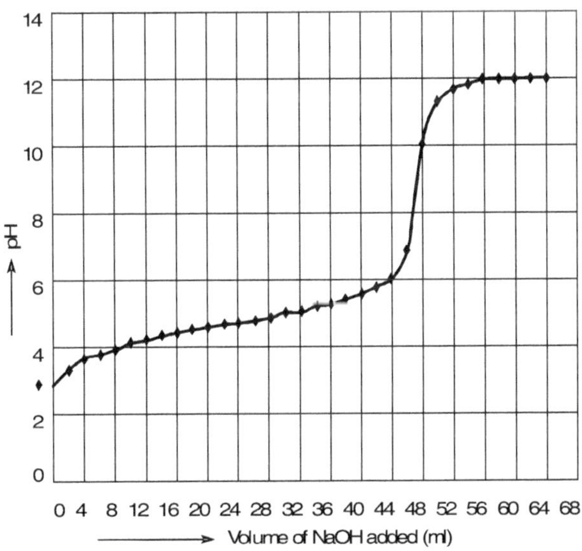

Figure 6.5 Titration between a weak acid and a strong base

III. Titration between a polyprotonic acid (O-phosphoric acid) and a strong base

Depending on the number of ionizable protons per molecule of an acid, the acids are classified as mono-, di- and polyprotonic acids. Orthophosphoric acid is a polyprotonic acid as every molecule gives out 3 protons, sequentially. Therefore, this acid shows 3 pK_a values as shown below.

$$H_3PO_4 \rightleftharpoons H_2PO_4^- + H^+ (pK_a = 2.1)$$

$$H_2PO_4^- \rightleftharpoons HPO_4^- + H^+ (pK_a = 7.2)$$

$$HPO_4^- \rightleftharpoons PO_4^- + H^+ (pK_a = 12.3)$$

1. Prepare 50 ml of 0.1M phosphoric acid.
2. Prepare 300 ml of 0.1M sodium hydroxide.

3. Take 50 ml of the acid in a 250 ml beaker and find out the initial pH (after calibrating the pH meter with a standard buffer, say at 4.01).
4. Now add 2.0 ml of 0.1M NaOH, mix with a glass rod or a magnetic stirrer. Note down the pH.
5. Repeat the above step by adding 2.0 ml portions of 0.1M NaOH until the pH reaches about 13.
6. Plot the readings on a graph sheet (pH vs volume of 0.1M NaOH added). Connect all the points by a smooth line and find out the pK_a values of phosphoric acid (Figure 6.6).

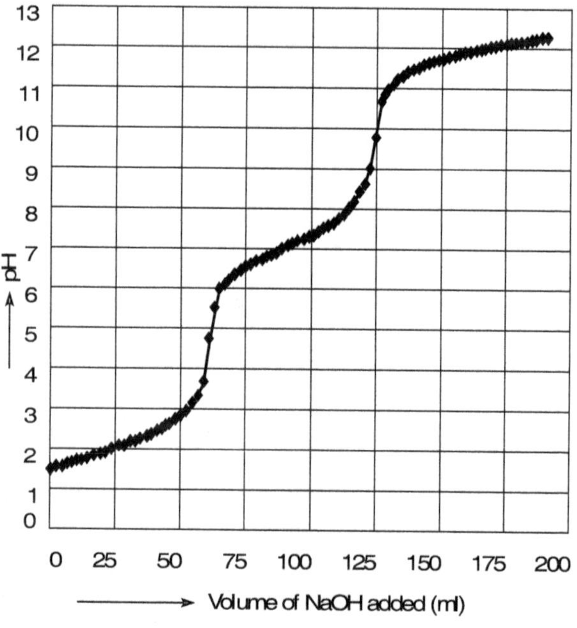

Figure 6.6 Titration between a polyprotonic acid and a strong base

IV. Titration curve for an amino acid

In acid solution, amino acids carry positive charges and hence they move towards cathode in an electric field. In alkaline solution, amino acids carry negative charges and therefore move towards anode. But at certain reactions around neutrality, they are found to be electrically neutral. This occurs as a result of dissociation of H^+ ion, which passes from the COOH to the NH_2 group. The amino acid is said to exist as zwitterion under such conditions.

$$\begin{array}{cc} NH_2 & NH_3^+ \\ | & | \\ H-C-COOH & H-C-COO^- \\ | & | \\ H & H \end{array}$$

Glycine   Glycine as Zwitterion

In this stage, the amino acid will not migrate to either electrode in an electrical field. The pH at which the amino acid has no tendency to move either to the positive or negative electrode is called its **isoelectric point**. At this pH, the amino acid molecule bears a net charge of zero. The isoelectric point is symbolized by PI.

The zwitterion form of glycine in solution is $^+H_3N–CH_2–COO^-$. If HCl is added to this solution of glycine in water, it will behave as a base.

$$^+H_3N - CH_2 - COO^- + HCl \rightarrow {}^+H_3N - CH_2 - COOH + Cl^-$$

On the other hand if an alkali is added, the glycine solution behaves like an acid.

$$^+H_3N - CH_2 - COO^- + NaOH \rightarrow H_2N - CH_2 - COO^- + Na^+ + H_2O$$

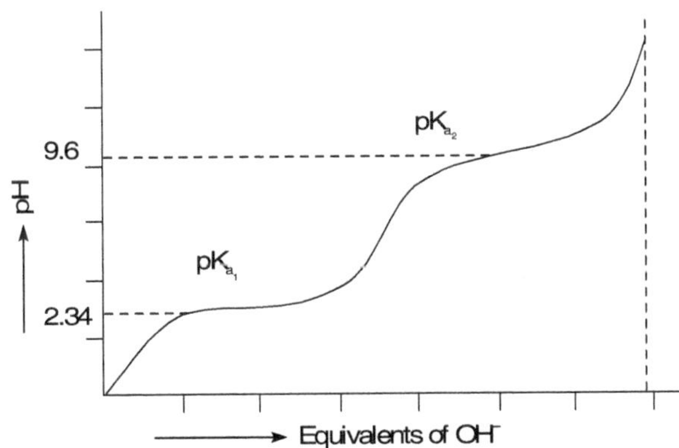

Figure 6.7 Titration curve for glycine

Figure 6.7 shows that the titration curve of glycine looks like that of a diproteinic acid. From the mid point of the first titration curve pK_{a_1} (for the dissociation of carboxyl group) and from the midpoint of the second titration curve pK_{a_2} (for the dissociation of amino group) the isoelectric point of glycine can be calculated. From these two pK_a values, the pH of the solution of glycine in its zwitterionic form can be calculated from the formula,

$$pH = \frac{pK_{a_1} + pK_{a_2}}{2}$$

Thus for glycine,

$$pH = \frac{2.34 + 9.6}{2}$$

$$pH = 5.97$$

Thus, the pH of glycine dissolved in pure water will be 5.97. At this pH, glycine will exist as zwitterion. This pH is known as the isoionic point of glycine, since glycine at this pH does not possess any net charge. This pH is also known as the isoelectric pH of glycine as at this pH glycine will be electrophoretically immobile.

REVIEW YOUR LEARNING

1. Define pH.
2. Derive Henderson–Hasselbalch equation.
3. Explain the principle, operation and applications of pH meter.
4. Explain the following:
   i. Electrolytic dissociation of water
   ii. Acids and bases
   iii. Buffers
   iv. Acid–base titration curves and pK_a values
   v. Titration curve for an amino acid

MANOMETRY

Manometers are used to measure pressure. Measurements involving gases are governed by three parameters, namely volume (V), pressure (P) and temperature (T). These are related to one another by the equation

$$\frac{PV}{T} = R$$

where R is a constant called gas constant.

It follows that at constant temperature, any change in the amount of a gas can be measured by a change in its volume if the pressure is kept the same or by a change in its pressure if its volume is kept constant.

Manometers measure a pressure difference by balancing the weight of a fluid column between the two pressures of interest. Large pressure differences are measured with heavy fluids, such as mercury (e.g. 760 mm Hg = 1 atmosphere). Small pressure differences, such as those experienced in experimental wind tunnels or venturi flowmeters, are measured by lighter fluids such as water (27.7 inch H_2O = 1 psi; 1 cm H_2O = 98.1 Pa). A simple manometer for pressure measurement is shown in the Figure 7.1.

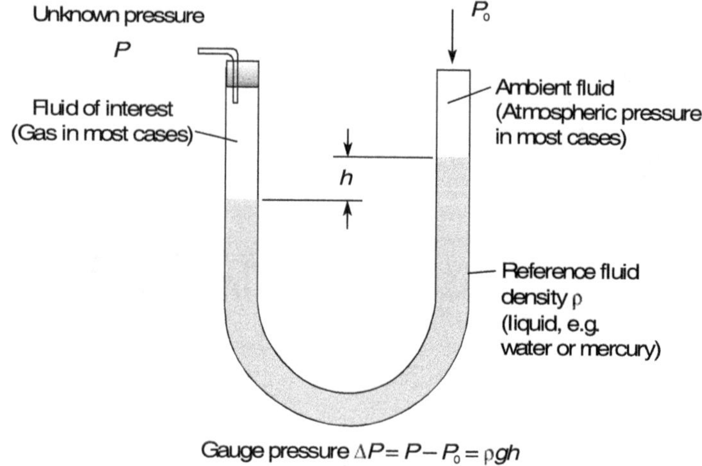

Figure 7.1 Pressure measurement by manometer

TYPES OF MANOMETERS

Piezometer Tube Manometer

The simplest manometer is a tube, open at the top, which is attached to the top of a vessel containing liquid at a pressure (higher than atmospheric) to be measured. An example can be seen in Figure 7.2. This simple device is known as a Piezometer tube. As the tube is open to the atmosphere, the pressure measured is relative to the atmospheric pressure and so is called gauge pressure.

Pressure at A = Pressure due to column of liquid above A

$$p_A = \rho g h_1$$

Pressure at B = Pressure due to column of liquid above B

$$p_B = \rho g h_2$$

This method can only be used for liquids (i.e., not for gases) and only when the liquid height is convenient to measure. It must not be too small or too large and pressure changes must be detectable.

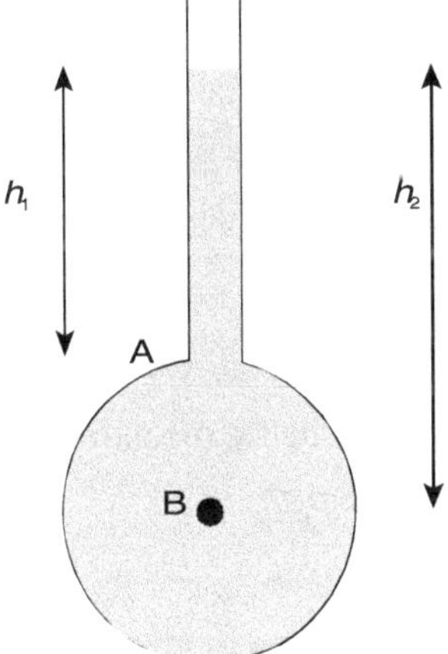

Figure 7.2 A simple piezometer tube manometer

U-tube Manometer

Using a U-tube enables the pressure of both liquids and gases to be measured with the same instrument. The U-tube is connected as in Figure 7.3 and filled with a fluid called the manometric fluid. The fluid whose pressure is being measured should have a mass density less than that of the manometric fluid and the two fluids should not be able to mix readily, i.e., they must be immiscible.

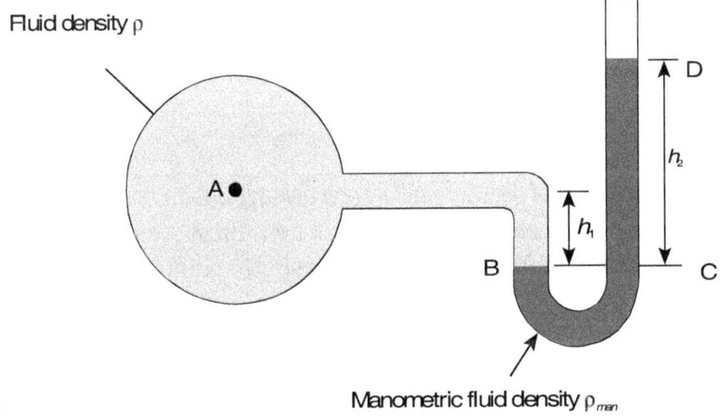

Figure 7.3 A U-tube manometer

Pressure in a continuous static fluid is the same at any horizontal level so,

$$\text{Pressure at B} = \text{Pressure at C}$$

$$p_B = p_C$$

For the left hand arm

$$\text{Pressure at B} = \text{Pressure at A} + \text{Pressure due to height } h_1 \text{ of fluid being measured}$$

$$p_B = p_A + \rho g h_1$$

For the right hand arm

$$\text{Pressure at C} = \text{Pressure at D} + \text{Pressure due to height } h_2 \text{ of manometric fluid}$$

$$p_C = p_{Atmospheric} + \rho_{man} g h_2$$

As we are measuring gauge pressure we can subtract $\rho_{Atmospheric}$

i.e., $$p_B = p_C$$

$$\therefore\ p_A + pg\,h_1 = \rho_{man}\, g h_2$$

$$p_A = \rho_{man} g h_2 - \rho g h_1$$

If the fluid being measured is a gas, the density will probably be very low in comparison to the density of the manometric fluid, i.e., $\rho_{man} >> \rho$. In this case the term $\rho g h_1$ can be neglected, and the gauge pressure is given by

$$p_A = \rho_{man} g h_2$$

Measurement of pressure difference using A U-tube manometer If the U-tube manometer is connected to a pressurized vessel at two points, the pressure difference between these two points can be measured.

If the manometer is arranged as in the Figure 7.4, then

$$\text{Pressure at C} = \text{Pressure at D}$$

$$p_C = p_D$$

$$p_C = p_A + \rho g h_a$$

$$p_D = p_B + \rho g(h_b - h) + \rho_{man} gh$$
$$p_A + \rho gh_a = p_B + \rho g(h_b - h) + \rho_{man} gh$$

Giving the pressure difference

$$p_A - p_B = \rho g(h_b - h_a) + (\rho_{man} - \rho)gh$$

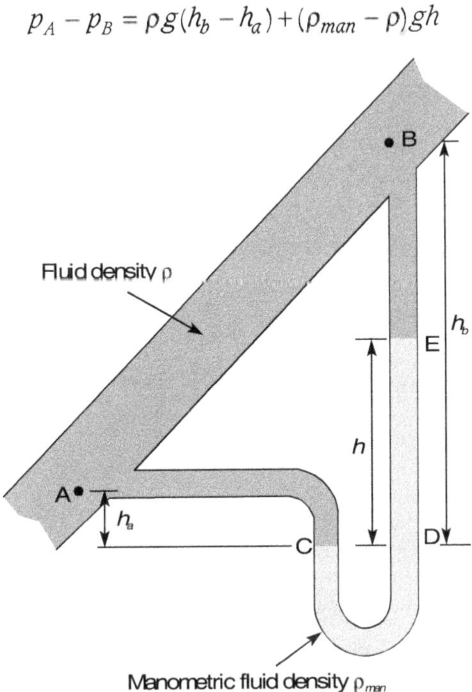

Figure 7.4 Pressure difference measurement using the U-tube manometer

Again, if the fluid whose pressure difference is being measured is a gas and

$\rho_{man} >> \rho$, then the terms involving ρ can be neglected, so

$$p_A - p_B = \rho_{man} gh$$

Advanced U-tube Manometer

The U-tube manometer has the disadvantage that the change in height of the liquid in both sides must be read. This can be avoided by making the diameter of one side very large compared to the other (Figure 7.5). In this case the liquid on the side with the large area moves very little whereas that on the small area moves considerably more.

Assume the manometer is arranged as in the Figure 7.5 to measure the pressure difference of a gas of negligible density and that pressure difference is $p_1 - p_2$. If the datum line indicates the level of the manometric fluid when the pressure difference is zero and the height differences when pressure is applied is as shown, the volume of liquid transferred from the left side to the right $= z_2 \times (\pi d^2/4)$

And the fall in level of the left side is

$$z_1 = \frac{\text{Volume moved}}{\text{Area of left side}}$$

$$= \frac{z_2(\pi d^2/4)}{\pi D^2/4}$$

$$= z_2\left(\frac{d}{D}\right)^2$$

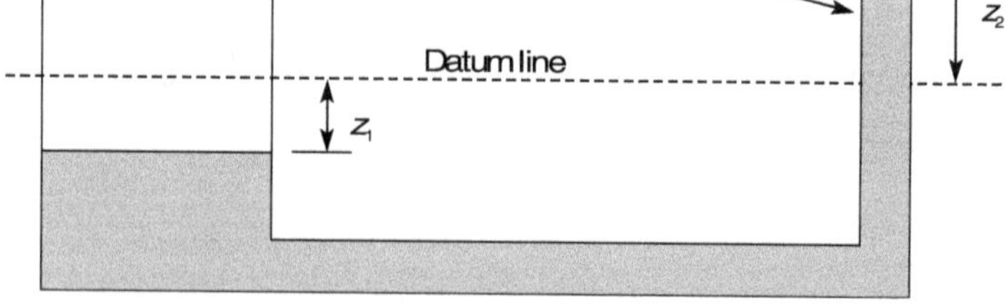

Figure 7.5 Advanced U-tube manometer

We know from the theory of the U-tube manometer that the height difference in the two columns gives the pressure difference so

$$p_1 - p_2 = \rho g\left[z_2 + z_2\left(\frac{d}{D}\right)^2\right]$$

$$= \rho g z_2\left[1 + \left(\frac{d}{D}\right)^2\right]$$

Clearly if D is much larger than d, then $(d/D)^2$ is very small and hence negligible; so

$$p_1 - p_2 = \rho g z_2$$

So only one reading need be taken to measure the pressure difference.

If the pressure to be measured is very small then tilting the arm provides a convenient way of obtaining a larger (more easily read) movement of the manometer. The above arrangement with a tilted arm is shown in the Figure 7.6.

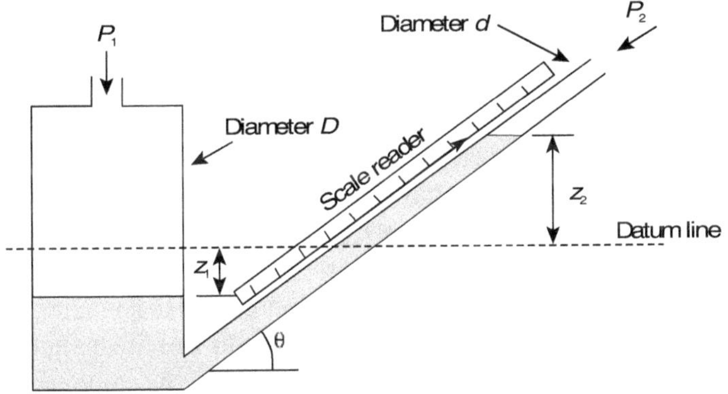

Figure 7.6 Tilted manometer

The pressure difference is still given by the height change of the manometric fluid but by placing the scale along the line of the tilted arm and taking this reading large movements will be observed. The pressure difference is then given by

$$p_1 - p_2 = \rho g z_2$$

$$= \rho g \times \sin\theta$$

The sensitivity to pressure change can be increased further by a greater inclination of the manometer arm. Alternatively the density of the manometric fluid may be changed.

MANOMETERS USED FOR MEASUREMENT OF OXYGEN CONSUMPTION IN A TISSUE OR AN ANIMAL

Manometry may be used to measure either uptake or evolution of O_2 and CO_2. Manometry also has the distinctive feature of allowing the simultaneous determination of O_2 and CO_2 exchange and has the advantage that the magnitude of the exchange is independent of the partial pressure of the gas at the beginning of the experiment.

The two principal types of manometers are the Warburg constant volume manometer (named after the discoverer of the technique) and Gilson constant pressure manometer.

Warburg Manometer

The apparatus consists of a detachable flask with a central well, equipped with one or more side arms attached to a manometer containing a liquid of known density (Figure 7.7). The flask is immersed in a water bath at a constant temperature. The system is shaken to promote a rapid gas exchange between the fluid and gas phase. The manometer has an open and enclosed end. By adjusting the position of three way stock cock, the end of the manometer can also be made to have contact with the atmosphere. The volume of air space in the flask and manometer connected to it is kept constant. This is done by adjusting the fluid in the arm of the manometer on the flask side to its original level. Since the volume and the temperature are kept constant, only the pressure varies as quantity of gas in the flask changes. This change could be measured by the difference between the heights of fluid in the two arms of the manometer. In the beginning, the height of the fluid in the two arms of the manometer is at the same level when the gas in the flask is at atmospheric pressure. As the experiment proceeds, the tissues kept in the flask consume oxygen and release CO_2; any change in the pressure will be due to net gaseous exchange.

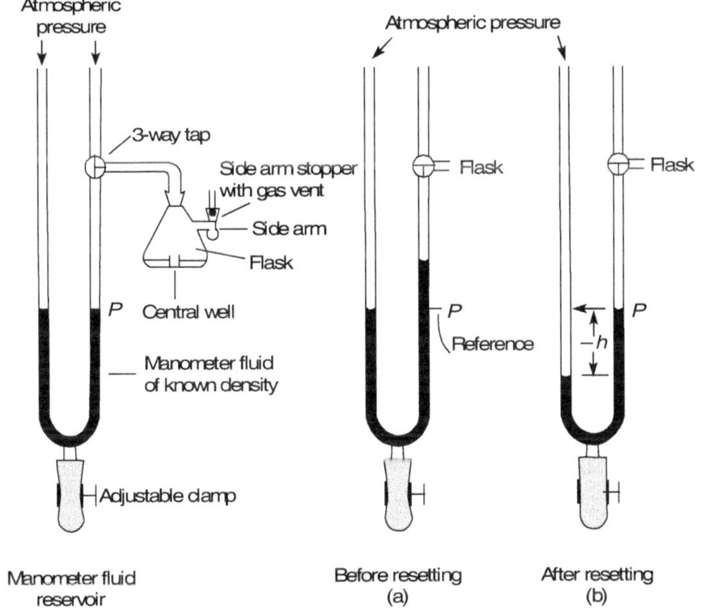

Figure 7.7 Warburg manometer

KOH in the central well of the flask facilitates the efficient absorption of CO_2 evolved. Hence, any change in pressure could be solely due to change in oxygen level. This change in pressure results in a change in the height of the fluid in the arm. At regular time course intervals, the meniscus of the fluid in the left-hand limb is returned to the reference point P by withdrawing fluid into the reservoir using the adjustable clamp, thereby measuring the resultant decrease in pressure at constant volume as h mm. The change h is related to the quantity X of gas evolved at standard pressure, which can be calculated by the following equation:

$$X = h\left(\frac{V_g \dfrac{273}{T} + V_t \alpha}{P_o}\right)$$

where,

V_g = volume of the gas space in the flask including that of the capillary from the flask to the reference mark P(mm^3),

T = temperature in water bath and
P_0 = standard pressure expressed in mm of manometric fluid.

Since, for a experimental flask being used to study a particular reaction under defined conditions, all values within the brackets in the equation are constant, $X = kh$ where k is the flask constant.

Since the Warburg respirometer is open to the atmosphere on one side, during the course of the experiment any change in the atmospheric pressure or in the ambient temperature will affect the reading. Hence for pressure correction, a thermobarometer is used which acts as a control. The experiment is conducted in the barometer without the tissue.

Gilson Constant Pressure Manometer

In Gilson constant pressure manometer, up to 14 experimental flasks and small U-tube capillaries are connected to the same reference flask by a gassing manifold (Figure 7.8). This arrangement allows

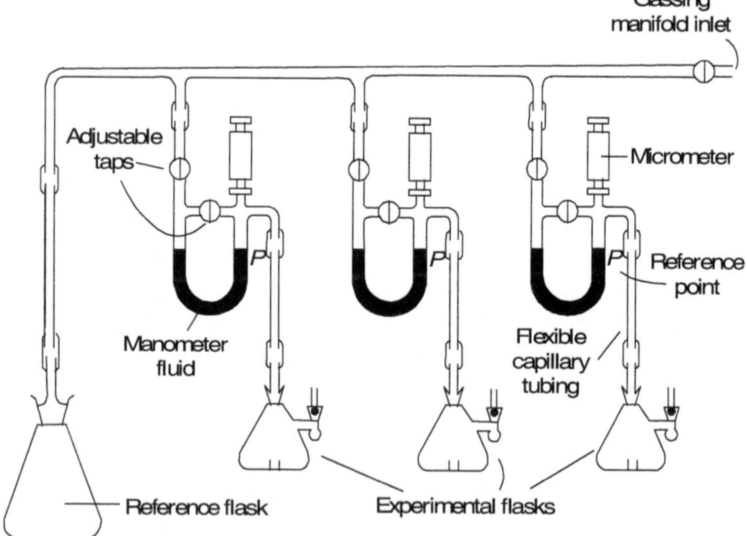

Figure 7.8 Diagrammatic representation of the Gilson constant pressure manometer

all flasks to be gassed simultaneously. The reference flask which is larger than the experimental flasks, eliminates errors due to the changes in the barometric pressure and water bath temperature. Attached to each tube is a micrometer calibrated in microlitres, which is used to adjust the level of fluid in the tube to a reference mark prior to taking readings, thereby ensuring that all gas volume changes are recorded at constant pressure. Changes in gas volumes within experimental flasks are thus measured directly thereby obviating the need to calibrate the flask and U-tube. Volume changes must however be adjusted to standard pressure allowing for the vapour pressure of water in the flask.

REVIEW YOUR LEARNING

1. Explain the principle of manometry.
2. Explain the measurement of pressure using a piezometer tube.
3. Explain the use of U-tube manometer.
4. Explain the structure and usage of tilted manometer.
5. Describe the manometers used for measurement of oxygen consumption in a tissue or an animal.

OSMOMETRY

Osmosis is the net diffusion of water molecules from a dilute solution or pure water itself to a more concentrated solution across a semipermeable membrane, which permits the diffusion of water but not of a solute. Thus, if an aqueous solution of sucrose is separated from water by a semipermeable membrane, water flows into the sucrose solution across the membrane. Thus, osmosis requires (a) semipermeability of the membrane separating the two solutions, so that the membrane is permeable to water but not to the solute and (b) a difference in concentration of the solute on the two sides of the membrane. In fact, water molecules diffuse in both directions across the semipermeable membrane, but a net diffusion or osmosis of water from the dilute to the concentrated solution results from a larger number of water molecules diffusing in that direction than in the reverse direction. Water continues to flow into the more concentrated solution across the membrane in this way until the hydrostatic pressure rises so high on the concentrated side of the membrane as to cause a transmembrane diffusion of water in the opposite direction at the same rate as the osmotic inflow. This excess of hydrostatic pressure, which exactly balances the osmotic influx of water from pure water to the concentrated solution, is called the **osmotic pressure** of that solution.

Osmotic pressure (π) may also be defined as the pressure that has to be exerted on the concentrated solution, separated from pure water by a semipermeable membrane in order to counteract and stop the osmotic inflow into that solution. It equals the difference between the hydrostatic pressures on the two sides of the semipermeable membrane at osmotic equilibrium.

Osmotic pressure is a colligative property of a solution. Properties whose magnitudes depend on the number of molecules or ions of the solution, but not on their nature or shape are called colligative properties of solutions. **Osmolality** (osmoles per kg solution) and **osmolarity** (osmoles per litre water) are measures of the total concentration of dissolved particles in a solution, without regard for the homogeneity or non-homogeneity of the molecular species, or their molecular weights, size, or density. Any substance dissolved in a solvent affects four interrelated colligative properties of the solvent–solute mixture. These colligative properties are:

- Decrease in vapour pressure,
- Decrease in freezing point,
- Increase in boiling point, and
- A change in osmotic concentration.

All of these properties are interrelated and are mathematically convertible. Therefore, the accurate determination of any one of these properties allows estimation of the other three, and is a measure of the osmotic concentration of the solution.

The solute molecules in a dilute solution behave almost like gas molecules, tending to occupy the entire volume of the solution in a gas vessel. Vant Hoff proposed the law of osmotic pressure, which resembles gas laws.

VANT HOFF'S LAWS OF OSMOTIC PRESSURE

Law 1 The osmotic pressure (π) of a so lution is directly proportional to the molar concentration (C) of the solute as long as the temperature is maintained constant.

$$\pi = k_1 C$$

where k_1 is constant.

If V litres of solution contain 1 mole of the solute, then

$$C = \frac{1}{V}$$

In that case, the expression for law 1 becomes:

$$\pi = \frac{k_1}{V}$$

or

$$\pi V = k_1$$

This is a relation comparable with Boyle's law

i.e., $PV = K$

If V litres of solution contain n mole of the solute, then

$$C = \frac{n}{V}$$

In that case, the expression for law 1 becomes

$$\pi = \frac{nk_1}{V}$$

or

$$\pi V = nk_1$$

Vant Hoff's first law implies that the more concentrated a solution, the higher is its osmotic pressure.

Law 2 The osmotic pressure of a solution is directly proportional to the absolute temperature (T) as long as its concentration remains constant.

$$\pi = k_2 T$$

where k_2 is constant.

This relation is analogous to the following expression for gas law when V is constant.

$$\frac{P}{T} = K$$

The Vant Hoff's equation is a combination of his two laws:

$$\pi V = nRT$$

$$\pi = \frac{n}{V} \times RT$$

$$\pi = CRT$$

where,

n = the number of moles of the solute in V litres of the solution,
C = the molar concentrations of the solute,
R = the molar gas constant (0.082 litre atm. K^{-1} mol^{-1}) and
T = the temperature in K.

DETERMINATION OF OSMOTIC PRESSURE

There are several different techniques employed in osmometry.

Simple Osmometer

A simple osmometer is a device used for measuring the osmotic strength of a solution, colloid or compound. A simple osmometer consists of a wide steel chamber with its bottom end closed by a semipermeable cellophane membrane supported on a perforated metal plate. The top of the chamber communicates directly with a calibrated upright capillary tube. The chamber is filled with the test solution up to the lower end of the capillary tube and placed in a larger vessel containing the pure solvent. The osmotic influx of the solvent through the cellophane membrane raises the fluid volume in the steel chamber and the fluid level in the capillary tube is read from its calibration to estimate the osmotic pressure.

Mercury Osmometer

This apparatus consists of a glass tube with a porous bottom coated with copper ferrocyanide, which serves as a semipermeable membrane. The glass tube is connected to a mercury manometer. The glass tube is filled with the test solution and the porous bottom is immersed in distilled water contained in a large vessel. Now, water enters into the tube through the semipermeable copper ferrocyanide coating by osmosis. This produces excess pressure in the porous bottom and this raises the mercury level in the manometer (Figure 8.1). The raise in the mercury level marks the osmotic pressure of the sample.

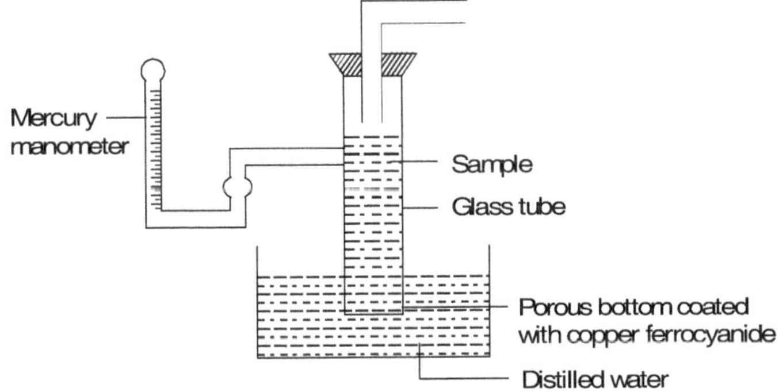

Figure 8.1 Mercury osmometer

Berkely–Hartley Method

The apparatus consists of an inner porcelain tube (I) enclosed in an outer and wider metal tube (O). The walls of the inner tube are made semipermeable by impregnation with copper ferrocyanide. The inner and outer tubes are filled with water and the test solution respectively (Figure 8.2). Pressure is applied

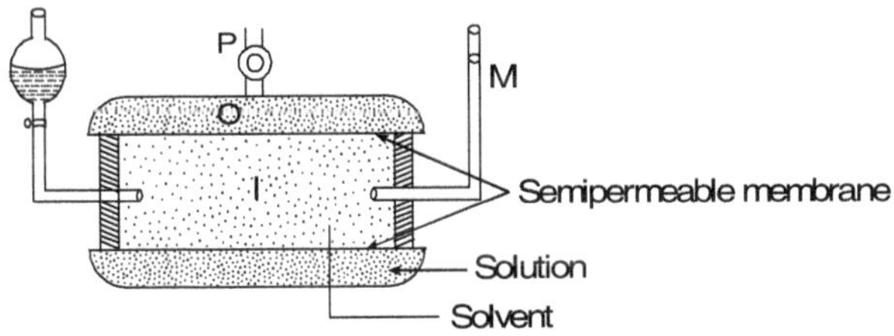

Figure 8.2 Berkely–Hartley apparatus

by a pump (P) on the solution in the outer tube until the solvent meniscus remains at its initial level in the manometer (M). This pressure, read from a pressure gauge gives the osmotic pressure. This method avoids the dilution of the solution due to the osmotic inflow of water.

Freezing Point Method

Freezing point depression is one of the oldest and easiest methods for determination of the osmotic concentration of biological fluids. In this method, the specimen is placed in a cooling chamber that is maintained at a temperature well below the freezing point of the solution. During the analysis, the samples are supercooled, initiating crystallization, a process called "seeding". Seeding can be achieved by mechanical vibration. The crystal formation results in release of the heat of fusion of water (80 cal/g water) causing the sample to warm to a point at which ice and solution exist in equilibrium, and the temperature remains constant for a period of time.

Osmotic concentration, or osmolality, is expressed in units of milliOsmoles (mOsm) per kg of water, where one mOsm is equivalent to one mM of dissolved solute particles. A solution containing 1 Osmole (1000 mOsm) of dissolved solute per kg of water lowers the freezing point of water by 1.858°C. Therefore, the freezing point depression of the sample can be converted to units of osmolality, or osmotic concentration by dividing by 1.858.

$$\text{Molar concentration}(C) = \frac{\text{Depression of freezing point (d)}}{\text{Molar depression constant of water (K)}}$$

The molar concentration thus obtained can be substituted in Vant Hoff's equation and the osmotic pressure of test solution can be calculated.

Freezing-point depression osmometry (Figure 8.3) is capable of measuring all osmotically active solutes within the range of

0–4000 mOsmol/kg H_2O, including volatile substances, making it the most sensitive.

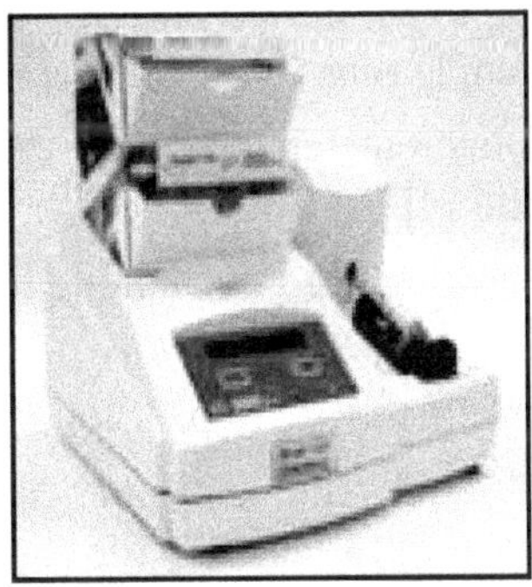

Figure 8.3 Freezing-point depression osmometer

However, the cryoscopic osmometer (Figure 8.4) does this calculation directly by converting the thermistor readings by direct comparison with readings obtained by using standard aqueous salt solutions of known osmolality.

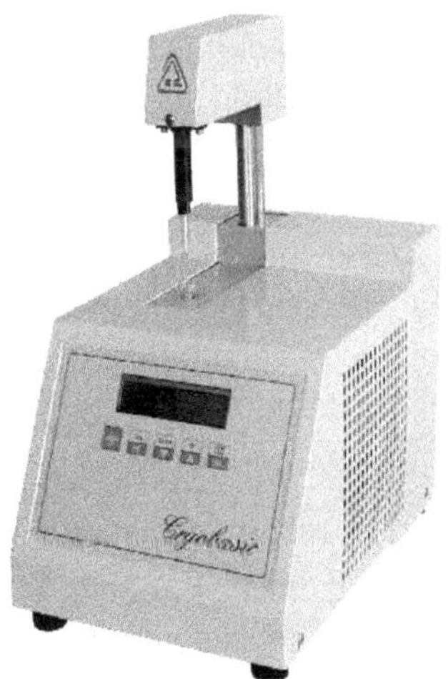

Figure 8.4 Cryoscopic osmometer

Measuring with a Cryoscopic Osmometer

1. When you get to the osmometer, the machine should be on, the operating head should be down and there should be a clean, dry sample tube in the refrigerator well.
2. Be sure the osmometer is cooled: the cool light should be cycling on and off.
3. Raise the operating head by pushing the head release button.
4. The **Range Switch** should be in the 0–2 position and the **Mode Switch** in the **RUN** position.
5. Place 50 μl of sample into an osmometer tube using the pipette provided. Avoid capturing air in the bottom of the sample tube, as this will adversely affect the reproducibility of readings.
6. Place the sample tube into the refrigerator well.
7. Lower the operating head so that the seed wire and probe enter the tube, and the head latches in the down position, without forcing it.
8. The digital display will be reading negative numbers, which will start to become more positive, and will count up to almost 1000. When the number hits 1000, the sample is seeded with a high amplitude vibration of the seed wire (and the osmometer makes a loud, obnoxious noise).
9. After the sample is seeded, the numbers on the digital display will decrease until they read the osmolality of the solution. The read light will come on and the head will pop up when the reading is complete. The number on the digital display is the osmolality of the solution.
10. Gently wipe off the probe and seed wire with a clean kimwipe (do not use gauze pads).

Whenever the instrument is not in use, there should be an empty sample tube in the refrigerator well, to prevent frost buildup in the well. After using the osmometer, turn it off and lower the operating head into the empty tube to protect the temperature probe. Do not force a sample tube into a frosted refrigerator well. If the well has accumulated frost, turn off the osmometer for several minutes to defrost, and remove the accumulated moisture from the well with a "Q-tip".

Vapour Pressure Osmometry

Vapour pressure osmometry is a technique to measure the number average molecular weight of a polymer. It is based upon Raoult's law that governs change in vapour pressure of a solution based on the mole fraction of the solute. Raoult's law states that the vapour pressure of each component in an ideal solution is dependent on the vapour pressure of the individual component and the mole fraction of the component present in the solution. Vapour pressure is the pressure of a vapour in equilibrium with its non-vapour phases.

The vapour pressure osmometer is comprised of two chambers: one for pure solvent and the other to contain solution, where the solute is the polymer whose molecular weight is unknown. Thermistors in each chamber provide an electrical signal (the acutal measurement) of differential heating to achieve a vapour equilibrium in each chamber.

By measuring solutions of different concentrations of solute with a known molecular weight standard, a plot of concentration versus difference in the electrical signal can be prepared. Similarly, the unknown is then prepared and measured with the number average molecular weight derived from the standard plot.

In the physiology of a whole plant, the osmotic potential determines the ability of the plant to take up water from the environment and to generate turgor pressure, although the freezing point can be critical to the survival of cells at low temperatures. Any of these properties can be measured directly and used to calculate the total concentration of the solutes in the solution (osmolality). Once that calculation is made, the other colligative properties can be estimated. The Wescor osmometer measures the vapour pressure (relative humidity) by measuring the dew point (temperature at which water condensate is in equilibrium with the water vapour in the overlying atmosphere).

The osmometer has a small chamber to which is sealed a thermocouple hygrometer. A thermocouple measures temperature by the voltage between two dissimilar metals that are joined together. Actually, there are two thermocouples—the difference in their temperatures is measured by the difference in the voltages across them (Figure 8.5).

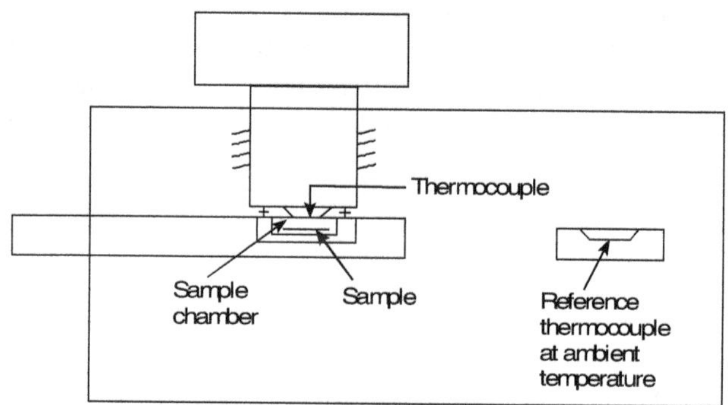

Figure 8.5 Vapour pressure osmometer

The relationship between temperature and voltage is reversible; running a current across the interface between the metals can raise or lower the temperature of the thermocouple. A small amount (8 μl) of solution placed on a piece of paper and inserted into the chamber quickly equilibrates with the chamber atmosphere. The instrument then runs a program to determine the dew point of the atmosphere.

The program runs as follows:

1. Once the sample is introduced and the chamber is sealed, the instrument allows time for thermal and vapour equilibration. The electronic circuit measures the thermocouple voltage under "quiescent" conditions.
2. An electrical current is fed through the thermocouple junction to cool it (by the Peltier effect) below the dew point. Water condenses from the air in the chamber and forms a thin liquid film on the junction surface.

3. The cooling current is reduced, and the temperature of the thermocouple tends toward an equilibrium that reflects the evaporation or condensation of water on its surface. Since it starts cold, condensation tends to warm the junction to the equilibrium temperature (dew point). When the temperature of the junction reaches the dew point, it stabilizes.
4. The electronics read the temperature of the junction (relative to the room temperature) by the voltage across the junction (relative to the voltage across a junction at room temperature). The temperature depression is proportional to the osmolality of the solution.

Figure 8.6 describes the steps of the program by the temperature of the sample chamber thermocouple as a function of time.

The final reading is calibrated in milliOsmoles/kg.

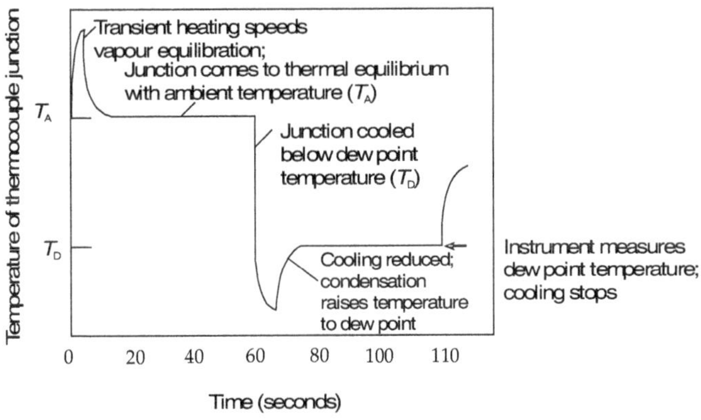

Figure 8.6 Measurement of dew point temperature in vapour pressure osmometer

APPLICATIONS OF OSMOMETRY

The maintenance of adequate body fluid volume and the correct distribution of this fluid between the body compartments is a critical part of homeostasis. The process of osmosis plays an important role in the movement of fluid within the body and the use of osmometry is an important part of the management of many patients. In addition to the application of osmometry to the measurement of body fluids, most commonly plasma and urine, osmotic action plays a part in the therapeutic actions of some drugs and its strength needs to be quantified in fluids administered to patients.

Osmolality of blood increases with dehydration and decreases with overhydration. In normal people, increased osmolality in the blood will stimulate secretion of antidiuretic hormone (ADH). This will result in increased water reabsorption, more concentrated urine and less concentrated blood plasma. A low serum osmolality will suppress the release of ADH, resulting in decreased water reabsorption and more concentrated plasma. Normal osmolality in plasma is about 280–296 milliosmoles per kilogram. This is contributed mainly by chloride, potassium, urea, and glucose, and additionally by other ions and substances in the blood.

Osmometry is also useful in determining the molecular weight of unknown compounds and polymers.

REVIEW YOUR LEARNING

1. Define or explain the following:
   - i. Osmosis
   - ii. Osmotic pressure
   - iii. Osmolality
   - iv. Osmolarity
   - v. Osmometry
   - vi. Osmometer
   - vii. Raoult's law
2. Explain Vant Hoff's laws of osmotic pressure.
3. Explain the techniques employed in osmometry.
4. Explain the use of mercury osmometer in the determination of osmotic pressure.
5. Explain Berkely–Hartley method of determination of osmotic pressure.
6. Explain freezing point method for determination of the osmotic concentration of biological fluids.

7. Explain the use of cryoscopic osmometer in the determination of osmotic pressure.
8. Explain the principle, instrumentation and use of vapour pressure osmometry.

CHROMATOGRAPHY

The term chromatography was originally applied by Michael Tswett, a Russian botanist in 1906 to a procedure where a mixture of different coloured pigments (chlorophylls and xanthophylls) was separated into its constituents. He used a column of $CaCO_3$ to separate the various components of petroleum ether chlorophyll extract into green and yellow zones of pigments. He termed such a preparation as chromatogram and the procedure as chromatography.

Chromatography is defined as a technique in which the components of a mixture are separated based upon the rates at which they are carried through a stationary phase by a liquid or gaseous mobile phase. Separation begins to occur when one component is held more firmly by the stationary phase than the other, which tends to move on faster in the mobile phase. Thus, the underlying principle of chromatography is first to adsorb the components of a mixture on an insoluble material and then suitable liquid solvents. The adsorbent can be in the form of sheet (paper and thin layer chromatography) or it can be packed into a column (column chromatography). A third form of chromatography is obtained with columns containing ion exchange resins (ion exchange chromatography).

In all chromatographic techniques, difference in affinity involves the process of either adsorption or partition. In adsorption, the binding of a compound to the surface of the solid phase takes place; whereas, in partition the relative solubility of a compound in two phases results in the partition of the compound in two phases. Thus, all types of chromatography known so far have been grouped in either of the two mentioned forms:

Partition chromatography	1. Paper chromatography 2. Thin layer chromatography 3. Gel filtration chromatography 4. Partition column chromatography	Liquid as mobile and solid as stationary phase
	5. Gas–liquid chromatography or vapour phase chromatography	Gas as mobile and liquid as stationary phase
	6. Liquid–liquid chromatography	Liquid as both stationary & mobile phase
Adsorption chromatography	1. Adsorption column chromatography 2. Thin layer chromatography	Liquid as mobile and solid as stationary phase
	3. Ion exchange chromatography	Electrolyte solution as mobile and some ionic polymer as stationary phase

PAPER CHROMATOGRAPHY

Paper chromatography was first used by Martin Consden and Gordon to separate a mixture of amino acids. The stationary phase is the filter paper on which the mixture of amino acids is made to migrate. The filter paper is made of cellulose molecules that are hydrated. The moving phase consists of a mixture of an immiscible solvent system of which one is usually water, which has a stronger affinity to the supporting medium. The other solvent has lower affinity to the stationary phase and carries the solute across the stationary phase. The most commonly used solvent mixtures for different type of substances for paper chromatography are given in the Table 9.1.

Table 9.1 Some mobile phases used in paper chromatography

Mobile phase	Type of substances to be separated
n-Butanol–acetic acid–water Isopropanol–acetic acid–water	Hydophilic substances
Formamide–chloroform Formamide–benzene Formamide–chloroform–benzene	Slightly hydophilic substances
Formamide–benzene–cyclohexane Dimethyl formamide–cyclohexane	Hydrophobic substances

As the solvent passes through an area of the paper containing a solute, the solute will begin to partition itself between the aqueous and the organic phases in proportion to its relative solubility in the two phases. The more soluble the solute is in the organic phase, the faster will be the solute carried along by the organic phase. Conversely, the greater the affinity for water, the slower the solute will move with respect to the solvent front. Thus, if several compounds possess different solubility characteristics, theoretically each will progress across the paper at a specific rate, which is different from that of any of the other compounds. The distance the solute moves in relation to the distance the solvent moves serves as a convenient means for identifying the solute. This relative rate of flow is the R_f value for the compound under the specified conditions of the experiment.

$$R_f = \frac{\text{Distance travelled by solute (measured to centre of the spot)}}{\text{Distance travelled by solvent}}$$

There are two techniques which may be employed for the development of paper chromatogram: (1) ascending and (2) descending method.

Ascending Paper Chromatography

This is the simplest of all the methods. It consists of a reservoir and the paper is held in position by means of clamps. The lower end of the paper is dipped into the solvent. The sample is spotted in a position just above the surface of the solvent so that, as the solvent moves vertically up the paper by capillary action, separation of sample is achieved (Figure 9.1). This method is preferable for quick analysis of a large number of substances.

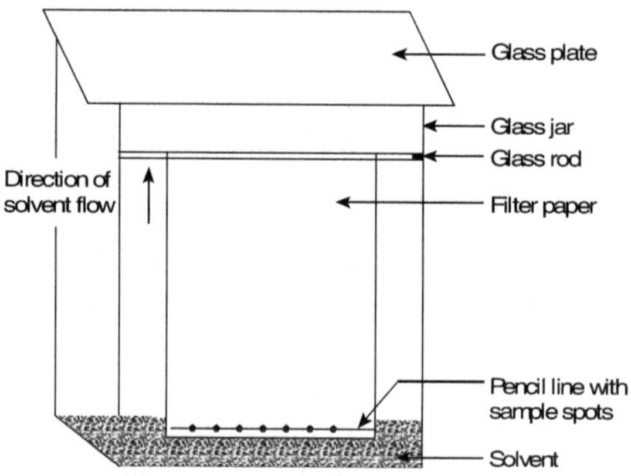

Figure 9.1 Ascending chromatography

Descending Paper Chromatography

In this method, the solvent is kept in a trough at the top of the chamber. The paper is then suspended in the solvent and the lid is placed at the top (Figure 9.2). The liquid moves down by capillary action as well as by the gravitational force. In this case, the flow is more rapid as compared to the ascending method. Because of this rapid speed, the chromatogram is completed in a comparatively shorter time.

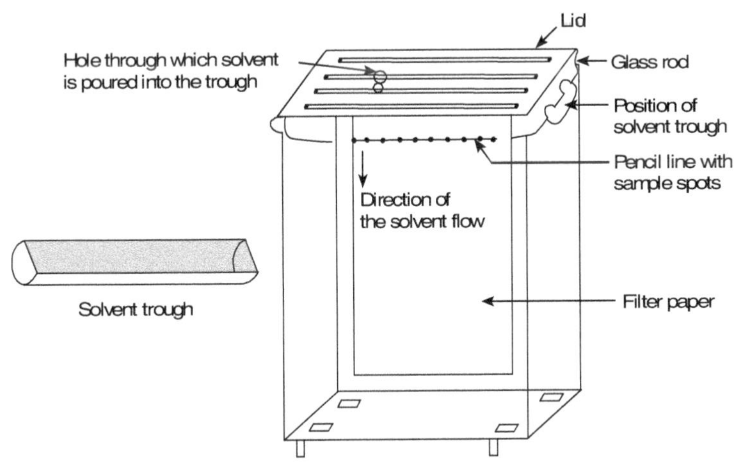

Figure 9.2 Descending chromatography

The method of paper chromatography may be one-dimensional or two-dimensional depending on the type of complexity involved in the analysis. The method given above is one-dimensional chromatography. Several compounds may have the same R_f value in a particular solvent system; running more than one chromatogram, each with a different solvent system, can separate these compounds. For obtaining better separation of components, a two-dimensional paper chromatography is applied. In this technique a square sheet of filter paper is taken. The test sample is applied to the upper left corner and chromatographed for few hours with one solvent mixture, e.g. butanol : acetic acid : water. The paper is dried and it is turned at 90° and again chromatographed in a second solvent mixture, e.g. collidine : water. This technique thus allows both vertical and horizontal separation of the amino acids (Figure 9.3).

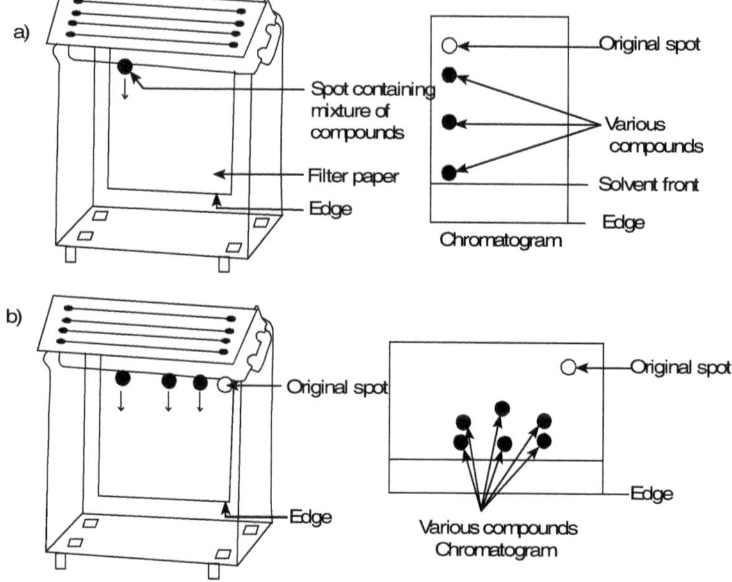

Figure 9.3 Two-dimensional paper chromatography

Forces in Operation

The movement of the solute molecules on the chromatogram depends on the net result of a number of forces operating in the system. These forces are of two types namely, propelling and retarding.

Propelling forces These include the capillary force and the solubility force of the solvent. The Whatman paper is made up of numerous fibrils, which are placed very close to each other, thus forming a network of capillaries. The solute rises through these capillaries as a result of capillary force. The smaller the pore of the capillaries, the greater is the height to which the solute rises.

The solubility force of the solvent refers to the capacity of the solvent to dissolve the solute. The rise of the solute also depends on its solubility in the solvent being used. The greater the solubility of the solute in the solvent, the greater is the height to which it rises in the chromatogram.

Retarding forces Concurrent with the forces of propulsion, certain retarding forces also operate in the system, which tries to drag the solute molecules from moving in either direction. These retarding forces include the gravitational force and the partition force. Under the ascending chromatographic technique, the solute molecules have to move against the gravitational force, which acts from below and tends to retard the movement of solute molecules in the upward direction. In the descending chromatography, the gravitational force does not figure as a retarding force, rather, it assists in propulsion and henceforth, acts as a propelling force.

The partition force refers to the force between the liquid and liquid molecules. The solvent molecules occupy the interstices between the fibrils on the Whatman paper. The solute molecules, to move on the Whatman paper, should displace the solvent molecules. If the space were not filled with the solvent molecules, the movement of the solute molecules through them would be much more facilitated. Thus the movement of the solute molecules on the Whatman paper in either direction is the net result of the interaction between various forces of propulsion and retardation. Movement of the solute molecules is exhibited only when the propelling forces exceed the retarding forces in magnitude.

Detection Various methods are adopted to identify the substances. The components can be identified by spraying chemical reagents imparting specific colour reactions. Amino acids can be identified by spraying ninhydrin which gives purple colour. Other methods of detection are i) UV and infrared absorption ii) fluorescence and, iii) radio activity.

Applications

Paper chromatography is used in the separation of a variety of compounds of biochemical interest including fatty acids, carbohydrates and amino acids. In clinical chemistry, chromatography is applied to separate and identify amino acids, sugar and various other compounds of clinical interest present in urine and other biological fluids.

In general, paper chromatography is the most widely used procedure for the separation and identification of amino acids. This method is so common that almost every type of compound is separated in this manner. Fields like biology, medicine and other intelligence departments use this method.

THIN LAYER CHROMATOGRAPHY (TLC)

Thin layer chromatography is used mainly for the separation of low molecular weight compounds. This technique is more or less similar to paper chromatography.

Thin layer chromatography may be either carried out by the adsorption principle (if the thin layer is prepared by an adsorbent such as Kieselguhr or alumina) or by the partition principle (if the layer is prepared by a substance such as silica gel which holds water like the paper).

Table 9.2 Solvent systems used in thin layer chromatography

Solvent systems	Stationary phase	Compounds to be separated
Butanol–acetic acid–water (4 : 1: 1)	Silica gel G	Amino acids
Petroleum ether–Diethyl ether–Acetone (90 : 10 : 1)	Silica gel G	Triglycerides
Chloroform–methanol–water (65 : 25 : 4)	Silica gel G	Phospholipids
Petrolem ether–propanol (99 : 1)	Kieselguhr G	Plant pigments
Butanol–acetone–phosphate buffer pH 5	Kieselguhr G	Mono- and disaccharides

A slurry of the stationary phase is made in a solvent. The slurry is poured onto the glass plate and spread evenly. The plates are then dried by keeping in an oven at 40°C. The sample is applied using a micropipette and dried. The plate is placed in a chamber containing the solvent and developed by ascending chromatography (Figure 9.4). Table 9.2 shows some of the solvent systems used in thin layer chromatography. After the solvent front has almost reached the top, the plate is dried. It can be rechromatographed at right angles with a second solvent for two-dimensional work.

Spots are located by natural colour, by fluorescence or by spraying various reagents. Materials can be eluted from the chromatogram by scraping off the stationary phase and by eluting the powder with a suitable solvent.

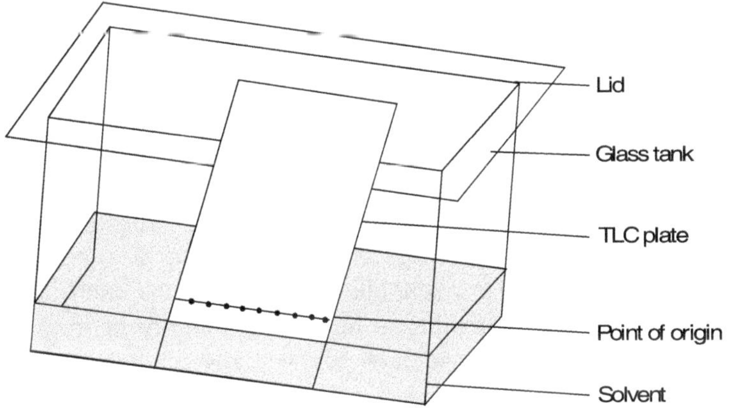

Figure 9.4 Thin layer chromatography

Applications

Thin layer chromatography is used almost exclusively for the separation of small molecules. Amino acids, nucleic acids, lipids, steroids, terpenoids, hydrocarbons and carbohydrates can be separated using thin layer chromatography. It is often used to identify drugs, contaminants and adulterants. It is also widely used to resolve plant extracts and many other biochemical preparations. TLC is commonly used in studying the incorporation of ^{14}C–labelled compounds into metabolites and for the analysis of urine and blood in pathological laboratories.

COLUMN CHROMATOGRAPHY

Column chromatography is defined as a separation process involving uniform percolation of a liquid through a column packed with finely divided material. The separation in the column is effected either by direct interaction between the solute components and the surface of the stationary phase or by adsorption of solute by the stationary phase.

Matrices Used in Column Chromatography

In column chromatography, the matrix is the material supporting the stationary phase. A matrix material should have high mechanical stability to facilitate good flow rate, good chemical stability, should be available in different particle sizes, and it should be chemically inert.

The most commonly used matrices are the following:

Cellulose It is a polysaccharide consisting of β-1, 4-linked D-glucose units. For matrix purpose, it is cross-linked with epichlorohydrin. The pore size is controlled by the extent of cross-linking. It is available in various forms such as bead, microgranular and fibrous forms. It is highly hydrophilic in nature.

Dextran This is a polysaccharide consisting of α-1, 6-linked D-glucose units. For matrix purpose, it is cross-linked with epichlorohydrin. But the product is less stable to acid hydrolysis. However, it is stable up to pH 12 and is hydrophilic in nature. Commercial examples include Sephadex and Sephacryl.

Agarose This is also a polysaccharide made up of 3,6-anhydro-1-galactose and D-galactose units. For the preparation of matrix material, it is cross-linked with 2,3-dibromopropanol to give gels that are stable in the pH range 3 to 14. It has good flow property and is highly hydrophilic. Agarose matrix materials should always be suspended in water or a suitable buffer solution. They should never be allowed to dry out. Otherwise, they undergo irreversible changes. Commercial examples include Sepharose and Bio-gel A.

Polyacrylamide It is a polymer of acrylamide monomer, cross-linked with N, N′-methyl-bisacrylamide. It is stable in the pH range 2 to 11, e.g. Bio-gel P.

Polystyrene It is a polymer of styrene, cross-linked with divinylbenzene. Polystyrene matrices have good stability over all pH ranges. They are mainly used in ion exchange and exclusion chromatography.

Silica It is a polymeric material produced from orthosilicic acid. Since this molecule possesses numerous silanol (Si–OH) groups, it is hydrophilic in nature. For matrix purpose, excess silanol groups are removed by treatment with trichloromethyl silane. They are stable in the pH range 3 to 8.

Packing of Column

The column (2.5 × 100 cm) is held in a vertical position and a glass wool plug is placed at the end of the chromatographic column. Before packing the column, the gel is allowed to swell in a particular solvent for a convenient period. If swelling is completed by heating, the resulting slurry is allowed to cool before packing. If slurry is prepared in cold, it should be deaerated under vacuum. Now the gel is allowed to settle and half of the supernatant solution is removed. The gel is again mixed with the solvent to form a suspension or slurry. After constant stirring, it is poured carefully to the column by making use of a glass rod. The entire amount of the slurry is added in one step to avoid uneven packing. After the settling of the gel, the eluent (buffer) is allowed to run through the column for a longer period before applying the sample, to get fine packing.

Application of the Sample

Application of the sample to the top of a prepared column should be done very carefully. A simple way is to remove most of the mobile phase from above the column by suction and to allow the remainder to drain into the column bed. The sample is then carefully applied by a pipette, and also allowed to run into the column slowly. After complete entry of the sample into the column bed, a small volume of mobile phase is applied very slowly a few times, and then it is added to a height of 2 to 5 cm in the column. The column is then connected to a big reservoir and the flow is adjusted so that the height of the mobile phase in the column is maintained constant.

The second method of sample application involves preparation of the sample in a high-density solution (e.g. in 1 per cent sucrose solution). When this dense solution is layered from a height of about 2 cm from the column bed, it automatically sinks to the surface of the column and passes quickly into the column.

A third method of sample application involves the use of capillary tube of syringe to pass the sample prepared in a high dense solution directly to the column surface. This is the most satisfactory method of the three discussed above.

Factors Affecting Efficiency of Separation

The following factors affect drastically the efficiency of separation of column chromatography:

1. **Dimension of the column** Column efficiency is improved by increasing the length/width ratio of the column. For preparative separations, length/width ratios of 10 : 1 to 100 : 1 have been found to be most satisfactory. In general, long columns are used for the separation of closely related compounds and large column diameters are used to accommodate large amount of samples.
2. **Particle size of the column packing** Particle size of the column packing material plays an important role in column chromatography. Decreasing the particle size of the adsorbent improves the separation. The recommended

particle size used for both adsorption and partition is 100 to 200 mesh range.

3. **Quality of solvents** Solvents having low viscosity are generally employed in high efficiency separations. Since the rate of flow of a solvent is inversely proportional to its viscosity, it is important to select a solvent with very low viscosity.
4. **Temperature of the column** The speed of elution increases at higher temperatures because adsorption is mostly reduced at higher temperatures. However, column chromatography is usually done at room temperature or low temperature (about 5 to 10°C).
5. **Flow rate** A medium flow rate is usually preferred. Very slow flow rate may sometimes cause zone spreading. Too fast flow rate often leads to non-equilibrium system and extensive tailing.
6. **Packing of the column** Packing of the column is very important in column chromatography. Poorly packed columns may lead to either zone spreading or extensive tailing.

Elution Techniques

The removal of the sample from the solid matrix using a suitable solvent is called **elution**. The total volume of material, both solid and liquid, in the column is called the **bed volume**. The volume of the mobile phase is known as **void volume**. The amount of liquid that must be added to produce a peak of a particular solute in the effluent is the **elution** or **effluent volume**. The time taken for each material to emerge from the column is referred to as its **retention time,** t_R. Elution volume and retention time are related to the flow rate of the mobile phase through the column.

There are three methods by which columns are eluted in column chromatography.

Simple method or Isocratic elution Column development using a single liquid as the mobile phase is known an **isocratic elution**.

Single solvent flows continually through the column. This is the most common method used in gel permeation chromatography.

Stepwise or batch elution method For preparative purposes, stepwise or **batch elution** method is commonly employed. In this procedure, the column is eluted with one solvent until a predetermined volume has been applied. Then a second solvent is added to elute the components in the column. This method is used in ion exchange and affinity chromatography.

Gradient elution method In order to increase the resolving power of the mobile phase, it is necessary to change continuously its pH, ionic strength or polarity. This is known as **gradient elution**. In gradient elution method, the column is eluted with a solution in which the concentration of one of the components is gradually increased. This is the most common way of eluting components in adsorption, affinity and ion-exchange columns.

Chromatogram

In column chromatography, if a detector is placed at the end of the column and its signal is plotted as a function of time or volume, a series of symmetric peaks is obtained. Such a plot is called **chromatogram**. The positions of the peaks on the time axis can be used to identify the components of the sample. The areas under the peaks provide a quantitative measure of the amount of compound.

Column chromatography involves **ion exchange**, **molecular sieve**, and **adsorption** or **partition** phenomenon. In adsorption column chromatography the substances are preferentially adsorbed by the adsorbent (stationary phase) packed in the column, while in partition column chromatography, the components of a mixture distribute them and then get separated.

The most commonly used column chromatography methods are discussed below.

ADSORPTION CHROMATOGRAPHY

It was first developed by the American petroleum chemist D.T. Day in 1900. Later, M.S. Tswett, the Polish botanist, in 1906 used adsorption columns in his investigations of plant pigments. It was not until about 1930 that the method was used extensively by chemists. The rate of adsorption varies with a given adsorbent for different materials. This principle of selective adsorption is used in column chromatography. In this method, the mixture to be separated is dissolved in a suitable solvent and allowed to pass through a

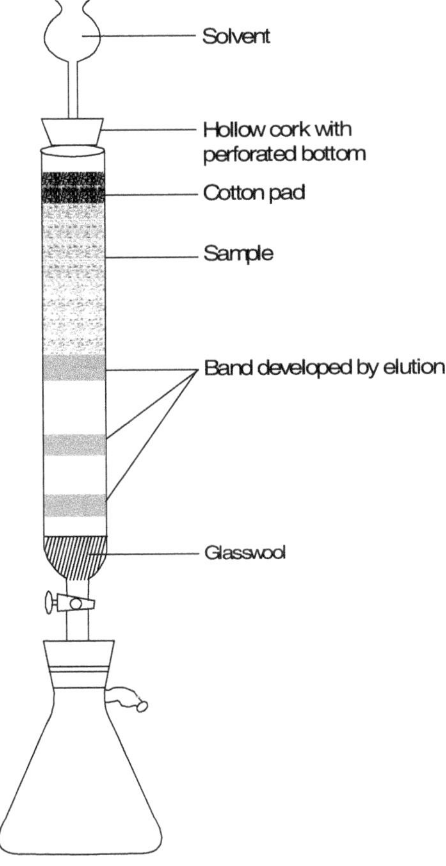

Figure 9.5 Adsorption column chromatography

tube containing the adsorbent. The component, which has greater adsorbing power, is adsorbed in the upper part of the column. The next component is adsorbed in the lower portion of the column, which has lesser adsorbing power than the first component. This process is continued. As a result, the materials are partially separated and adsorbed in the various parts of the column. The initial separation of the various components can be improved by passing either the original or some other suitable solvent slowly through the column. The various bands present in the column become more defined (Figure 9.5). The banded column of adsorbent is termed as a chromatogram. The portion of a column, which is occupied by a particular substance, is called its zone.

Principle

Adsorption chromatography is a technique in which separation of components of a mixture takes place by the adsorption efficiency of the substances to the solid stationary phase. The most strongly adsorbed component forms the topmost band while the least strongly adsorbed material forms the lowermost band on the adsorbent column. Since adsorption is a surface phenomenon, the degree of separation depends upon the surface area of the adsorbent. The distribution coefficient (K) is given by

$$K = \frac{\text{Amount of solute per unit of stationary phase}}{\text{Amount of solute per unit of mobile phase}}$$

The distribution coefficient depends mainly upon the temperature and the concentration of the substance.

Types of Adsorbents (Stationary Phase)

There are a number of adsorbents commercially available for use in adsorption chromatography. The most commonly used adsorbents are Fuller's earth, powdered charcoal, polystyrene beads, hydroxyapatite, silica or silica gel, and alumina or aluminium oxide gel.

Fuller's earth It consists of a mixture of minerals obtained from certain clay deposits. The minerals include Kaolinite, Montmorillonite and Halloysite. It is widely used to decolorize food oils and for removal of pigments from wine samples. The only advantage is its very low cost. It is sold under the trade name of Florisil.

Powdered charcoal It is a powerful adsorbent but its properties are markedly dependent on its origin and on the process by which it is prepared. Some of the resource materials are bone, blood and coconut. Reproducibility of results is always a problem with charcoal. Finely divided charcoal is very difficult to wet and preparing a column packed with charcoal is very troublesome.

Polystyrene beads Polystyrene polymers or polystyrene-divinyl benzene copolymers can be modified as tiny beads, without any charged groups on their surface. By controlled cross-linking, beads can be made with known porosity and swelling properties. As they lack ionized groups, these beads function by adsorption only. They are useful for the separation of alkaloids, steroids, etc. They are commercially available under the trade name "XAD resins".

Hydroxyapatite It is a basic calcium phosphate with the probable formula of $3Ca_3(PO_4)_2 \cdot Ca(OH)_2$. The preparation of hydroxyapatite is tedious and the quality of the product frequently varies from batch to batch. The gel must not be allowed to dry out and freeze. Drying or freezing alters the degree of hydration grossly and leads to a change in the gel structure. It is useful in protein separation.

Silica or silica gel Silica is a very popular adsorbent because it binds a diverse range of substances with moderate strength. Silica gel is prepared by precipitating hydrated silica from sodium silicate on addition of acid under carefully controlled conditions. It is then exhaustively washed, dried and sized precisely. It is less stable than alumina and so should be handled carefully to avoid any change in size by mechanical damage. It can be dried at relatively modest temperatures. Silica is used in the form of crushed and powdered glass, which has very strong adsorptive property for some proteins.

Alumina or aluminium oxide gel Alumina is a powerful adsorbent that is widely used. It is prepared as a precipitate of aluminium hydroxide by addition of sodium hydroxide to some soluble aluminium salts. The gelatinous precipitate is then washed, dried and sized. Depending on the later treatment given, it may be made acidic, basic or neutral, which show the presence of certain groups on the surface. It must be strongly heated to bring it to the anhydrous state (Al_2O_3). Alumina is mechanically quite strong. It has the advantage of not crumbling to still finer particles when manipulated for packing into columns.

Preparation of the Column

1. The powdered substance is poured inside the tube in small lots and the tube is tapped until the necessary length of the column is built up. This method is called dry packing.
2. In the second method, the bottom of the tube is closed and the column is filled with the solvent. The powdered material is added in small quantities and it gently settles down against the viscosity of the solvent. This method of packing is known as wet method. Interlocking of air bubbles in the powder must be avoided in both methods.

Separation Procedure

The different components can be separated from one another as distinct bands across the length of the adsorbent column. The elution is carried out by two different methods.

Extrusion method In this method, the running of the solvent is stopped when reasonable separation of bands has occurred. Then the column is drained, extracted carefully and kept horizontally on a table. The bands are then separated by cutting with the help of a knife. The scrapings are extracted with a solvent and then estimated by a suitable technique.

Liquid chromatography This involves a continuous elution with the solvent so that the components are washed out one after another.

Applications of Adsorption Chromatography

Adsorption chromatography is used for the separation of

i. Polycyclic aromatic compounds, phenols, amines, etc.
ii. Urinary 17-ketosteroids and their glucuronides.
iii. Plasma cortisol.
iv. Aliphatic hydrocarbons from aromatic hydrocarbons.
v. Geometrical isomers.
vi. Technical products in a highly purified state.

ION-EXCHANGE CHROMATOGRAPHY

Ion exchange may be defined as a process concerned with the reversible exchange of ions in solution with ions electrostatically bound to an inert support medium. Ion-exchange separations are carried out usually in columns packed with an ion-exchanger. An ion-exchanger is a solid material that has charged groups to which ions are electrostatically bound. Ion-exchangers can exchange these ions for ions in aqueous solution. They can be used in column chromatography to separate molecules according to the charge.

Principle

The principle of ion-exchange chromatography is based on the attraction between oppositely charged particles. If a mixture of charged and uncharged molecules is passed through a column of ion-exchanger, charged molecules adsorb to ion exchangers reversibly. The column is eluted with buffers of different pH or ionic strength. The components of the eluting buffer compete with the bound material for the binding sites and eventually displace the charged particle originally present in the mixture.

Properties of Ion-Exchangers

An ion-exchanger contains covalently linked charged groups. If a group is negatively charged, it will exchange positive ions and it is called a **cation exchanger**. A commonly used group in cation exchanger is the sulphonic group, SO_3^-. If H^+ is bound to the group, it can exchange one H^+ for one Na^+ or two H^+ for one Ca^{2+}. The sulphonic acid group is called a strongly acidic cation exchanger. Phenolic hydroxyl and carboxyl groups belong to weakly acidic cation exchangers. If the charged group is positive, e.g. quaternary amino groups, it is a strongly basic **anion exchanger**. The most common weakly basic anion exchangers are aromatic or aliphatic amino groups.

The matrix can be made of various materials such as dextran, cellulose, agarose and copolymers of styrene and vinyl benzene. The total capacity of an ion-exchanger describes its ability to take up exchangeable ions and is usually expressed as milliequivalents of exchangeable groups per milligram of dry weight. This number is supplied by the manufacturer and it is an important property because if the capacity exceeds, ions will pass through the column without binding.

Choice of Ion-Exchanger

The decision to choose an ion-exchanger depends on the sample's nature. If the sample has a single charge, the choice is very clear. However, many substances, e.g. proteins carry both negative and positive charges and the net charge depends on pH. If a protein is stable at pH values above the isoelectric point, an anion exchanger should be used. If it is stable at values below the isoelectric point, a cation exchanger should be used.

The choice between strong and weak exchangers is based on the effect of pH on charge and stability of the constituents to be separated. If the sample contains a weakly ionized substance, a strong ion-exchanger can be used. However, if the substance is liable, a weak ion-exchanger is preferable because strong exchangers are capable of denaturing the molecules. Weak ion-exchangers are excellent for the separation of molecules with a high charge, because the weakly charged ions usually fail to bind. In general, weak exchangers are more useful than strong exchangers. The various types of ion-exchangers and few examples for each type is given in the Table 9.3.

Table 9.3 Types of ion-exchangers

Ion exchanger	Matrix	Functional group	Trade name
Strong cationic Exchanger	Dextran	Sulphopropyl	Sephadex
	Styrene-divinyl benzene	Sulphonic acid	AG 50
	Polystyrene	Sulphonic acid	Dowex 50
Weak cationic exchanger	Dextran	Carboxy methyl	CM-Sephadex
	Cellulose	Carboxy methyl	CM-Cellulose
	Acrylic	Carboxylic	Bio-Rex 70
Strong anionic exchanger	Dextran	Diethyl (-2 hydroxy propyl)-aminoethyl	QAE-Sepahdex
	Styrene-divinyl benzene	Tetramethyl ammonium	AG 1
Weak anionic exchanger	Dextran	Diethyl aminoethyl	DEAE-Sephadex
	Cellulose	Diethyl aminoethyl	DEAE-Cellulose
	Expoxyamine	Tertiary amino	AG 3

Buffer and Ionic Conditions

Usually, cationic buffers and anionic buffers are used for anionic exchangers and cationic exchangers respectively. Since ionic strength is a factor in binding, a buffer that has a high buffering capacity must be chosen but the ionic strength need not be too high.

Mechanism of Ion-exchange Chromatography

The exchanger is prepared in a way that it is fully charged. The sample containing the ionic species to be separated is allowed to percolate through the exchanger slowly so that a new equilibrium is achieved.

$$E^+y^- + x^- \rightarrow E^+x^- + y^-$$

where,

E^+ is the charged anion exchanger,

y^- is the counter-ion of the opposite charge associated with the exchanger matrix and

x^- is the charged molecule in the sample to be separated.

This molecule can now exchange sites with the counter-ion. The neutral and cationic molecules will not bind at all. Once the exchange of counter-ion is achieved, x^- can be eluted by decreasing the pH of the solvent. Finally, the anion exchanger is washed with higher concentration of y^- so that the counter-ions are attached again to the anion exchanger, which gets fully charged (Figure 9.6).

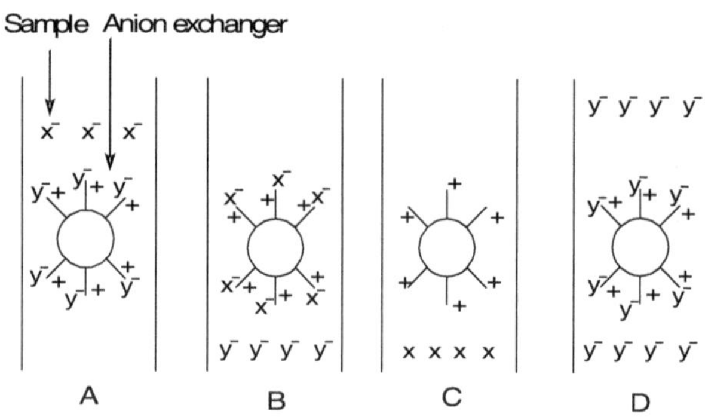

A = Sample applied to an anion exchanger column.

B = Anion x^- has bound to column displacing y^-.

C = Elution by decreasing the pH of the solvent. x^- is converted to an uncharged species and is displaced.

D = Recharging the column with increased concentrations of y^- in the solvent.

Figure 9.6 Mechanism of ion-exchange chromatography

Applications of Ion-exchange Chromatography

i. Resin exchangers are most useful for small organic molecules and metallic ions (e.g. to separate Ca^{2+} from Mg^{2+}).

ii. Metal-ion-free reagents are prepared by ion-exchange chromatography.

iii. Cellulose, dextran and polyacrylamide exchangers are used for the separation of proteins and polysaccharides.

iv. Nucleotides, amino acids, vitamins and other biologically important small molecules are separated using dextran and polyacrylamide exchangers.

v. Ion-exchange chromatography is also used in desalting. A special type of ion-exchange, called ion-retardation resin, is available for this purpose.

vi. A special type of resin, called mixed-bed resin, is used to prepare deionized water. These resins contain equivalent amounts of groups in the H^+ and OH^- forces. As salt ions are taken up, H^+ and OH^- ions, which are just the components of water, are released.

AFFINITY CHROMATOGRAPHY

This technique is useful for the isolation of proteins, polysaccharides, nucleic acids and other classes of naturally occurring compounds. It exploits the functional specificities of biological systems. It can rapidly achieve separations that are time-consuming, difficult or even impossible using other techniques. A complex mixture of substances in solution can be applied to an affinity chromatography column and purification of the substances of interest can be accomplished in a single step.

Principle

Affinity chromatography makes use of a specific affinity between a substance to be isolated and a molecule that can specifically bind (a ligand). The column material is synthesized by covalently coupling a binding molecule (which may be a macromolecule or a small molecule) to an insoluble matrix. The column material is then specifically able to adsorb from the solution, the substance to be isolated. Elution of the desired substance from the column is accomplished by changing the conditions to those in which binding does not occur.

Properties of Matrix

An ideal matrix must possess the following properties:

i. It must contain suitable groups for ligand binding.
ii. It should exhibit good flow properties.
iii. It must be stable during binding of the macromolecule and the subsequent elution.
iv. It should have minimum interaction with other macromolecules.

The commonly available matrix materials for affinity chromatography are given in the Table 9.4.

Table 9.4 Matrix materials for affinity chromatography

Material	Commercial name
Cross-linked dextrans	Sephacryl S
Agarose	Sepharose or Bio-Gel A
Polyacrylamide gel	Bio-Gel P
Polystyrene	Bio-Beads

Selection and Attachment of Ligand

A prior knowledge of the biological specificity of the compound to be purified is essential to select an appropriate ligand. One can select a ligand that shows absolute specificity to the compound to be isolated. Alternatively, a ligand that shows group selectivity can be selected. Such ligands will bind to a closely related group of compounds. An example to this type of ligand is 5′-AMP, which can bind reversibly to many NAD^+-dependent dehydrogenases because 5′-AMP is structurally similar to a part of the NAD^+ molecule.

A ligand can be attached directly to the matrix. The optimum length of this spacer arm is 6–10 carbon atoms. Spacers should have two reactive functional groups at their ends. One reactive functional group is necessary to bind with the matrix and the other reactive functional group is necessary for combining with the ligand. Spacers are required only for small-immobilized ligands. They are not necessary for macromolecular ligands. A number of spacer arms are commercially available. Few examples are 1, 6-diaminohexane, 6-aminohexanoic acid and 1,4-bis (2, 3-epoxypropoxy) butane.

The most common method of attachment of the ligand to the matrix involves the preliminary treatment of the matrix with cyanogen bromide (CNBr). By altering the reaction conditions and the relative proportion of the coupling agents, the number of ligand molecules that are attached to each matrix couple can be changed.

Experimental Procedure

The experimental procedure for affinity chromatography is similar to that used in other forms of liquid column chromatography. The column is packed with the ligand-treated matrix. The buffers used must contain cofactors such as metal ions that are necessary for ligand-macromolecule interaction. The sample is then applied and fractionated. The macromolecule binds with the ligand and gets immobilized. The column is eluted with more buffer to remove non-specifically bound contaminants.

The purified or immobilized compound is recovered from the ligand by either specific or non-specific elution (Figure 9.7). Non-specific elution is achieved by a change in either pH or ionic strength.

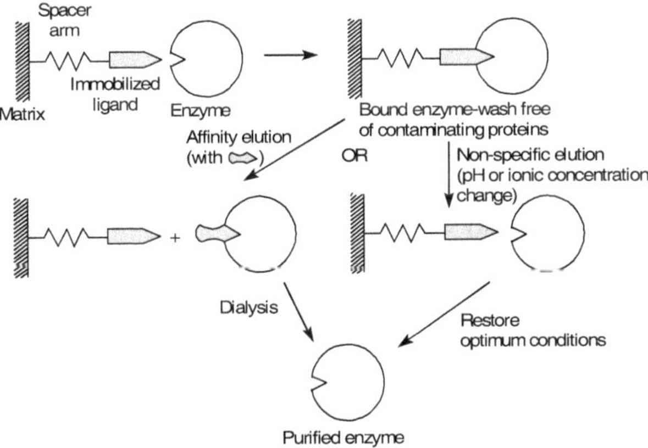

Figure 9.7 Diagram of purification of an enzyme by affinity chromatography

Elution using dilute acetic acid or ammonium hydroxide produces a change in the state of ionization of groups present in the ligand and macromolecule. The pH of the collected fractions is readjusted to the optimum value to reduce the chances for protein denaturation. A change in ionic strength causes elution of macromolecules due to a disruption of the ligand-macromolecule interaction. 1M NaCl is usually employed for this purpose. Finally, the protein is purified from low molecular species by exclusion chromatography. Examples of ligands used in affinity chromatography are given in the Table 9.5.

Table 9.5 Ligands used in affinity chromatography

Ligand	Affinity to macromolecule
Avidin	Biotin-containing enzymes
5′-AMP	NAD^+-dependent dehydrogenases
2′, 5′-ADP	$NADP^+$-dependent dehydrogenases
Fatty acids	Fatty acid binding proteins
Heparin	Lipoprotein, lipases, DNA polymerases
Concanavalin A	Glycoprotein containing α-D-glucose and α-D-mannose
Proteins A and G	Immunoglobulins
Cibacron blue	Nucleotide-requiring enzymes, coagulation factors
Soybean lectin	Glycoprotiens containing N-acetyl-α (or) β D-galactopyranosyl residues

Applications of Affinity Chromatography

Affinity chromatography is used for the purification of a wide range of enzymes and other proteins, including receptor proteins and immunoglobulins.

1. Lectin-sepharose is used for the isolation of glycoproteins, separation of different lymphocytes and for fractionating pure T cells.

2. 5´-AMP- and 2´, 5´-ADP-sepharose are used for the purification of many NAD^+, $NADP^+$- and ATP-dependent dehydrogenases and kinases.
3. Protein A-agarose is used to purify human Immunoglobulin G (IgG) from serum or cell extracts.
4. Cibacron-blue agarose binds to any enzyme having an affinity with nucleotides since it is structurally similar to nucleotides. It is used for the isolation of enzymes such as kinases, dehydrogenases, albumin, DNA polymerase and coagulation factors.
5. Poly(A)-agarose binds nuclear mRNA molecules having poly(U) sequences. It is used to fractionate ribonucleoproteins, RNA polymerases and many viral mRNA molecules. Lysine-agarose binds ribosomal RNA and the protein, plasminogen.
6. Heparin-agarose is used for the purification of bone collagenase, hepatitis B surface antigen, uterine estradiol receptor, murine myeloma DNA polymerase, ribosomes, adrenal tyrosine hydroxylase and several androgen receptors.

GEL PERMEATION CHROMATOGRAPHY OR GEL FILTRATION CHROMATOGRAPHY OR MOLECULAR SIEVE CHROMATOGRAPHY

Gel permeation chromatography is also known as gel filtration chromatography or molecular sieve chromatography. It is a special type of partition chromatography in which the separation of macromolecules is based only on molecular size. A column is prepared of tiny particles of an inert substance that contains small pores. If a solution containing different molecular weight substances is passed through the column, molecules larger than the pores move only in the space between the particles and hence are not retarded by the column material. However, molecules smaller than the pores diffuse in and out of particles. In this way, they are slowed in their movement down the column. Hence, molecules are eluted from the column in the order of decreasing size or decreasing molecular weight (Figure 9.8).

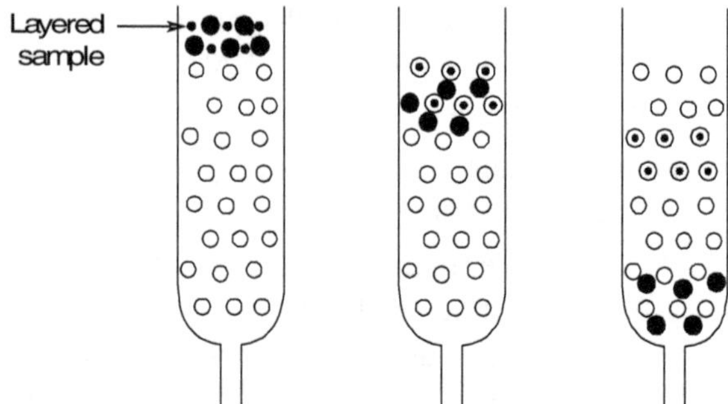

Figure 9.8 Mechanism of gel filtration chromatography

Properties and Types of Gels

The gels used as molecular sieves consist of cross-linked polymers that are generally inert. They are uncharged molecules and they do not bind or react with the sample being analysed. The space present within the gel is filled with liquid. Three types of gels are currently used. They are dextran, agarose and polyacrylamide.

Dextran is a polysaccharide composed of glucose residues. It is produced by the fermentation of sucrose and the fermentation is caused by the microorganism, *Leuconostoc mesenteroides*. It is prepared with various degrees of cross-linking to control pore size and is supplied as dry beads. It is commercially available under the trade name Sephadex.

Polyacrylamide gels are prepared by cross-linking acrylamide with N, N′-methylene-bisacrylamide. The pore size is determined by the degree of cross-linking. These gels contain a polar carboxylamide group on alternate carbon atoms. They are commercially available under the trade name Bio-gel.

Agarose gels are obtained from certain seaweeds. Agarose is a linear polymer of D-galactose and 3,6-anhydro-1-galactose. It is dissolved in boiling water and forms a gel when cooled. The concentration of the material in the gel determines the size of the pores. Agarose is supplied in the market as wet beads called Sepharose. Table 9.6 gives a list of gels that can be used for the separation of substances of different molecular weight.

Table 9.6 Gels used for the separation of substances of different molecular weights

Material	Trade name	Fractionation range (Mol. wt.)
Dextran	Sephadex G-10	0 – 100
	Sephadex G-25	1,000 – 5,000
	Sephadex G-75	3,000 – 70,000
	Sephadex G-100	4,000 – 1,50,000
	Sephadex G-200	5,000 – 8,00,000
Agarose	Sepharose 2B	$2 \times 10^6 - 25 \times 10^6$
	Sepharose 6B	$1 \times 10^4 - 20 \times 10^6$
Polyacrylamide	Biogel P-2	100 –1,800
	Biogel P-60	3,000 – 60,000
	Biogel P-150	15,000 –1,50,000
	Biogel P-300	60,000 – 4,00,000

Advantages of Gel Chromatography

1. The separation of all substances is independent of pH, temperature, ionic strength and buffer composition.
2. Since there is no adsorption, very labile substances are not affected by this chromatography method.
3. There is less zone spreading than with other chromatographic techniques.
4. The elution volume is directly related to molecular weight.

Estimation of Molecular Weight of a Protein

One of the most important uses of gel permeation chromatography is the determination of the molecular weight of proteins. To determine the molecular weight of a protein, the elution volumes of different proteins with known molecular weight are first measured.

A plot of the elution volume against the logarithm of the molecular weight gives a straight line. The molecular weight of unknown protein is calculated by interpolating elution volume as shown in the Figure 9.9a.

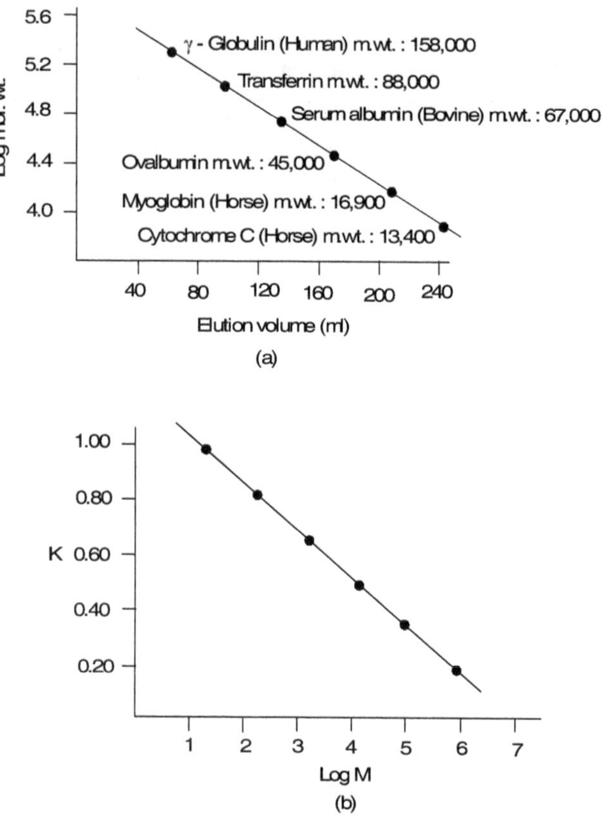

Figure 9.9 Molecular weight determination

Alternatively, a plot of parameter K versus log M (molecular weight) yields a straight line (Figure 9.9b).

$$K = \frac{V_e - V_o}{V_s}$$

where,

V_e is the elution volume required to elute a particular substance from the column.

V_o is the void volume required to elute a molecule that never entered the stationary phase. It is measured usually by passing the excluded high molecular weight material, dextran blue, through the column.

V_s is the volume of the stationary phase. It is measured as $V_t - V_o$ in which V_t is the total volume of the column.

The molecular weight of a protein can be calculated by interpolating the K as shown in the Figure 9.9b. The precision of these determinations is about 90%. The error is primarily a result of the breadth of the band obtained during elution.

Applications of Gel Permeation Chromatography

1. It has widespread use in purifying enzymes and other proteins and in fractionating nucleic acids.
2. It is used in the determination of molecular weight of proteins.
3. It is used for desalting low molecular weight compounds during protein purification.
4. Using agarose gels various types of RNA and viruses have been successfully fractionated and purified.
5. In the assay of enzymes for the determination of cofactor requirements, the enzyme preparation sometimes contains inhibitors of small molecular size of the cofactors themselves. Such molecules are easily removed with the dextran of polyacrylamide gels.
6. Solutions of high molecular weight substances can be concentrated by the addition of dry Sephadex G-25. Water and low molecular weight substances are absorbed by the swelling gel while the high molecular weight substances remain in solution.
7. It is used to study the binding of ligands with proteins or receptors.

LIQUID–LIQUID CHROMATOGRAPHY

In liquid–liquid chromatography, the liquid stationary phase is attached to a supporting matrix by purely physical means, and in bonded-phase liquid chromatography, the stationary phase is covalently attached to the matrix. An example of liquid–liquid chromatography is one in which a water stationary phase is supported by a cellulose, starch or silica matrix, all of which have the ability to bind physically as much as 50% (w/v) water and remain free-flowing powders. The advantages of this form of chromatography are that it is cheap, has a high capacity and has broad selectivity. Its disadvantage is that the elution process may gradually remove the stationary phase, thereby altering the chromatographic conditions. This problem is overcome by the bonded-phase liquid chromatography. Most bonded-phases use silica as the matrix, which is derivatized to immobilize the stationary phase by reaction with an organochlorosilane. This process is called **silanization**. There are three commonly used modes of liquid–liquid partition chromatography that differ in the relative properties of the stationary and mobile phases. They are normal phase liquid chromatography, reversed-phase chromatography, and countercurrent chromatography.

Normal-Phase Liquid Chromatography

In this form of chromatography, the stationary phase is polar and the mobile phase is relatively non-polar. The most popular stationary phase is an alkylamine bonded to silica. The mobile phase is generally an organic solvent such as hexane, heptane, dichloromethane or ethylacetate. The mechanism of separation exploits the ability of the analyte to displace molecules of the mobile phase adsorbed as a monolayer on the stationary phase, as well as the ability of the analytes to compete with molecules of the mobile phase in the formation of a bilayer on the stationary phase surface. The order of elution of analytes is such that the least polar is eluted first and most polar last. Indeed, polar analytes generally require gradient elution with a mobile phase of increasing polarity, generally achieved by the use of methanol or dioxane. The main advantage of normal phase liquid chromatography is its ability to separate analytes that have low water solubility.

Reversed-Phase Liquid Chromatography

In this form of liquid chromatography, the stationary phase is non-polar and the mobile phase relatively polar. By far the most commonly used type is the bonded-phase form, in which alkylsilane groups are chemically attached to silica. The mobile phase is commonly water or aqueous buffers, methanol, acetonitrate, or tetrahydrofuran or mixtures of them. Reversed-phase liquid chromatography differs from most other forms of chromatography in that the stationary phase is essentially inert and only non polar (hydrophobic) interactions with analytes are possible. Chromatographic separation of analytes is determined principally by the characteristics of the mobile phase and probably involves a combination of adsorption and partition mechanisms.

In reversed-phase chromatography, polar analytes elute first and non-polar analytes last. Non-polar analytes may need a gradient using increasing proportions of a polarity solvent such as hexane.

Reversed-phase HPLC is probably the most widely used form of chromatography, mainly because of its flexibility and high resolution. It is widely used to analyse drugs and their metabolites, insecticide and pesticide residues, and amino acids. In non-polar form, reversed-phase chromatography can be used to separate lipophilic compounds such as fats.

Countercurrent Chromatography (CCC)

This separation process is based upon the distribution of a compound between two immiscible liquid phases. These phases may be mixtures of organic solvents, buffers, salts and various complexing agents. The ratio of the concentration of the compound in the two phases at equilibrium is called the partition coefficient. A mixture of substances with different partition coefficients can be quantitatively separated by a technique known as countercurrent distribution.

The apparatus most commonly used is the Craig's Countercurrent Distribution apparatus (Figure 9.10). It consists of 30–1000 interconnected H-shaped vessels (the so-called train), each of which retains a fixed volume of the stationary liquid phase. The solute mixture is introduced into the first vessel in the train and equilibrated with the immiscible and less dense mobile phase by the repeated rocking of the vessel through 90°. After equilibration is complete (1 to 2 min), the mobile phase is transferred to the next vessel as a result of the complete tipping of the first vessel. When this returns to its original position, the fresh upper phase is introduced automatically. The whole process is repeated so that the mobile phase is transferred progressively along the series of lower phases. The solutes are transferred at a rate determined by their distribution coefficients and the relative volumes of the two solvents in each vessel. Each solute eventually accumulates in a specific group of vessels, the resolution being determined by the total number of transfers and their differences in partition coefficient. A number of miniaturized versions of the technique are now available commercially.

CCC is one of the few forms of chromatography that have been used successfully for cell organelle fractionation. It has also been useful for cell fractionation and membrane receptor isolation.

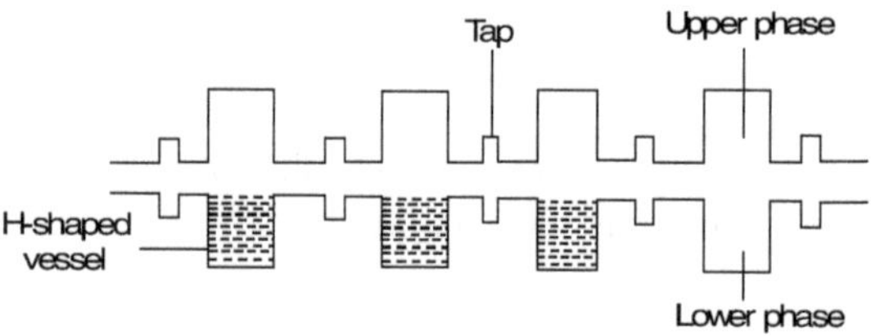

Figure 9.10 Craig's countercurrent distribution apparatus

GAS–LIQUID CHROMATOGRAPHY

Gas–Liquid Chromatography (GLC) or simply gas chromatography (GC) is a type of partition chromatography in which the mobile phase is a carrier gas, usually an inert gas such as helium or nitrogen, and the stationary phase is a microscopic layer of liquid on an inert solid support. A gas chromatograph uses a thin capillary fibre known as the column, through which different chemicals pass at different rates depending on various chemical and physical properties. As the chemicals exit at the end of the column, they are detected and identified electronically (Figure 9.11). The function of the column is to separate and concentrate different components in order to maximize the detection signal.

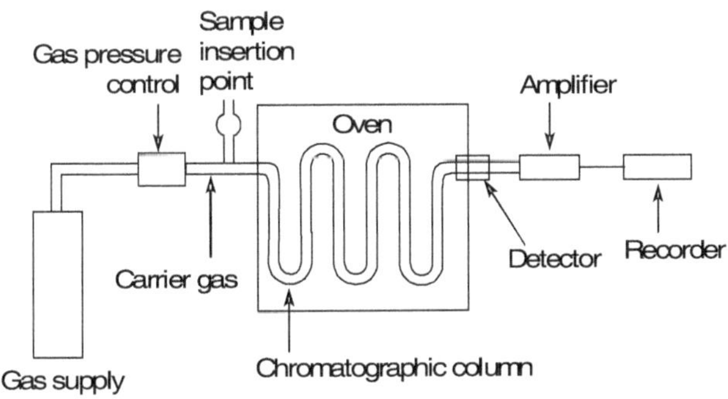

Figure 9.11 Gas–liquid chromatography

Principle

A stationary phase of a high-boiling liquid material is supported on an inert granular solid. This material is packed into a narrow coiled glass or steel column, through which an inert carrier gas such as nitrogen, helium or argon is passed. The column is maintained in an oven at an elevated temperature that enables the compounds to be separated in the evaporating state. The basis for the separation is the difference in the partition coefficients of the volatilized compounds between the liquid and gas phases. As the compounds are carried through the column by the carrier gas, they pass through a detector that is linked via an amplifier to a chart recorder. The chart recorder marks a peak as each separated compound passes through it.

Instrumental Components

Carrier gas The carrier gas must be chemically inert. Commonly used gases include nitrogen, helium, argon, and carbon dioxide. The choice of a carrier gas is often dependent upon the type of detector used. The carrier gas system also contains a molecular sieve to remove water and other impurities.

Sample injection port For optimum column efficiency, the sample should not be too large, and should be introduced onto the column as a "plug" of vapour. Slow injection of large samples causes band broadening and loss of resolution. A microsyringe is used to inject the sample through a rubber septum into a flash vaporizer port at the head of the column. The temperature of the sample port is usually about 50°C higher than the boiling point of the least volatile component of the sample. For packed columns, sample size ranges from tenths of a microlitre up to 20 microlitres. Capillary columns, on the other hand, need much less sample, typically around 10^{-3} μl. For capillary GC, split or splitless injector is used.

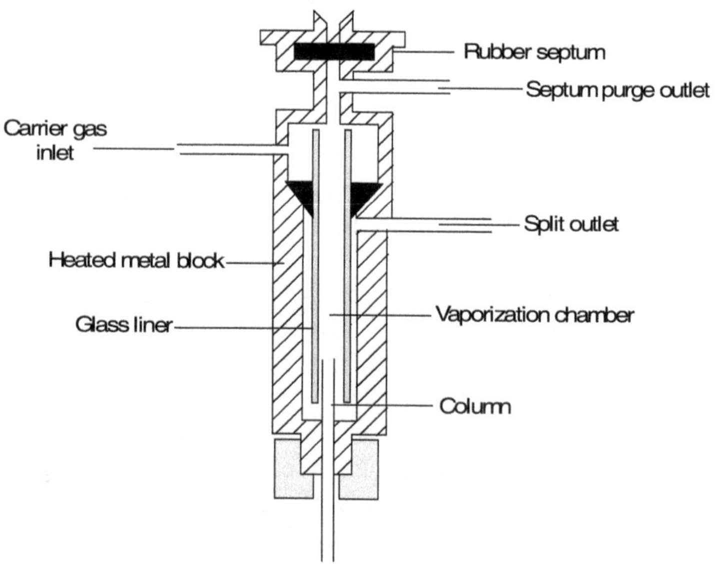

Figure 9.12 Split/splitless injector

Split or splitless injector contains a heated chamber containing a glass liner into which the sample is injected through the septum. The carrier gas enters the chamber through the carrier gas inlet. The sample vaporizes to form a mixture of carrier gas, vaporized solvent and vaporized solutes. A proportion of this mixture passes onto the column, but most of it exits through the split outlet. The septum purge outlet prevents septum bleed components from entering the column (Figure 9.12).

Columns There are two general types of column, **packed** and **capillary** (also known as **open tubular**). Packed columns contain a finely divided, inert, solid support material (commonly based on diatomaceous earth) coated with liquid stationary phase. Mostly packed columns are 1.5–10 m in length and have an internal diameter of 2–4 mm. Capillary columns have an internal diameter of a few tenths of a millimetre. They can be one of two types—**wall-coated open tubular** (WCOT) or **support-coated open tubular** (SCOT) or **porous layer open tubular column** (PLOT). Wall-coated columns consist of a capillary tube whose walls are coated with liquid stationary phase. In support-coated columns, the inner wall of the capillary is lined with a thin layer of support material such as diatomaceous earth, onto which the stationary phase has been adsorbed. SCOT columns are generally less efficient than WCOT columns. Both types of capillary columns are more efficient than packed columns. In 1979, a new type of WCOT column was devised—the **fused silica open tubular** (FSOT) column (Figure 9.13). These have much thinner walls than the glass capillary columns, and are given strength by the polyimide coating. These columns are flexible and can be wound into coils. They have the advantages of physical strength, flexibility and low reactivity.

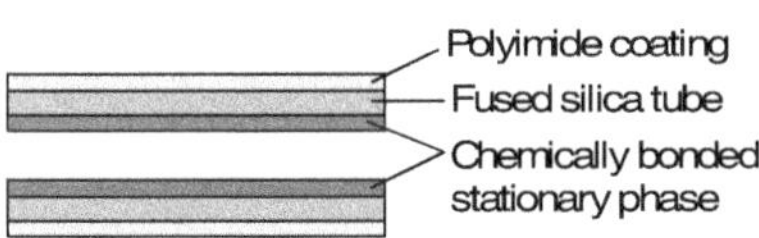

Figure 9.13 Cross-section of a fused silica open tubular column

Matrix or support material Support material is present only in PLOT columns. This is used to provide a supporting surface for stationary phase. The most commonly used support material is diatomaceous silica (Celite). To avoid direct exposure of the support material to the sample, the hydroxyl groups present in the celite are modified. This is done by silanization of the support material with hexamethyl disilazane. Besides the support material, the glass column and the glass wool plug located at the base of the column are also silanized.

Stationary phase Stationary phase chemicals are high-boiling organic compounds. Examples are polyethylene glycols, methyl phenyl silicon gum and esters of adipic, succinic and phthalic acids. Depending upon the analysis, the concentration of the stationary phase coated onto the support material varies from 1% to 25% loading. The choice of stationary phase for analysis depends on the mixture of compounds to be separated. There are two types of stationary phase organic compounds. The first type, known as selective phase organic compounds, is used where separation occurs on the basis of different chemical characteristics of components. The second type, known as a non-selective phase organic compounds, is used where separation is based on the differences in boiling points of the sample components.

The columns are dry-packed under a slight positive gaseous pressure. The column is then conditioned for about 48 hours by heating near to the upper working temperature limit. To avoid contamination of detector during conditioning, the column is disconnected from the detector. The operating temperature for the analysis should not exceed the boiling point of the stationary organic phase chosen. Very high temperature results in volatilization of stationary phase itself and contamination of the detector, which produces unstable baseline by the recorder.

Preparation of Sample

Polar compounds are retained in the column for excessive periods and this leads to poor resolution and peak tailing. Therefore, they are not directly applied. Conversion of these polar groups (e.g. –OH, $–NH_2$, –COOH) into non-polar derivatives increases the volatility and effective distribution coefficients of these compounds. Silanization, methylation and perfluoroacylation are common conversion methods for carbohydrates, fatty acids and amino acids respectively. Non-polar organic compounds do not need any such conversions and they can be directly applied.

Application of Sample

The sample for GLC is dissolved in a suitable solvent such as acetone or methanol and is injected into the column using a microsyringe through a septum in the injection port. The injection area of the column is kept usually at a slightly higher temperature than the column. This helps in a rapid and complete volatilization of the sample.

Column Temperature

For precise work, column temperature must be controlled to within tenths of a degree. The optimum column temperature is dependent upon the boiling point of the sample. As a rule, a temperature slightly above the average boiling point of the sample results in an elution time of 2–30 minutes. Minimal temperatures give good resolution, but increase elution time. If a sample has a wide boiling range, then temperature programming can be useful. The column temperature is increased (either continuously or in steps) as separation proceeds.

Separation Procedure

The most commonly used carrier gases are nitrogen, helium and argon. These gases are passed at a flow rate of 40 to 80 ml per minute through the column. The column temperature is maintained according to the stationary phase chosen. Since partition coefficients are sensitive to temperature, two types of temperature control techniques are employed. Maintaining a constant temperature throughout the column length is known as **isothermal analysis**.

Increasing the temperature gradually throughout the column length is referred to as **temperature programming**.

Detectors

A number of detectors are used in gas chromatography. The most common one is the **thermal conductivity detector** (TCD), which monitors changes in the thermal conductivity of the effluent. The main advantage of the TCD is that it can detect any substance (except the carrier gas). Some of the other detectors are sensitive only to specific types of substances. Other detectors include the flame ionization detector (FID), electron capture detector (ECD), **flame photometric detector** (FPD), **photo-ionization detector** (PID), and **Hall electrolytic conductivity detector**. Some of the commonly used detectors are given in the Table 9.7.

Table 9.7 GC detectors

Detector	Type	Support gases	Selectivity	Detectability
Flame ionization (FID)	Mass flow	Hydrogen and air	Most organic compounds	100 pg
Thermal conductivity (TCD)	Concentration	Reference	Universal	1 ng
Electron capture (ECD)	Concentration	Make-up	Halides, nitrates, nitrites, peroxides, anhydrides, organometallics	50 fg
Nitrogen–Phosphorus	Mass flow	Hydrogen and air	Nitrogen, phosphorus	10 pg
Flame photometric (FPD)	Mass flow	Hydrogen and air, possibly oxygen	Sulphur, phosphorus, tin, boron, arsenic, germanium, selenium, chromium	100 pg

(Contd.)

Table 9.7 (Continued)

Detector	Type	Support gases	Selectivity	Detectability
Photo-ionization (PID)	Concentration	Make-up	Aliphatics, aromatics, ketones, esters, aldehydes, amines, heterocyclics, organosulphurs, some organometallics	2 pg
Hall electrolytic conductivity	Mass flow	Hydrogen, oxygen	Halide, nitrogen, nitrosamine, sulphur	

In FID (Figure 9.14), the effluent from the column is mixed with hydrogen and air, and ignited. Organic compounds burning

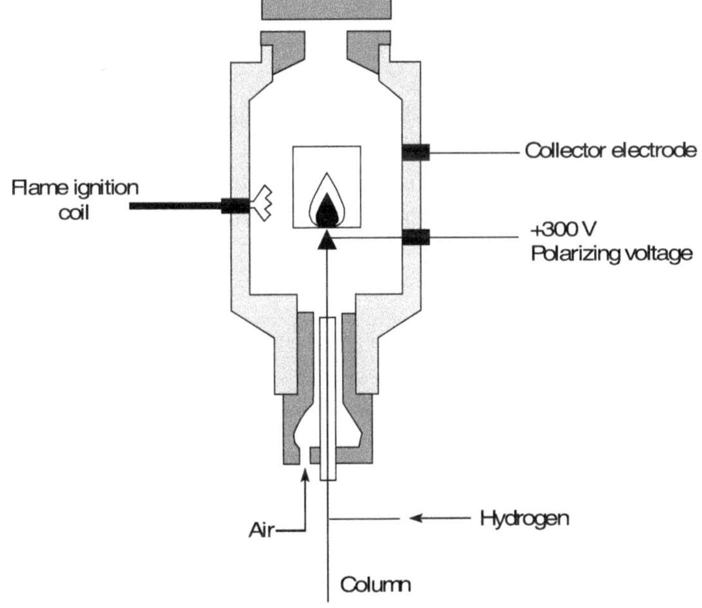

Figure 9.14 Flame ionization detector

in the flame produce ions and electrons, which can conduct electricity through the flame. A large electrical potential is applied at the burner tip, and a collector electrode is located above the flame. The current resulting from the pyrolysis of any organic compound is measured. FIDs are mass-sensitive rather than concentration-sensitive; this gives the advantage that changes in the flow rate of mobile phase do not affect the detector's response. The FID is a useful general detector for the analysis of organic compounds; it has high sensitivity, a large linear response range, and low noise. It is also robust and easy to use, but unfortunately, it destroys the sample.

Identification of Compounds

A detector indicates the amount of material emerging from the column. It does not indicate directly the identity of the compounds that produces a particular peak. There are various means of identifying peaks. For example, if the substances present in a mixture are known in advance and the purpose of the investigation is to measure the amount of components present in the mixture, a duplicate sample containing a small amount of an added known substance to the mixture is rechromatographed. If the known substance is the same as one of the components in the mixture, the size of that peak will increase. If the identification of unknown substances is still difficult, the most useful method of identification is mass spectroscopy.

Advantages of GLC

1. Due to the great length of the column, the separation of low molecular weight compounds is much easier.
2. Sensitivity is extraordinary. A concentration of 10^{-12} gram is detectable for many substances.
3. Since separations are frequently achieved in less than a minute, the speed of separation is also extraordinary.
4. Using a non-destructive detector, GLC can be used for preparative purposes.

Uses of GLC

1. It is widely used for the qualitative and quantitative analyses of a large number of compounds because this technique provides a high speed of resolution, very good reproducibility and high sensitivity. It is based on the partitioning of compounds between a liquid and gas phase.
2. Thousands of volatile organic compounds can be separated by GLC.
3. Non-volatile substances can also be separated if converted into volatile ones by oxidation, acylation, alkylation, etc.
4. Alcohols, esters, fatty acids and amines present in biological samples are often separated by GLC.
5. Concentration of individual elements such as carbon and hydrogen can be determined very accurately.
6. Functional groups can also be identified. For example, alkoxy groups are detected by iodination to form an alkyl iodide, which is easily identified.
7. The position of a double bond can be determined by cleaving the bond by oxidation or ozonolysis. The products are identified by chromatography.

HIGH PERFORMANCE LIQUID CHROMATOGRAPHY (HPLC)

A wide variety of separation techniques have been described earlier. Gas chromatographic technique could successfully separate and analyse components in a relatively shorter time and with more precision as compared to other techniques of chromatography. However, gas chromatographic technique has certain limitations as only those substances which can be evaporated in the temperature range of 300–400°C could be analysed. There are approximately 20% of the known organic compounds, which have the stability and volatility to be excited into a gaseous phase at such a temperature range. This property has imposed a limitation, and only a small portion of organic compounds can be analysed by gas chromatography. The technique of liquid chromatography on the other hand is considered to be a more versatile technique as it can successfully analyse the remaining 80% of the organic compounds.

Liquid chromatography is a slow separating technique which is performed in vertical columns and under gravitational flow.

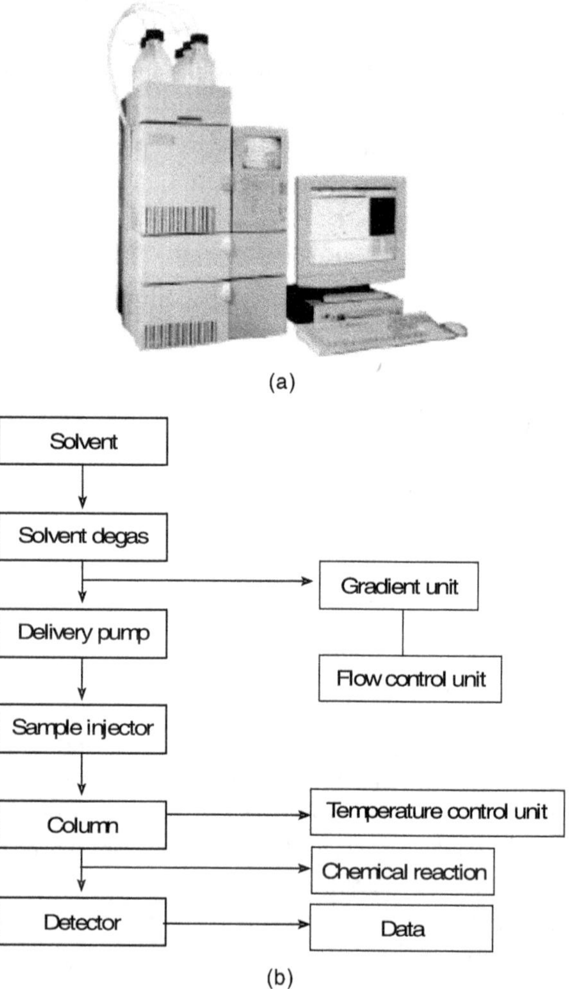

Figure 9.15 High performance liquid chromatography a) Experimental unit b) Flow diagram

Recent developments have improved its speed and versatility. The increase in speed has been achieved by pumping the solution through a column at a pressure of up to more than 10,000 psi. The versatility has been achieved by using column particles of smaller diameter and surface area and by using suitable packing structure. These improvements have provided about seven times more resolving power than a normal liquid chromatographic technique. Such an improved technique is known as high performance liquid chromatography (HPLC). The characteristic features of HPLC are sensitivity, ready adaptability to accurate quantitative determination, and suitability for separating non-volatile species or thermally fragile species. HPLC allows separation and measurements to be made in a matter of minutes. This technique has also been given a few other names, such as high pressure liquid chromatography and high speed liquid chromatography. HPLC system consists of four major components. These are (i) pump (ii) injector (iii) column and (iv) detector. The essential components of HPLC are shown in the Figures 9.15a and 9.15b.

HPLC is equipped with one or more glass or steel reservoirs which are equipped with a means to remove dissolved gases, usually oxygen and nitrogen. It is essential to remove them as they interfere by forming bubbles in the column and detector system. The presence of these bubbles leads to spreading of bands. The methods for removing the bubbles consist either of a vacuum pumping system, or a device for heating and stirring the solvents. The bubbles are best removed by passing the solvent through a multipore filter under vacuum before introducing into the solvent reservoir.

Separation by HPLC, which employs a single solvent of constant composition, is called an **isocratic elution**. When two or more solvent systems of significant differences in polarity are used, the method is called **gradient elution**. Accordingly the techniques are called isocratic or gradient techniques. Modern HPLC equipment is often provided with devices which introduce solvents from two or more reservoirs into a mixing chamber.

Different Components of HPLC

Delivery pumps These pumps deliver a steady stream of solvent from the reservoir to the detector through the column. This is one of the most important components of HPLC. Delivery pumps should have sufficient pressure, resistance, less pulsation and discharge of constant volume. The pump can deliver solvent at a pressure up to 10,000 psi with a flow rate of over 50 millilitres per minute. About 80% of separations done by HPLC require pressure less than 1200 psi. There are three types of pumps, which are commonly used. These are (i) reciprocating pump (ii) displacement pump and (iii) pneumatic pump.

i. **Reciprocating pumps** This type of pump is widely used. The pump consists of a small chamber in which the solvent is pumped by a back and forth motion of a motor-driven piston (Figure 9.16). The advantage of this pump is that it facilitates replacement of solvents. The disadvantage is that this type of pump causes pulsed flow.

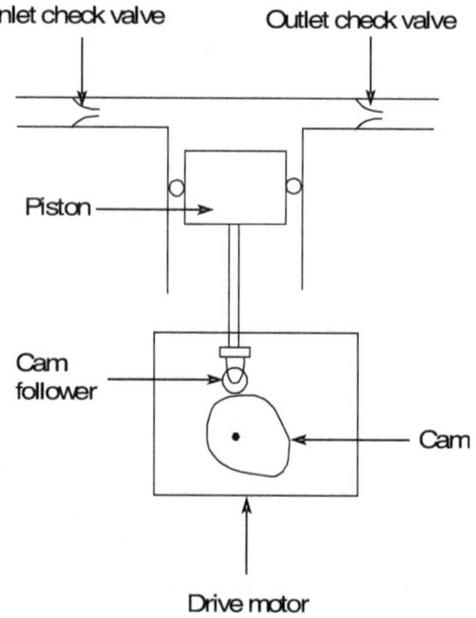

Figure 9.16 Reciprocating pump

ii. **Displacement pump** This pump consists of large syringe-like chambers equipped with a plunger, which is activated by a screwdriver mechanism through a motor (Figure 9.17). This type of pump produces a flow which is quite independent of viscosity and backpressure and with no pulsed flow. The disadvantage is that it has limited solvent capacity.

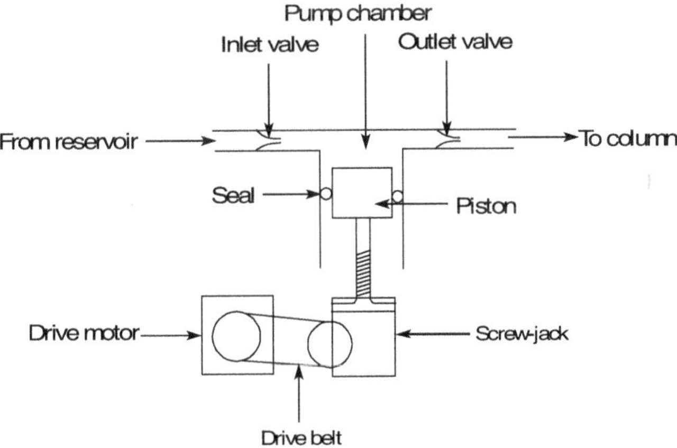

Figure 9.17 Displacement pump

iii. **Pneumatic pump** This pump contains a mobile phase which is contained in a collapsible container and placed in a vessel. This can be pressurized by a compressed gas. The advantage is that these pumps are inexpensive and pulse-free. The disadvantage is that it depends on solvent viscosity and column backpressure.

Sample injection unit An unknown sample is introduced into the flowing stream of solvent with an injector. The volume of sample used ranges from 0 to 500 μl and since the system operates at high pressures, the injection process is complicated. Accordingly various injection methods have been designed. These are described below:

i. **Syringe injection** The simplest means of sample introduction is done by using microsyringes which are designed to withstand the pressures up to 1500 psi. The injection is done through a self-sealing elastomeric septum.

ii. **Stop flow injection** In this type of syringe injection no septum is used. The flow of the solvent is momentarily stopped and the sample is directly injected onto the head of the column. This method is very convenient and simple.

iii. **Sampling valve** This sampling device is most widely used. In this method, a sample loop of a fixed capacity is connected to a high-pressure valve and the sample is filled into the sample loop through a syringe. Sampling loops permit the introduction of sample at pressures up to 7000 psi.

Columns The most important part of conducting separation is column. The column is made of stainless steel, which differs in length and inside diameter depending on their application (Figure 9.18). The column size ranges in length from 10–30 cm with the inside diameter being 4 to 10 mm. Microcolumns as short as 3.0 to 7.5 cm with an inside diameter ranging from 1.0 to 4.6 mm are most widely used. In most of the cases, the columns are often used at room temperature. However, the temperature of the column affects speed of affinity and diffusion between the sample and the column. In addition, temperature also affects the solubility of the sample and viscosity of the solvent. Thus use of columns in the thermostatic ovens is recommended. The method used for the column in thermostatic ovens is either jacket method or air-circulating method. Most instruments are equipped with ovens that permit the use of controlled temperature from a few degrees above ambient to 100–150°C.

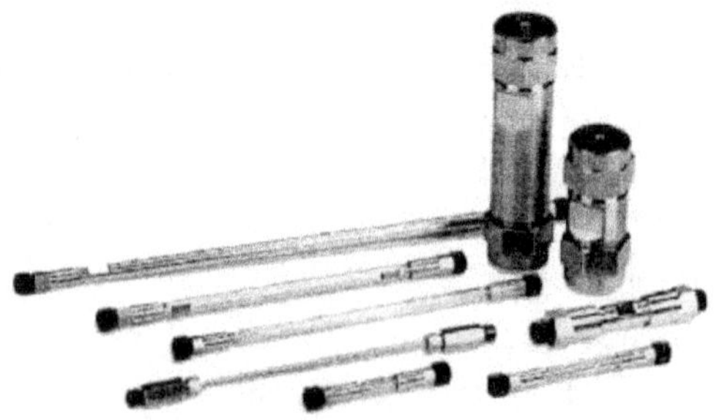

Figure 9.18 HPLC column

Column packing Several kinds of column packing are used. There are essentially three types of particles which are used for packing. These are given below.

i. **Pellicular particles** These consist of glass beads whose surfaces are coated with a thin silica layer. Interaction for the separation is produced on the stationary phase supported by thin silica layer. These particles have a diameter of 30–35 μm.
ii. **Microporous particles** These particles are composed of silica, alumina, or ion exchange resins and have diameters ranging from 3 to 10 μm. In this case only small solute molecules reach the pores where they interact with the stationary phase.
iii. **Macroporous particles** These contain macropores in addition to micropores, the particles of which have diameter of several hundred angstroms. These particles have high porosities.

Although there are a great variety of packing materials, all are based on the principles of separation by size, absorption, solubility and ion exchange.

The technique of HPLC is used for the following

a. Liquid–solid separation chromatography
b. Liquid–liquid partition chromatography
c. Gel permeation chromatography

The particles are thus selected according to the nature of substances to be separated. The selected particles are first dried in an oven before coating. They are then packed dry, after which the mobile phase is passed through the column for about two hours to remove air and allow the absorbent to equilibrate with the mobile phase.

Successful separation could be achieved by establishing proper balance between the attraction of the solvent and solid support for the sample molecules. In liquid chromatography, solvent and solid packing both compete for the sample. If the sample is chemically more like the solvent, then it is necessary to make the solvent less like the sample by changing the solvent polarity. Alternatively the packing may be made similar to the sample. Better sample separation is achieved by matching the polarities of the sample and packing and by using a solvent that has very different polarity. The relative polarities of commonly used solvents are graphically represented as the "eluotropic series". Solvents in the eluotropic series with comparable polarity rankings may have quite different solubility parameters. Although polarity is a useful basis of ranking solvent, other solvent characteristics like hydrogen bonding also affect the solvent strength. These factors are taken into account in "solubility parameter".

Selection of solvent One of the major problems is the selection of mobile phase. In general, highly polar materials are separated using partition chromatography while non-polar materials are separated using adsorption chromatography, and very high non-polar materials are separated using partition chromatography.

Detectors The column retains each component for a different length of time. The individual components leave the column one after another. The relative concentration of each component is recorded with respect to time with the help of detectors. All detectors used in

Table 9.8 Detectors used in HPLC

Detector basis	Type	Max. sensitivity
UV absorption	Selective	2×10^{-10}
IR absorption	Selective	1×10^{-5}
Fluorometry	Selective	1×10^{-11}
Refractive index	Non-selective	1×10^{-7}
Conductometric	Selective	1×10^{-8}
Mass spectrometry	Non-selective	1×10^{-10}
Electrochemical	Selective	1×10^{-12}
Radioactivity	Selective	–

HPLC are classified into two main types—a selective detector which detects only a part of the components of a sample and a non-selective detector which detects almost all the components. The characteristics of the detectors used in HPLC are given in Table 9.8.

i. **Ultraviolet absorbance photometric detector** Initially a fixed wavelength detector was used, which used low pressure mercury lamp as light source and its strongest bright line located at 254 nm as measuring wavelength. In recent times, a variable wavelength type based on photometric system is used. The most powerful ultraviolet

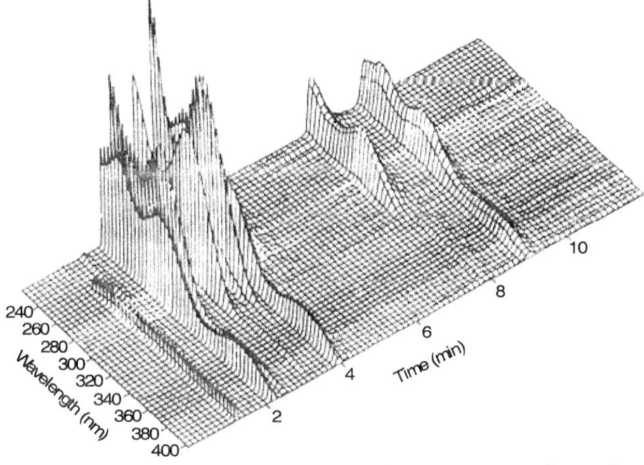

Figure 9.19 Simultaneous measurement of absorbance at all wavelengths by diode array technique

spectrophotometric detectors are diode array instrument UV detector whose sensitivity may go up to 0.06 ppm for compounds with high absorption. Diode array technique allows the simultaneous measurement of absorbance at many or all wavelengths within 0.01s (Figure 9.19).

ii. **Fluorometric detector** This detector is used to detect only fluorescent substances. Fluorometric compounds are frequently encountered in the analysis of such materials as pharmaceuticals, natural products, clinical samples and petroleum products. In case substances do not possess fluorescence, a colouring reagent is added to convert them into a fluorescent compound.

iii. **Refractive index detector** Refractive index detectors have the significant advantage of separating all solutes. They are the general detectors like flame detectors in gas chromatography. These detectors have some disadvantages such as low sensitivity, tendency to be affected by temperature and inability to conduct gradient elution.

iv. **Electrochemical detectors** Electrochemical detectors of several types are used. The devices are based upon the methods such as amperometry, polariography, coulometry and conductometry. The electrochemical detector appears to have a potential for fulfilling a long time need to work as universal detector. The electrochemical detector has been particularly successful in the assay of catecholamines, vitamins and antioxidants.

Recorder The signals from the detectors are recorded as deviation from a baseline. Two-open recorders are used with instruments having two detectors. The peak position along the curve relative to the starting popping, denotes the particular component. With proper calibration, the height or area of peak is a measure of the amount of the component present in the sample.

Qualitative and Quantitative Analysis by HPLC

In HPLC method the separated components are recorded as a peak. The resulting plot of detector response versus time is known as chromatogram. The qualitative and quantitative analysis of the separated components is done by comparing the chromatogram with a standard substance. The analysis to detect what substances each peak contains is termed as qualitative analysis whereas the analysis of the volume of such substances is termed as quantitative analysis. The greatest advance in detection using HPLC has been made by coupling of the HPLC to a mass spectrometer. More recently, HPLC has been interfaced with NMR spectroscopy to give structural information about the analytes.

Applications

All types of biological molecules can be assayed or purified using HPLC.

HPLC has a number of advantages over gas chromatographic method, which is apparent in the followings points.

1. Gas chromatographic method is not applicable to thermally unstable substances. On the other hand, there is no decomposition of substances in HPLC method, which is operated at room temperature.
2. The separation of each of the components of a substance is easily done by HPLC method whereas this is not true with gas chromatographic method.
3. The non-volatile substances are converted into volatile substances by chemical means in gas chromatographic method such as methyl esterification of fatty acids and trimethyl silanization of steroids. Such a kind of chemical treatment is not needed in HPLC.
4. HPLC performs more difficult separations than gas chromatography. This is because HPLC controls a greater number of separating variables. Besides, a greater variety of packing materials contributes to wider separating possibilities by HPLC.

REVIEW YOUR LEARNING

1. Define/explain the following:
    i. Chromatography
    ii. Column chromatography
    iii. Chromatogram
    iv. Retention time
    v. Effluent volume
    vi. Ion-exchangers
    vii. Adsorption chromatography

viii. Partition chromatography
ix. Gel permeation chromatography
x. Affinity chromatography
xi. Liquid–liquid chromatography
xii. Countercurrent chromatography
xiii. Elution techniques

2. Explain the principle, types and applications of paper chromatography.
3. Explain the principle and applications of thin layer chromatography.
4. Discuss the types of matrices used in column chromatography.
5. Describe the methods of application of sample in column chromatography.
6. Explain the principle, mechanism and applications of ion exchange chromatography.
7. Describe gel permeation chromatography and its applications.
8. Describe affinity chromatography and its applications.
9. Explain the principle, operation and application of gas–liquid chromatography.
10. Describe HPLC and its operation.
11. Write a brief note on detectors used in GLC and HPLC.

ELECTROPHORESIS

Electrophoresis is an important analytical tool used in all branches of biology. Electrophoresis is a Greek word meaning, "borne by electricity". A Russian physicist, **Alexander Reuss**, made the first electrophoretic observation of colloidal particles in 1807. He filled up a glass tube with water and clay. On passing electric current through the glass tube, he was excited to see some colloidal particles moving. **Michael Faraday** and **DuBois Reymond** also confirmed this discovery and they continued the study and found that positively charged particles moved towards the negative pole and negatively charged particles moved towards positive pole. These particles moved at different speeds depending on the number of excess charges they carried on their surface. The greater the number of charges, the faster the migration.

The movement of the charged particles in an electric field depends upon time, electric current, conductivity of the solvent and charge of the molecule to be separated. The **electrophoretic mobility** is defined as the distance travelled by the particles in one second under the potential gradient of one volt per centimetre. The different compounds in a mixture will have different electrophoretic mobilities and hence they can be separated. A mixture of amino acids, proteins, and nucleic acids can be separated by electrophoresis.

The two main types of electrophoretic methods are: (i) Moving boundary electrophoresis in which separation is carried out in the absence of a supporting medium. (ii) Zone electrophoresis in which paper or gel is used as supporting medium for separation.

MOVING BOUNDARY ELECTROPHORESIS

Tiselius of Sweden first developed moving boundary electrophoresis in 1930. In this method, separation is carried out in the absence of the supporting medium. A large volume of sample is needed and this method is rarely used. In this method, buffered negatively charged macromolecules are placed in a U-shaped cell (Figure 10.1).

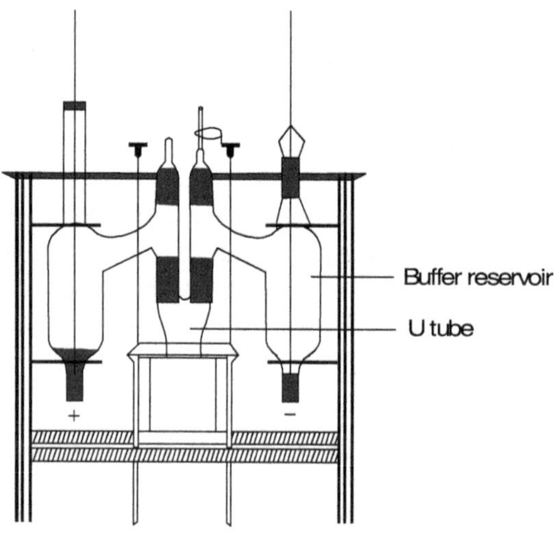

Figure 10.1 Tiselius apparatus

When an electrical current is applied, proteins migrate towards the anode. When they migrate from the macromolecule solution to the pure buffer, they form a boundary. As a result of this, there is a sharp change in the refractive index of the solution at this boundary. The changes in the refractive index are measured by Schlieren optical device. This yields electrophoretic patterns which show the direction and relative rate of migration of the molecules in the sample (Figure 10.2).

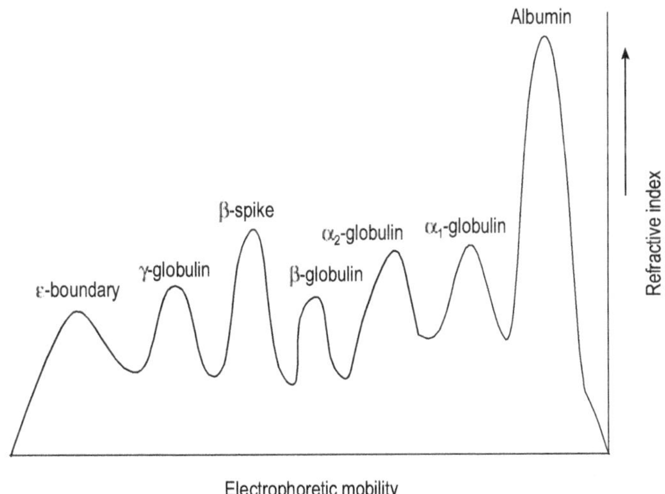

Figure 10.2 Electrophoretograph of human serum proteins recorded by Schlieren scanning method

ZONE ELECTROPHORESIS

This is the most commonly used type of electrophoresis. This method requires very small quantity of the sample. Paper or gel is used as the supporting medium for the separation of complex mixtures.

PAPER ELECTROPHORESIS

Paper electrophoresis is a type of zone electrophoresis (Figure 10.3). A great advantage of this technique is that only small volume of the sample is required. Upon separation, the molecules are immobilized by fixation at different zones. The molecules are then detected by the following methods:

- Staining them on a supporting medium
- Visualizing by ultraviolet light
- By virtue of enzymatic reaction
- By radioactivity

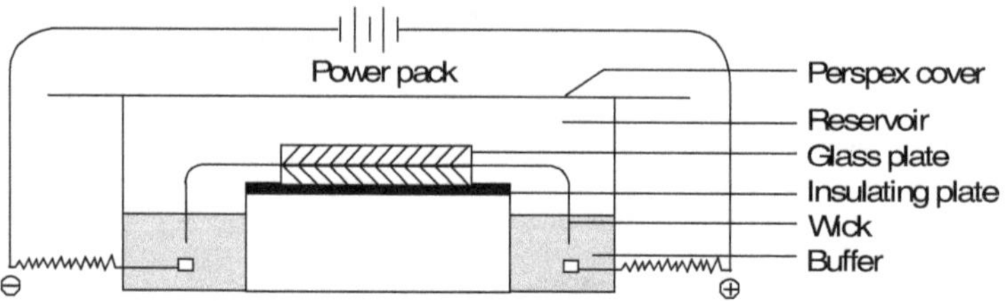

Figure 10.3 Paper electrophoresis

Paper electrophoresis is very useful in the study of normal and abnormal plasma proteins. The equipment required for electrophoresis consists of two units, a power pack and an electrophoretic cell. The serum under investigation is mixed with bromophenol blue (a blue coloured stain) and spotted at the centre of a strip of a special filter paper saturated with barbitone buffer of pH 8.6. When an electric current of proper amperage and voltage is passed through the paper, charged protein fractions bearing different charges migrate at different rates. After a run of about 5 to 6 hours, the paper is dried and stained with a solution containing bromophenol blue. In human serum, five different bands can be identified on paper electrophoresis (Figure 10.4). They are designated in the order of decreasing mobility as albumin, alpha globulin, α_2-globulin, beta globulin, fibrinogen and gamma globulin. Albumin, being the fastest moving fraction of the proteins of plasma, forms the last band of the paper. Gamma globulin which is the slowest moving protein, forms a band at the other end. The rest of the fractions take their positions in between these two bands.

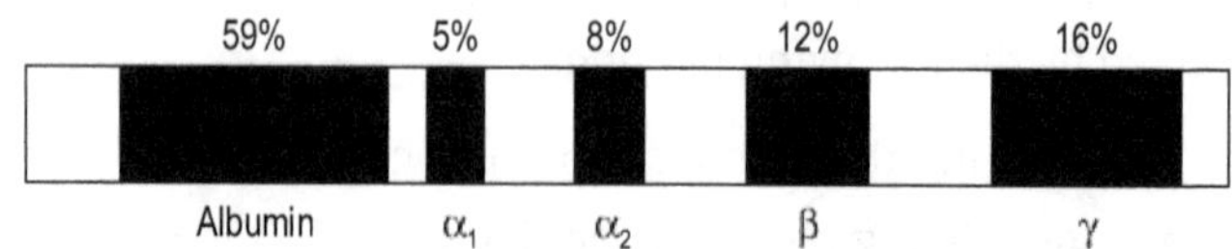

Figure 10.4 Electrophoretic pattern of human serum proteins

GEL ELECTROPHORESIS

Various types of gels are used as supporting media to separate complex mixtures. They are: (i) Starch gel, (ii) Agar gel, (iii) Polyacrylamide gel and (iv) Agarose acrylamide gel. The use of gels as supporting media has enhanced resolution, particularly for proteins and amino acids.

Gel electrophoresis is usually carried out through any one of the following methods.

1. Column electrophoresis
2. Slab gel electrophoresis

Column electrophoresis In this type, the gel is polymerized in a thin glass column of known diameter. Both ends of the column are not closed. After polymerization, this column is fitted between the upper and lower buffer reservoirs. (Figure 10.5a, b).

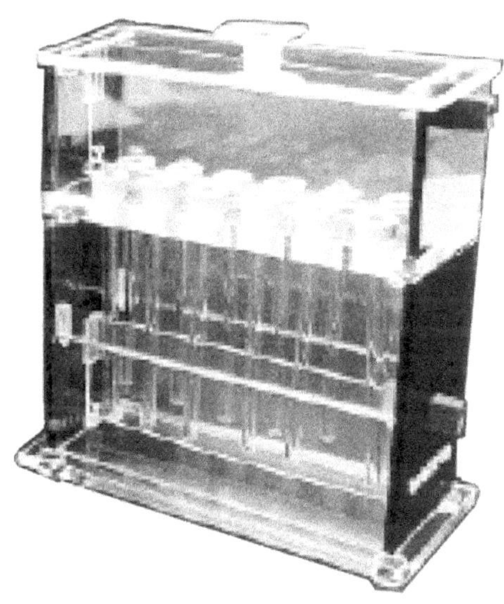

Figure 10.5 (a) Column electrophoresis apparatus

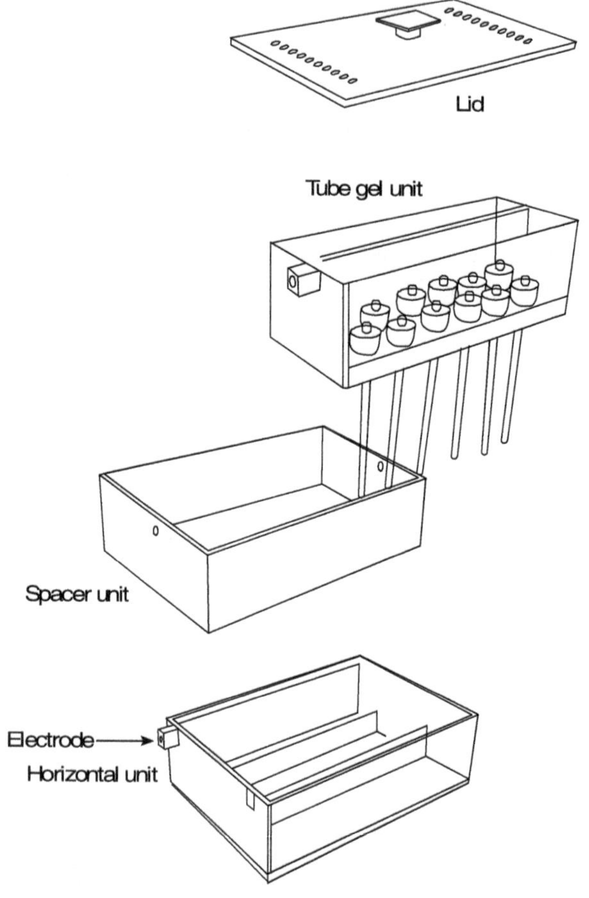

Figure 10.5 (b) Line diagram of column electrophoresis apparatus

Slab gel electrophoresis In this type, the gel is set or polymerized into a thin slab between two glass plates (Figure 10.6a, b). The thickness of the slab gel can be adjusted by increasing the thickness of two glass plates. The sample well is made by placing a comb-shaped jig into the glass before it is polymerized. After the gel has set, the comb is removed. Since a number of wells can be arranged side by side, a number of samples can be loaded simultaneously and compared under identical conditions. This technique is highly useful in the field of molecular biology.

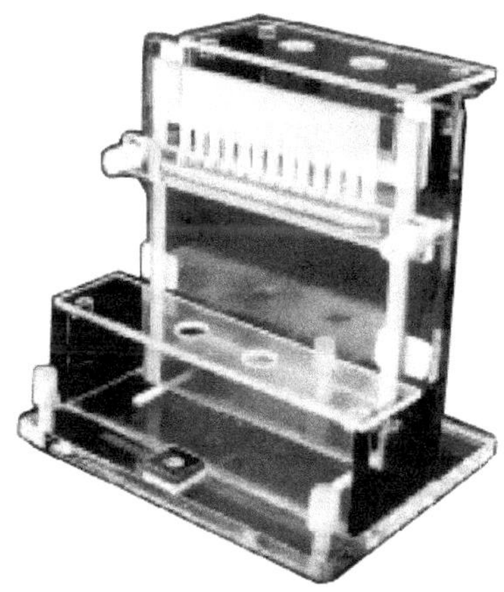

Figure 10.6 (a) Slab gel electrophoresis apparatus

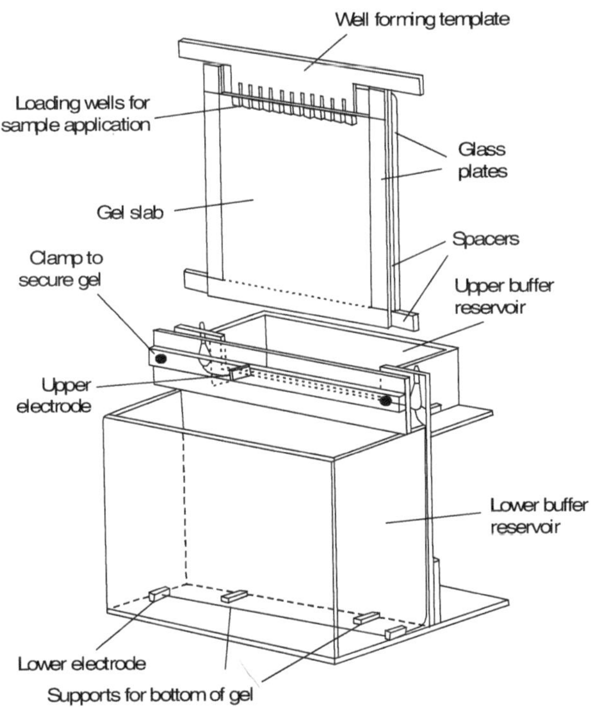

Figure 10.6 (b) Line diagram of slab gel apparatus

Starch Gel Electrophoresis

Smithies (1955) introduced zone electrophoresis using starch gel as a stabilizing medium. The starch gel apparatus consists of two buffer chambers, each compartmentalized and connected by paper wicks. The electrode is kept in one chamber and the other chamber is connected to the gel tray (Figure 10.7).

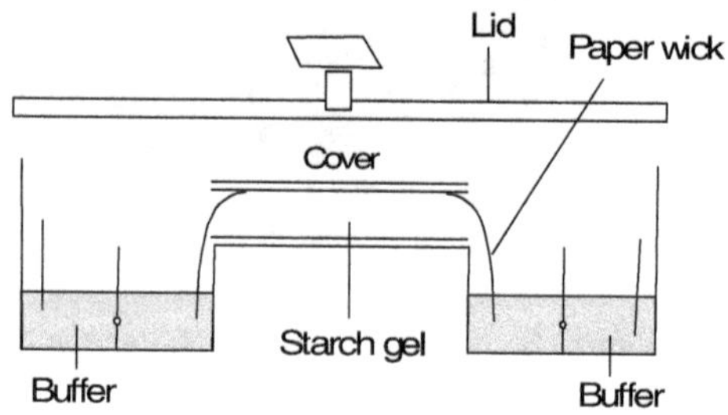

Figure 10.7 Starch gel electrophoresis

Starch is hydrolysed in acidified acetone at 37°C. The suspension is then neutralized with sodium acetate and washed with a large amount of distilled water and acetone. This hydrolysed starch when heated and cooled in an appropriate buffer sets as a gel. The sample is applied to the gel by soaking a small piece of filter paper and inserting it in the gel. The main drawback of starch gel is its unlimited pore size. Moreover, it is difficult to prevent contamination of starch gel by microorganisms. Another disadvantage of starch is that upon staining to detect the separated components, the starch gel turns opaque making direct photoelectric determination impossible. However, the resolving power of starch gel is very high. Starch gel does not absorb proteins; hence the separation of proteins gives sharp bands without tailing. The gels can be made transparent by soaking in glycerol, heated to 70–80°C for 5 minutes. Then the gel can be densitometrically scanned. Starch gel has no denaturing effect on enzymes. To analyse the presence of active enzymes, the specific substrate solution is poured in drops on the gel. The appearance of coloured band indicates the presence of active enzymes. The results can be immediately recorded by photography by using appropriate filters (Figure 10.8).

Figure 10.8 Starch gel electrophoretic pattern of acid phosphatase

Agar Gel Electrophoresis

Agar gel has been much utilized for immunoelectrophoresis by Graber and Williams (1959). Generally, a horizontal tank is used for agar gel electrophoresis. To prepare agar gel (1–2%) on microscopic slides, the agar solution is poured into a petri dish first. It is allowed to cool and a flat surface is obtained. Microscopic slides are placed on the gel and fresh agar solution about 2-mm thick is poured again. The resulting gel is of uniform thickness and sample is loaded in circular wells made in the gel (Figure 10.9).

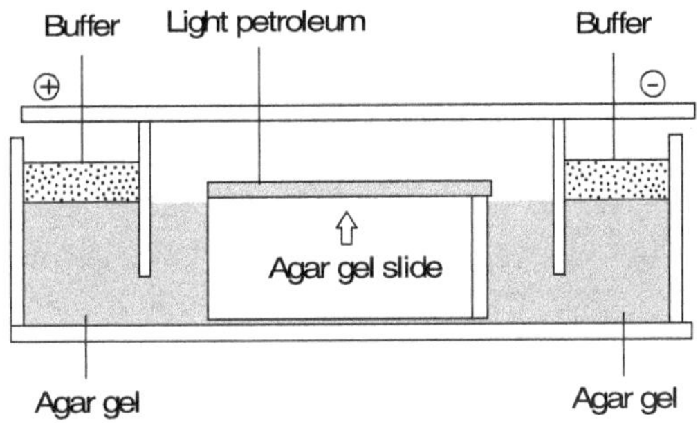

Figure 10.9 Agar gel electrophoresis

Agarose Gel Electrophoresis

Agarose is a natural product purified from red seaweed (Rhodophyta). It is a polysaccharide of alternate 1,4-linked 3, 6-anhydro-α-L-galactopyranose and 1,4-linked α-D-galactopyranose residue and arranged into a double helix. It dissolves on boiling and forms a gel when cooled.

In this electrophoresis technique, agarose is used at a concentration of 1 to 3%. Agarose gels are prepared by suspending dry agarose in aqueous buffer and then boiling the mixture until a clear solution is formed. It is allowed to cool to room temperature to form a rigid gel. Numerous inter- and intramolecular hydrogen bonds within and between the long agarose chain molecules are responsible for the gelling properties. This cross-linking gives the gel good anticonvectional properties. The pore size is controlled by the initial concentration of agarose. Low concentrations of agarose produce large pore size and high concentrations produce smaller pore size.

DNA molecules having high molecular weights cannot penetrate through even a weak cross-linked polyacrylamide gel. This problem is solved by agarose. A 0.8% agarose gel can accept DNA molecules with molecular weight as high as 59×10^6. Gels of different concentrations must be used according to the molecular weight of DNA samples. This is necessary because very large molecules cannot penetrate through the more concentrated gels and very small molecules will pass through the dilute gels easily.

In agarose gel electrophoresis technique, the glass slides on which agarose gels are prepared are kept on the bridges of the apparatus. The gel is connected to the anode and cathode buffer reservoirs through paper wicks (Figure 10.10). The sample is applied in the wells made in the gel. When electrophoresis is complete, the DNA is identified by immersing the gel in a dilute solution of the fluorescent dye, ethidium bromide. After some time, the gel is washed to remove unbound dye from the agarose and then it is illuminated with ultraviolet light to excite the fluorescence of ethidium

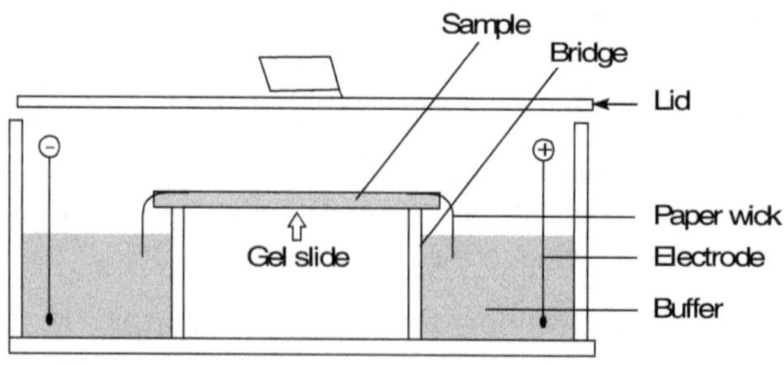

Figure 10.10 Agarose gel electrophoresis

bromide. The gel with its fluorescent bands is then photographed. Agarose gel can be used for immunoelectrophoresis technique.

Submarine Gel Electrophoresis

The most convenient method to study DNA and RNA fragments by electrophoresis is by agarose gels (Sharp *et al.,* 1973). Agarose gel of desired thickness can be cast in specially made trays called 'gel platform'. In order to make sample wells in gel, a suitable comb is placed in the gel platform during gel casting. After the gel is set, the comb is taken and rectangular wells are obtained. Agarose gel along with the gel platform is placed in the apparatus horizontally. Buffer is poured to submerge the gel and electrophoresis is carried out. This method of electrophoresing the gel under submerged condition is popularly known as submarine gel electrophoresis (Figure 10.11).

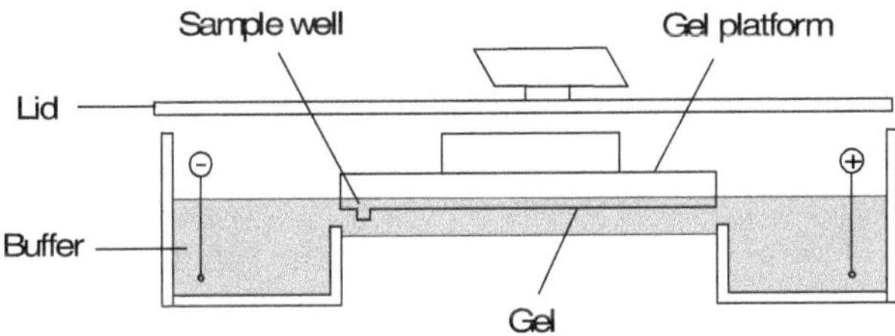

Figure 10.11 Submarine gel electrophoresis

After the electrophoresis, the gel is stained with ethidium bromide for detecting nucleic acids. Ethidium bromide binds with DNA/RNA and gives fluorescent bands when viewed in the transilluminator. Generally, Perspex sheets used to make gel platforms are opaque. So the gel to be viewed should be taken out of the gel platform carefully. This is very difficult and sometimes the gel breaks while handling. Also, since ethidium bromide is carcinogenic, it should be handled cautiously. To overcome these difficulties, gel platforms or the floor of the apparatus are made of UV-transparent Perspex sheet. In such cases bands can be viewed without taking the gel out and thus the handling process has become very convenient and safe with the UV-transparent gel platforms.

The submarine gel electrophoresis system has several advantages:

1. Handles thick agarose gels with a supporting gel platform comfortably.
2. Joule heat generated in the gel during electrophoresis is effectively dissipated by the surrounding buffer by convection current.
3. The pH changes in the vicinity of the anode and cathode is minimized by the decompartmentalization of the system. Anode and cathode reservoirs become a single compartment by the flooring of buffer to submerge gels.

Preparative Electrophoresis

Separation and enrichment of molecules of interest is the goal of preparative electrophoresis. For example, DNA or RNA of desired fragments are often prepared in pure form in terms of milligrams by using preparative submarine gel electrophoresis apparatus.

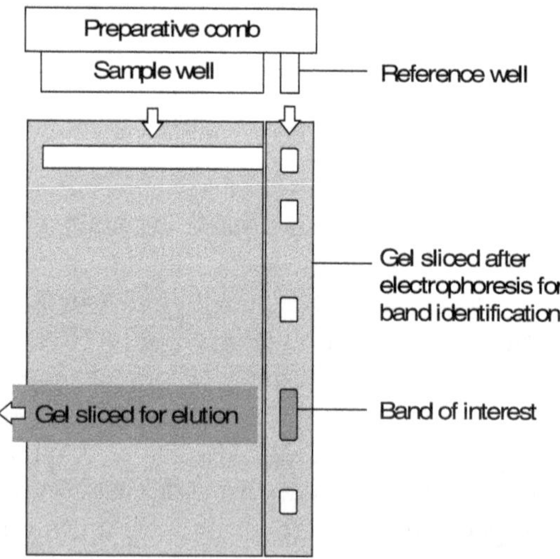

Figure 10.12 Preparative gel used for large-scale preparation of DNA or RNA

The simplest method to achieve this goal is electrophoresing large volumes of sample by using preparative comb (Figure 10.12). The preparative comb is made of thick sheet (6–10 mm) so that wells formed will have large sample capacity. The comb also has a small reference well. A known marker molecule is applied to it and after the electrophoresis, this lane is used for detecting and identifying the corresponding band in the sample zone. The separated band is then eluted and concentrated by other methods. Depending on the components present in the sample, the duration of electrophoresis is extended. In preparative models, a heat exchanger is also provided. Water from a thermostat bath is circulated in the gel bed. This arrangement is essential for prolonged electrophoresis. Submarine gel system easily resolves DNA up to 20 kb.

Pulse Field Gel Electrophoresis

In 1984, **Schwartz** and **Cantor** developed a new agarose gel electrophoresis technique called Pulse field gradient gel electrophoresis (PFGE) to separate larger DNA molecules. PFGE is based on the use of two electrical fields, which can be applied alternatively in different angles for a defined time duration which is called as pulse time.

Pulse field electrophoresis technique also employs agarose gels for separation as in submarine electrophoresis. But the apparatus is different. These are developed according to the arrangement of electrodes, gel and nature of electric field. The apparatus developed by **Swartz** and **Cantor** (1984) uses a gel and a pair of electrodes to provide an intersecting electric field in the plane of the gel as shown in the Figure 10.13. Although they were able to achieve separation of very large DNA molecules up to the size of yeast chromosomes, samples travel in distorted and non-linear lanes making further analysis difficult.

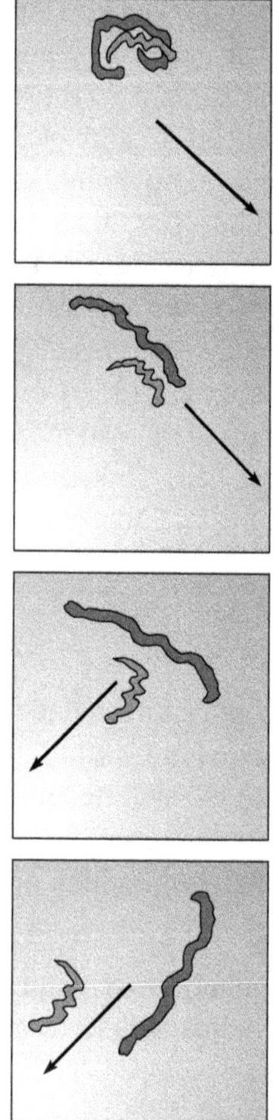

Figure 10.13 Separation of larger DNA fragments in PFGE (*Arrow indicates direction of electrical field*)

Hexagonal electrode apparatus The apparatus described by Chu *et al.* (1986) uses an array of electrodes arranged hexagonally around the gel. Hexagonal electrode arrangement mimics infinitely long pairs of electrodes intersecting at an angle of 120 degrees

(Figure 10.14). This unit separates DNA in straight lanes. But it requires a pulse generator with a complicated circuitry.

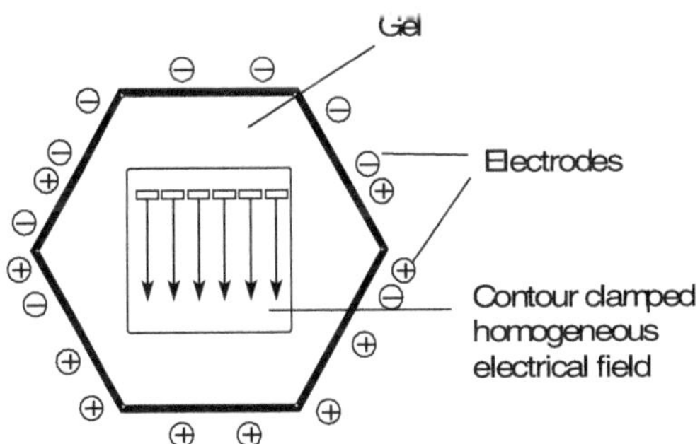

Figure 10.14 Hexagonal electrode arrangement for PFGE

Rotating platform apparatus Southern (1987) developed an apparatus for PFGE. In this unit, instead of altering the electrical field using different electrodes, the gel placed on a circular gel-running platform is periodically rotated back and forth at specified angles. The single pair of electrodes provide a uniform electric field for distortion-free separation and the unit ultimately requires only a sample circuitry.

Field Inversion Gel Electrophoresis (FIGE)

FIGE is another milestone in pulse field electrophoresis described by Carle *et al.* (1986). Unlike the PFGE techniques, in FIGE the polarity of the electrode is simply reversed periodically. The molecules migrating in the forward direction are subjected to reversed polarity and hence they tend to travel back. As the forward pulse is always kept longer than the reverse pulse, the net result of molecular migration is in thc forward direction.

FIGE has got several advantages over other techniques:

1. It can be carried out in any horizontal or vertical apparatus. Submarine gel apparatus or vertical slab gel apparatus with insufficient cooling provision can be employed.

2. Relatively thin gels can be used in vertical gel which are easily and quickly processed.
3. No extra pair of electrodes is needed.
4. By choosing an appropriate ramp (decrease in the pulse rate during the course of a run) molecules can be resolved within a limited size range. This sort of trick is very essentially required in techniques like Restriction Fragment Length Polymorphism (RFLP) analysis where separation and accurate measurement of fragment must be made.
5. FIGE is faster than other pulse field techniques; for example, by using FIGE, DNA fragments between 10–15 kb can be separated in 6 hours. In other pulse field techniques 5–10 kb fragments take about few days for separation.

Applications

Agarose gel electrophoresis is used to

1. isolate a large number of proteins
2. identify the purity of isolated proteins
3. determine the sequences of DNA
4. find out the presence of mutation in DNA or RNA (Southern and northern blotting)
5. detect the precursor molecules of tRNA, rRNA and mRNA
6. study the kinetics of the interconversions of conformation in many tRNAs
7. find out the number of subunits present in a protein
8. determine the molecular weight of proteins and DNA

Polyacrylamide Gel Electrophoresis

Electrophoresis in acrylamide gel is frequently referred to as PAGE (Polyacrylamide Gel Electrophoresis). The components used in PAGE are acrylamide, bisacrylamide and TEMED (Tetramethyl ethylene diamine). Cross-linked polyacrylamide gels are formed from the polymerization of acrylamide monomer in the presence of small amounts of bisacrylamide. The pore size in the gel can be varied by changing the concentration of both the acrylamide and bisacrylamide.

Several proteins of biological importance contain more than one polypeptide chain. These proteins are referred to as oligomeric proteins. The structure of these proteins is stabilized by hydrogen bonding, disulphide linkages or by hydrophobic interactions. The subunits of these proteins can be separated from each other by a class of compounds known as solubilizers, e.g. urea, beta-mercaptoethanol and sodium dodecyl sulphate (SDS).

Beta-mercaptoethanol breaks disulphide bridges present in the oligomeric proteins. SDS is an anionic detergent and disrupts macromolecules whose structure has been stabilized by hydrophobic interactions. SDS binding imparts a large negative charge to the denatured polypeptides.

PAGE combined with SDS is the most widely used method for analysing protein mixtures quantitatively. It is particularly useful for monitoring protein purification. In 1967, A. Shapiro, E. Vinuela and J. Maizel showed that the molecular weight of most proteins could be determined by measuring the mobility of most proteins in polyacrylamide gels containing SDS.

Initially, the sample to be separated is boiled for five minutes in a sample buffer containing SDS and beta-mercaptoethanol. This treatment completely denatures proteins present in the sample and imparts negative charge to the polypeptide chains. The sample buffer also contains an ionizable tracking dye (e.g. bromophenol blue), which allows to monitor the electrophoretic run.

SDS-PAGE is usually carried out as column electrophoresis (Figure 10.15). The gel is polymerized in a column. This column is then fitted between the upper and lower buffer reservoirs. An electrophoretic apparatus usually contains 8 to 12 columns.

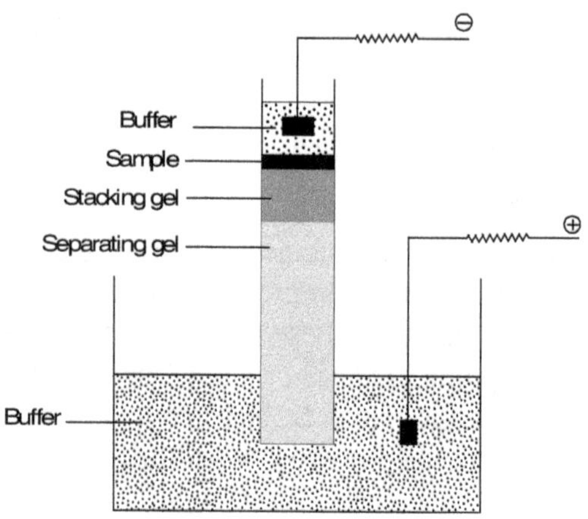

Figure 10.15 SDS-PAGE

In SDS-PAGE, the gel consists of two parts, namely, the main separating gel (about 10 cm long) and a shorter stacking gel (about 1 cm long). The main separating gel is first poured into a glass tube and allowed to set. Then 1 ml of stacking gel is poured on the top of the separating gel. The stacking gel has a very large pore size and the separating gel has comparatively small pore size.

Samples of proteins of known and unknown molecular weight are layered on the top of each column separately. The stacking gel allows the proteins to move freely and concentrate over the separating gel under the influence of the electric field. Proteins continue their movement towards the anode. Since proteins have the same charge per unit length, all proteins travel with the same mobility. However, as they pass through the separating gel, the proteins separate owing to the molecular sieving properties of the gel. The smaller proteins move fast as they can pass through the pores of the gel. But large proteins move slowly since they are retarded by frictional resistance due to sieving effects of gel. When the dye reaches the bottom of the gel, the current is turned off. The gel is removed from the glass tube and stained with appropriate stain solution (Figure 10.16).

A plot of the distance migrated versus log of the molecular weight gives a straight line. Hence, if a protein of unknown molecular weight is electrophoresed with two or more of proteins of known molecular weight, the molecular weight of unknown protein can be calculated to an accuracy ranging between 90 and 95 per cent (Figure 10.17). This is the most common way of estimating the molecular weight of protein subunits.

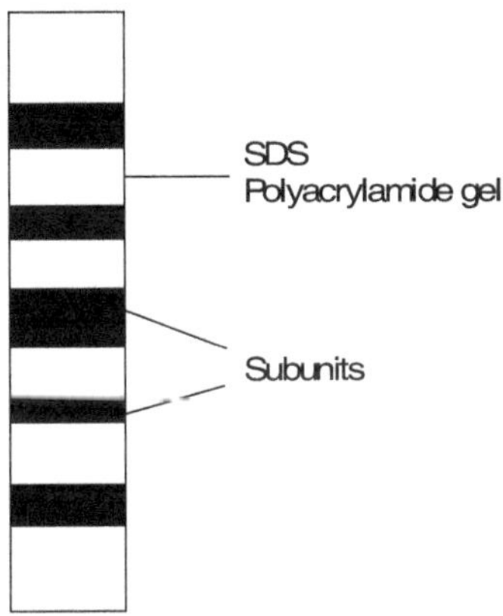

Figure 10.16 Separation of enzyme subunits by SDS-PAGE

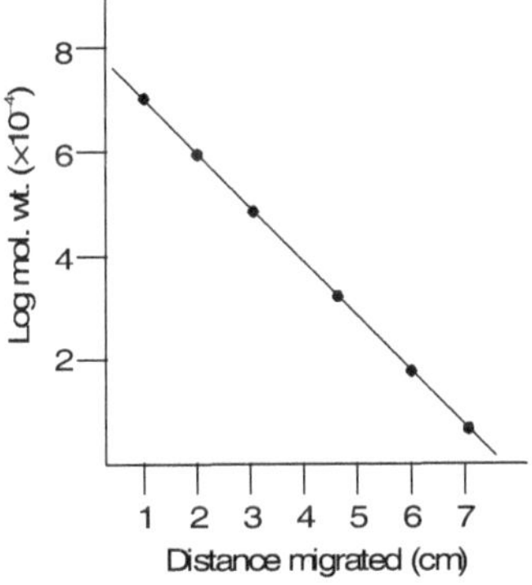

Figure 10.17 Determination of molecular weight

ISOELECTRIC FOCUSING

The method of separating proteins according to their isoelectric points in a pH gradient is called isoelectric focusing. This technique was discovered by H. Svensson in Sweden. This method has a high-resolution power because ordinary paper electrophoresis resolves plasma proteins into six bands whereas isoelectric focusing resolves them into 40 bands.

In conventional electrophoresis, the pH between anode and cathode is constant and the positively charged ions migrate towards the cathode and negatively charged ions migrate towards the anode. But in isoelectric focusing, a stable pH gradient is arranged. The pH gradually increases from anode to cathode. When a protein is introduced at a pH which is lower than its isoionic point, it will possess a net positive charge and will migrate in the direction of the cathode. Due to the presence of pH gradient, the net charge of the molecule changes due to ionization as it moves forward. When the protein encounters a pH where its net charge is zero, it will stop migrating. This is the isoelectric point of protein. Each protein present in the mixture will migrate to its isoelectric point and stop its migration at that point (Figure 10.18). Thus, once a final, stable focusing is reached, the resolution will be retained for a long time.

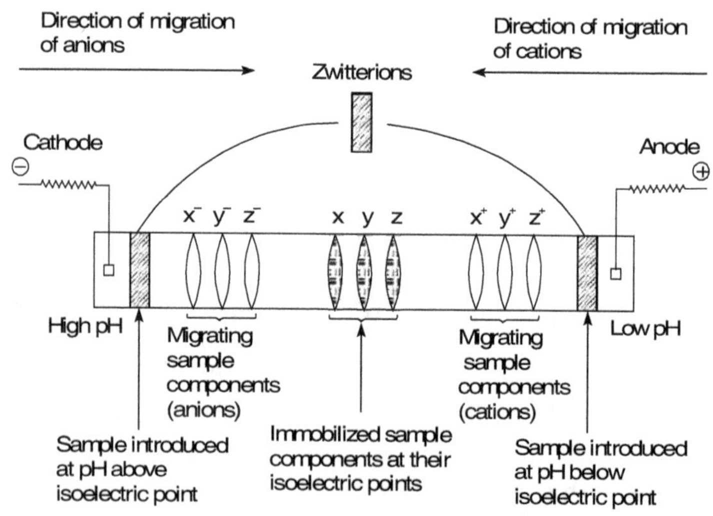

Figure 10.18 Isoelectric focusing

Isoelectric focusing is widely used for the separation and identification of serum proteins. It is used in the food and agriculture industry, forensic and human genetic laboratories, for research in enzymology, immunology and membrane biochemistry, etc.

CONTINUOUS FLOW ELECTROPHORESIS

The continuous flow electrophoresis is used for separation of particles in large-scale production. Electrophoresis takes place continuously as the separating material is carried upwards by flow of carrier buffer through annular space between two vertical concentric cylinders (Figure 10.19). The outer cylinder is rotated to maintain a stable laminar flow of the buffer solution. An electrical

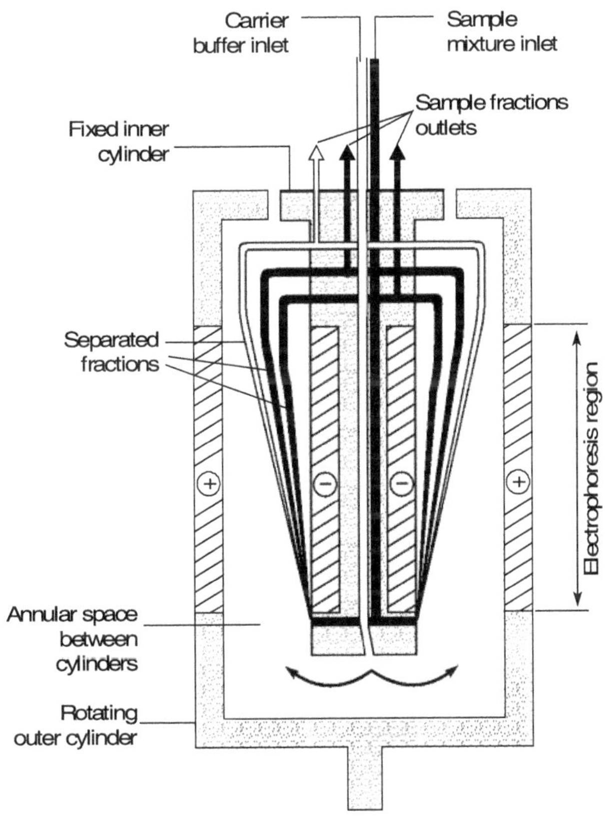

Figure 10.19 Continuous flow electrophoresis

field is applied between the two cylinders, causing the sample material to separate radially as it is carried upwards by the buffer flow. At the top of the inner cylinder, a series of radial slits enable the buffer stream to be separated into as many as thirty individual fractions. However, the resolution is not good. But this method is useful in the separation of particles and cells such as human erythrocytes.

CAPILLARY ELECTROPHORESIS

Capillary electrophoresis has been considered as one of the most practical advances in separation sciences. Capillary electrophoresis (CE) has variously been referred to as high performance capillary electrophoresis (HPCE), capillary zone electrophoresis (CZE) and free solution capillary electrophoresis (FSCE). Mikkers made the first publication of electrophoresis of macromolecules in capillary tubing in 1997. Capillary electrophoresis is a powerful nanotechnique characterized by high speed, high resolution, excellent mass sensitivity, low sample consumption and high efficiency.

Principle

Capillary electrophoresis involves a combined action of two major forces: (i) electroosmotic flow and (ii) electrophoretic separation. Under the influence of these forces, all compounds in the sample normally travel in one direction (towards cathode), allowing detection of positively charged, neutral and negatively charged molecules at one single location along the capillary tube. Once again, separation takes place in the buffer alone as in the moving boundary electrophoresis.

Components Involved in Capillary Electrophoresis

Capillary tube Generally a fused silica capillary column with internal diameter of 20–75 μ and external diameter ranging from 150–375 μ is used. Capillary lengths used also vary in the range from 30–100 cm and connect the anode and cathode reservoirs at the ends.

Detector According to the molecule under study, different detection schemes are available. But UV detector with a suitable modification in the capillary line is universally adopted.

Power source Capillary electrophoresis is carried out at 250–300 V/cm and a power source capable of delivering up to 30,000 V is required.

The basic apparatus for CE is shown in Figure 10.20. A small plug of sample solution (5 to 30 μ m^3) is introduced into the anode end of the fused silica capillary tube containing an appropriate buffer. Sample application is carried out in two ways: High voltage injection and Pressure injection.

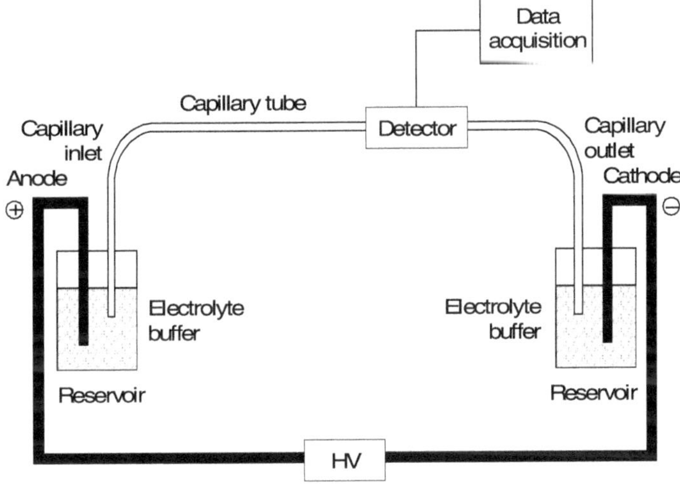

Figure 10.20 Capillary electrophoresis

1. *High voltage injection* With the high voltage switched off, the buffer reservoir at the positive electrode is replaced by a reservoir containing the sample, and a plug of sample (e.g. 5 to 30 μ m^3) is introduced into the capillary by briefly applying high voltage. The sample reservoir is then removed, the buffer reservoir is replaced, voltage again applied and the separation is then commenced.
2. *Pressure injection* The capillary is removed from the anodic buffer reservoir and inserted through an air-tight

seal into the sample solution. A second tube provides pressure to the sample solution, which forces the sample into the capillary. The capillary is then removed, replaced in the anodic buffer and a voltage applied to initiate electrophoresis.

A high voltage (up to 50 kV) is then put across the capillary tube and the component molecules in the injected sample migrate at different rates along the length of the capillary tube. Electrophoretic migration causes the movement of charged molecules in solution towards the electrode of opposite charge. Owing to this electrophoretic migration, positive and negative sample molecules migrate at different rates. However, although analytes are separated by electrophoretic migration, they are all drawn towards the cathode by electroendosmosis. Since this flow is quite strong, the rate of electroendosmotic flow usually being much greater than the electrophoretic velocity of the analytes, all ions are carried towards the cathode. Positively charged molecules reach the cathode first because the combination of electrophoretic migration and electroosmotic flow cause them to move fastest. As the separated

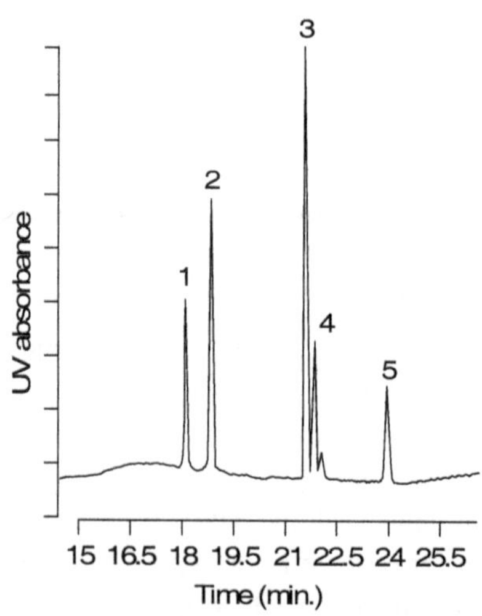

Figure 10.21 Capillary electrophoretograph of some structurally related peptides

molecules approach the cathode, they pass through a viewing window where they are detected by an ultraviolet monitor that transmits a signal to a recorder, integrator or computer. Typical run times are between 10 and 30 min. A typical capillary electrophoretograph is shown in Figure 10.21.

Applications

1. Capillary electrophoresis is used to separate a wide spectrum of biological molecules including amino acids, peptides, proteins, DNA fragments and nucleic acids, as well as a number of small organic molecules such as drugs or metal ions.
2. CE can identify point mutations which occur in DNA and which cause some diseases.
3. CE can be used to quantify DNA.
4. Chiral compounds can be resolved using CE.
5. A range of small molecules, drugs and metabolites can be measured in solutions such as urine and serum.

IMMUNOELECTROPHORESIS

Immunoelectrophoresis technique is based on the electrophoretic mobility and precipitation. In the precipitation reaction, an antigen combines with its specific antibody to form antigen–antibody complex. Since most of the antigen–antibody complexes are insoluble, they can be seen with the naked eye. Thus, immunoelectrophoresis exploits the specificity of the reaction between an antigen and antibody, and molecular sieving of the gel in which this reaction takes place.

Qualitative Immunoelectrophoresis

This analysis is carried out in agarose gel barbitone buffer on a microscope slide. As shown in Figure 10.22a, about 100 μg of a specific antigen [(human serum albumin (HSA)] and a mixture of antigens [(human serum (HS)] are placed in two wells. The slides are connected by thick wet filter paper wicks to the electrode wells and a direct electric current of about 8 mA per slide is passed for 1 to 2 hours. Charged particles are separated electrophoretically (Figure 10.22b). When the electrophoresis is over, the troughs are filled with appropriate antisera and incubated overnight at room temperature in a humid chamber. The antigens diffuse radially and the antibodies diffuse laterally resulting in the antigen–antibody precipitation arc as shown in Figure 10.22c.

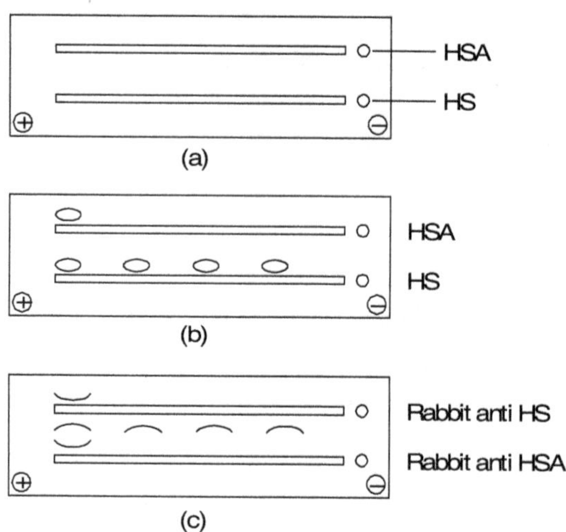

Figure 10.22 Qualitative immunoelectrophoresis

Counter Immunoelectrophoresis

This is a kind of immunoprecipitation technique originally described by Bussard (1959). The technique exploits the fact that during electrophoresis, antibodies tend to migrate towards the cathode while most antigenic proteins proceed towards anode. As a result, antigens and antibodies cross each other to form a precipitin line in a matter of few minutes. By using this technique, the presence of an antigen in a sample can be detected very quickly by immunodiffusion. In this technique, antigen and antibody solutions are added in the wells punched in agar gel at optimum distance (Figure 10.23a). Electrophoresis at pH 8.6, facilitates the antigens to move towards the anode and antibodies to move in the opposite direction

(Figure 10.23b), subsequently resulting in the formation of precipitin arc at the antigen–antibody crossing point (Figure 10.23c).

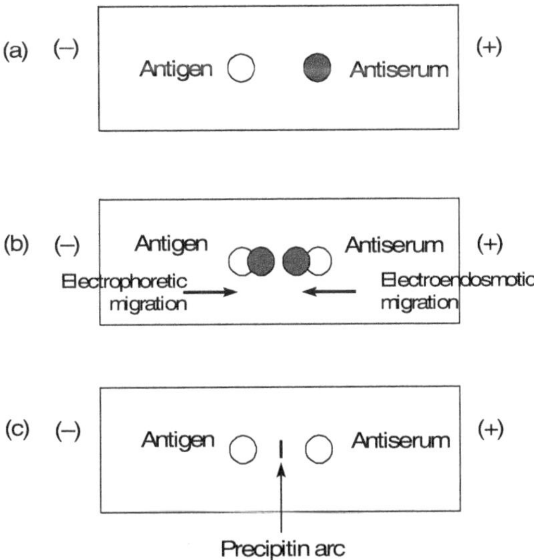

Figure 10.23 Counter immunoelectrophoresis

Crossed Immunoelectrophoresis (CIE)

This is a semi-quantitative two-dimensional immunoelectrophoresis technique, developed by Clarke and Freeman (1967). It involves a powerful combination of two electrophoretic methods.

1st dimension Simple electrophoresis of proteins (antigens) in agarose gel (Figure 10.24a) and

2nd dimension Electrophoretic migration of antigen (separated in the 1st dimension) through antibody containing gel (Figure 10.24b).

Antigenic proteins separated in the 1st dimension are allowed to migrate at right angles into the antibody-containing gel. A rocket-like immunoprecipitation peak is formed according to the amount of antigen present (Figure 10.24c). Such immuno peaks can be visualized during or immediately after electrophoresis and can be stained. The position of the peaks can be used for identification.

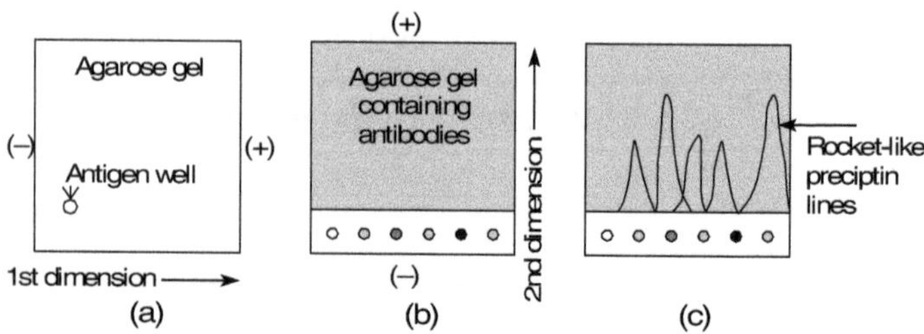

Figure 10.24 Crossed immunoelectrophoresis

Advantages

1. High resolution of proteins in the electrophoretic separation.
2. Proteins present in trace amounts are also detected as visible peaks.
3. Peaks can be identified by position and various specific staining methods.
4. Peaks can be quantified by using appropriate standards.
5. The technique requires no expensive instrumentation.
6. Suitable for studying various proteins present in complex biological mixtures like blood, membrane proteins and other body fluids.

ANALYSIS OF BANDS

After electrophoresis, the electrophoretic medium (paper or gel) is carefully taken out for qualitative or quantitative analysis. The following methods are followed to examine protein or nucleic acid bands.

1. Direct photometric scanning
2. Staining methods
3. Radiolabelling and autoradiography
4. Enzyme assay

5. Immunological methods
6. Blotting and detection

Direct Photometric Scanning

This method gives instant result. Proteins absorb UV maximally at 280 nm and nucleic acids at 260 nm. If a transparent gel containing separated bands of proteins or nucleic acids are allowed to move across the UV beam, they absorb UV and the resulting beam shows a decrease in intensity. This is monitored by a photocell as Optical Density (OD). According to the intensity of the band, the OD shows different values. The method of monitoring OD at different locations in gel to analyse bands is called **photometric** or **densitometric scanning** (Figure 10.25). Earlier, scanning was done manually and the OD measured at every point was plotted on a graph. It was drawn as peaks of different heights and widths corresponding to the bands. Recently, highly sophisticated densitometers are available with microprocessor to scan gels in UV or visible light ranges and analyse the data instantaneously.

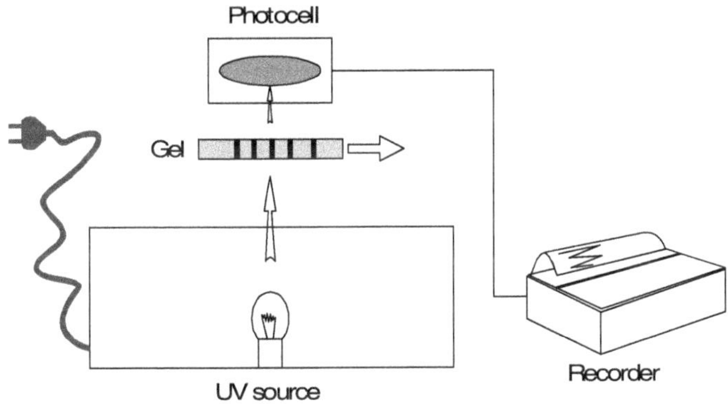

Figure 10.25 Densitometer

Scanning can be done for gels immediately after electrophoresis without any pretreatment. The only requirement for scanning gel in UV is to use a quartz trough or slide which passes UV through it. Although scanning is done in a matter of few minutes and results are obtained immediately, various UV-absorbing impurities present in the gel interfere with the results. As a result, sensitivity of the method declines. So this method is suitable only while dealing with large quantities of sample especially to locate bands in preparative procedures.

Staining Methods

Staining is the most common procedure. Proteins, glycoproteins, lipoproteins and nucleic acids are detected by various staining procedures. The ability of the staining method to detect these molecules (sensitivity) varies with the stain (dye). In order to achieve greater sensitivity of detection, a number of staining procedures are developed. Staining procedures involve several times of washing in staining and destaining solutions; so it is essential to fix the separated molecules with the electrophoretic medium. Otherwise, they are likely to diffuse out during the staining procedures. Table 10.1 gives a list of various dyes to detect several important biochemical components:

Table 10.1 Dye/chemical used to identify biomolecules

Stain	Sensitivity	Application
Amido black	1.0–5.0 μg	Proteins on membranes/gels
Coomassie blue	0.1–1 μg	Simple, fast and consistent protein stain
Copper stain		Simple, fast, reversible stain, SDS-PAGE
Zinc stain		Simple, fast, reversible stain Elution, blotting possible, SDS-PAGE only
Silver stain	1–10 ng	Proteins and nucleic acids on gels
Radiant red		Fluorescent RNA stain for denaturing gels
Ethidium bromide	1–1 ng	Fluorescent stain for DNA/RNA

(Contd.)

Table 10.1 (Continued)

Stain	Sensitivity	Application
Methylene blue	5 ng	Visual stain for nucleic acids
Indian ink	100 ng	Permanent protein stain for membranes
Biotin–avidin	30 ng	Permanent protein stain
Alkaline phosphatase	5–100 pg	Chromogenic detection of proteins
Colloidal gold	5–100 pg	To detect proteins on blot membranes.

Ethidium bromide (EB) is a fluorescent dye used to detect nucleic acids in agarose gels. It detects both single-stranded and double-stranded nucleic acids. It intercalates with double-stranded DNA and RNA to form UV-fluorescent complexes. When these complexes are excited by UV (300 nm), they give fluorescence at 590 nm (orange colour). Orange colour bands can be visualized in a gel containing DNA/RNA bands stained with EB (0.5 μg/ml) placed under UV illuminator. In order to view and photograph EB-stained DNA/RNA bands, different types of UV illuminators are used. They vary in their performance depending on the way they are designed, power, nature of UV source and filters used (Sealy and Southern, 1990).

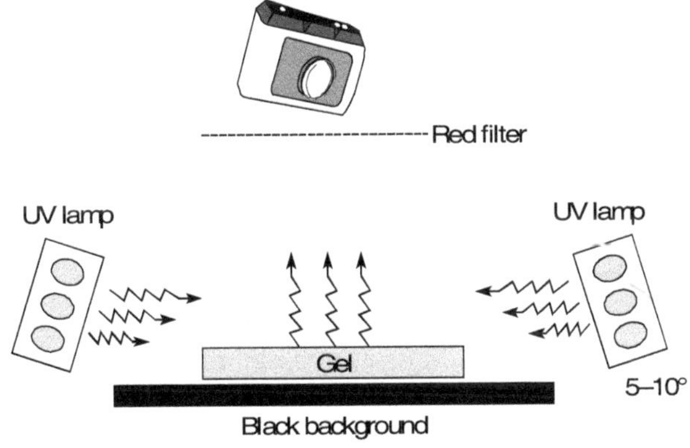

Figure 10.26 Incident illumination

i. Incident Illumination The UV lamps kept by the sides illuminate the gel placed on a black sheet. UV lamp is tilted to get maximum incident light. Fluorescent bands are viewed and photographed (Figure 10.26). Protective goggles are essential to protect the eyes. Red filter is used on the camera to get a picture. Since quartz filter is not used, it is less expensive but has low sensitivity.

ii. Transillumination In UV transilluminator, UV lamps are housed in a closed cabinet just below the viewing window. The viewing area is made of quartz filter (Figure 10.27). Gel placed on quartz filter is illuminated fully so that all the EB–DNA complexes are excited to give maximum fluorescence. As there is direct illumination, extreme safety measures should be taken to protect our eyes and skin. The camera also requires special interference filters to get a smooth image of the bands. This is very expensive, but it provides the maximum sensitivity.

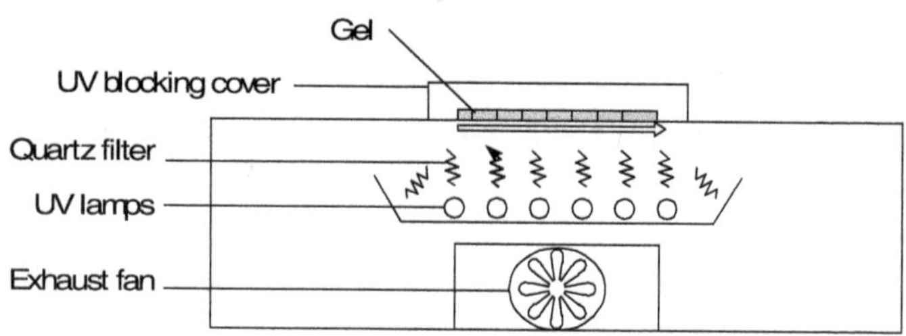

Figure 10.27 UV Transilluminator

In white light transilluminator, white fluorescent lamps and white translucent sheets are used instead of UV lamps and quartz filter. This arrangement offers uniform white light for visualizing and taking a photograph of coomassie blue or other visual dye-stained bands.

iii. Dark-field illumination In order to avoid direct exposure of the hazardous UV, Anbalagan (1992) designed a novel dark-field illuminator (Figure 10.28). It is based on illuminating the gel by the hidden UV source and visualizing the bands against a dark background. UV pass sheet replaces the expensive quartz filter and at the same time, gel placed on it is illuminated as in transillumination. As fluorescent bands are clearly viewed against a dark background, they can be photographed with minimal efforts. Simplicity makes the whole process safer, less expensive and offers satisfactory results.

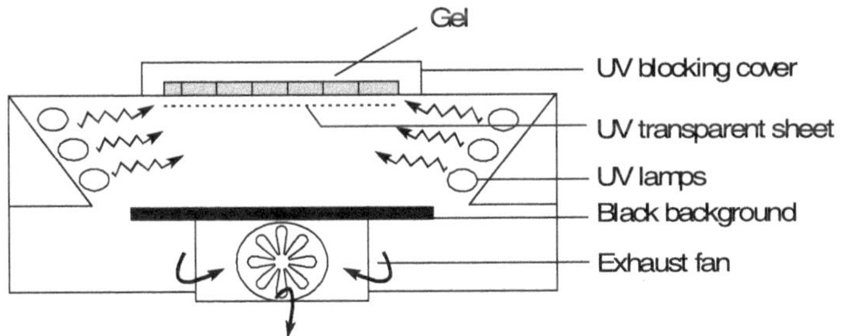

Figure 10.28 Dark-field UV illuminator

Gel Documentation System

Gel documentation or image analysing system is a modern way of studying results of electrophoretic patterns accurately. Gel, after staining, is placed on the illuminator and the image captured by a camera is then quantitatively or qualitatively analysed by a computer (Figure 10.29). It consists of the following components:

1. An illuminator—white light or UV light or combined

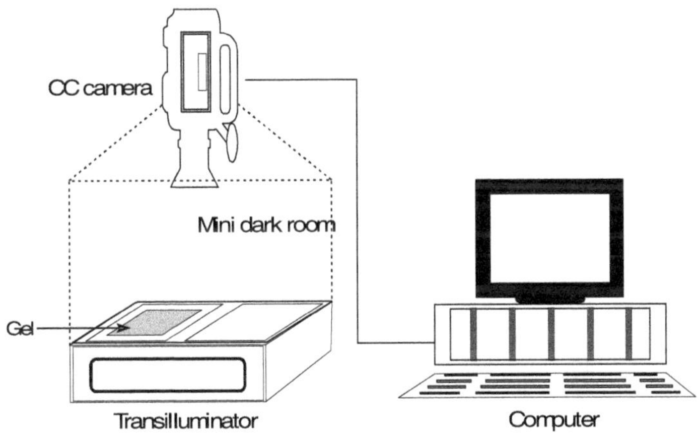

Figure 10.29 Gel documentation system for analysing gel images

2. Video camera/digital camera
3. Computer with suitable software

Illuminator Different types of illuminators are used. White-light illuminator can be used for viewing and taking a photograph of colour bands like coomassie-stained protein bands, UV transilluminator for detecting fluorescent bands and combined illuminator for both purposes. It is now available with a mini dark room facility, so that UV exposure is totally avoided.

CC-Camera This camera also known as digital camera captures live images of the gel and passes the image to computer for analysing. Different types of lens and filters are available to get the best image of the electrophoretic pattern. The software provided with the computer controls the camera function. Its resolution is at the pixel (768 × 620) level so the images are accessible to various functional analyses directed by the software. The camera has zoom lens (8 mm to 48 mm) so that even a mini size gel as small as 5 × 4 cm can be analysed. The camera height can be varied as it is mounted on an adjustable stand.

Computer A computer with a user-friendly software provides simplified acquisition, optimization and documentation of images on a single one. Different software packages are available for 2D gel analysis, RFLP, fingerprinting and DNA sequence analysis.

Once the image is acquired, the intensity, contrast of images and brightness can be optimized.

Advantages

1. Lanes can be selected manually or automatically and bands are scanned.
2. Area and quantity of the bands are determined. Molecular weight and pI are also estimated.
3. Standard curves and regression analysis can be carried out.
4. Spots in 2D gel are identified rapidly and quantified.

5. In RFLP and fingerprinting, using external or internal reference pattern, manual or automatic band identification, quantification, pattern recognition, and rapid similarity comparison are performed.
6. Powerful software packages are available which can process DNA images acquired from densitometry system or storage phosphor imaging system. Base sequences are generated quickly and displayed as base letters directly.

Autoradiography

Radiolabelling and autoradiography are sensitive methods to identify the macromolecules. Radioactive isotopes are unstable elements. They undergo decay and become chemically different elements as shown below.

$$C^{14} \rightarrow {}_7N + e^-$$

$$_{92}U^{238} \rightarrow {}_{90}Th^{234} +_2 He^4$$

This process is known as radioactivity. It is an irreversible process and the energy is released in the form of α, β and γ radiation.

The time required for radioactivity in an original sample to be reduced to half its original level is known as half-life. Radioactive elements exhibit characteristic half-life and emissions. Properties of some commonly used radioactive elements are shown in the Table 10.2.

Proteins and nucleic acids can be labelled with radioactive elements during their synthesis using radioactive amino acids/ nucleotides or post-synthetically by iodination, reductive methylation, etc. When such labelled molecules are electrophoresed, their presence in the gel can be detected by either monitoring the radiation using X-ray film or by measuring radioactivity.

Table 10.2 Properties of some commonly used radioactive elements

Radioisotope	Half-life	Radiation	Energy *(MeV)*
C^{14}	5570 years	β^-	0.155
H^3	12.3 years	β^-	0.0176
Ca^{45}	165 days	β^-	0.254
S^{35}	87 days	β^-	0.167
P^{32}	14.5 days	β^-	1.701
I^{131}	8.05 days	$\beta^- \gamma^-$	β^- 0.608
			γ^- 0.364

Monitoring the radiations using X-ray film is a direct method (Laskey and Mills, 1975). In this method the gel containing radioactive bands is placed next to an X-ray film. The radiation converts silver halide in the film to metallic silver, which can be seen as black spots on developing. The intensity of the film image depends on the nature of radiation, quantity of the radioactive molecules present, time of exposure, thickness of the gel, etc. About 6000 dpm/cm^2 of C^{14} or 1600 dpm/cm^2 of ^{35}S/ ^{125}I or 500 dpm/cm^2 of ^{32}P present in dried polyacrylamide gel will give film image absorbance of 0.02 A_{540} in 24 h exposure. As, ^3H emits very low energy radiation, it fails to penetrate the gel matrix and is not detected. The direct method of recording the presence of radioactive molecules in X-ray film is known as **autoradiography**. As there exists a linear relationship between the intensity of the spots and quantity of radioactive molecules, autoradiograms can be studied by densitometry also.

Fluorography Fluorography enhances autoradiographic images several-fold. Fluorography is a modification of autoradiography and is based on impregnating the gel with a scintillator, which converts low-energy particles to visible light to form an image on X-ray film. This method becomes very sensitive when carried out at –70°C. By this method low-energy β particles from ^3H (8000 dpm/cm^2) produces an image in 24 h. This image enhancement is more than 1000 times when compared to direct autoradiography (Laskey, 1980).

Phosphor imaging Phosphor imaging is an ultra-sensitive system. Storage phosphor screen is highly sensitive to β, γ, UV and X-rays. It traps ionizing radiation and stores it as phosphor particles. When the screen is scanned with laser beam, the screen releases the stored energy as blue light. This visible light is collected and converted to digital images by a computer and it can be analysed with the help of suitable software (Figure 10.30).

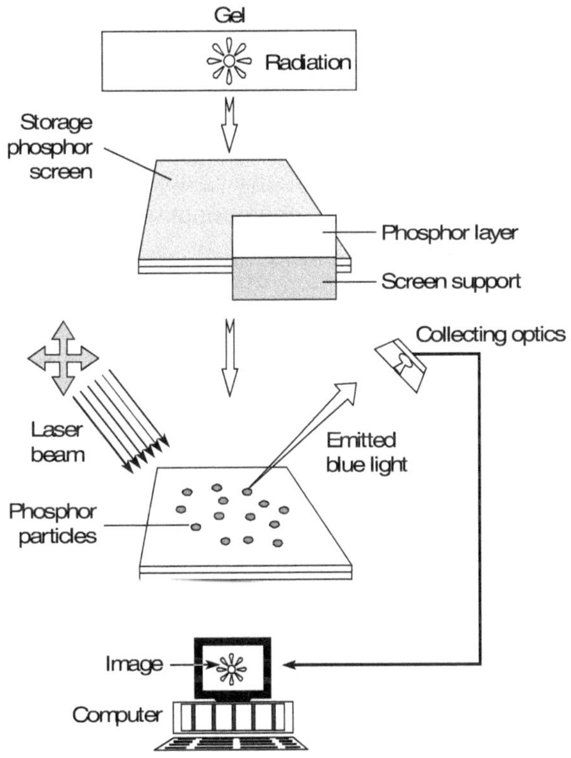

Figure 10.30 Phosphor imaging system

Enzyme Assay

Enzyme assay detects proteins rapidly and more specifically. When enzymes are separated by electrophoresis, they can be identified in the gel medium by carrying out the specific enzymatic reaction. **Gabriel** (1971) and **Rothe** (1994) surveyed a number of enzymes that can be identified in electrophoresis. As an example, the assay of lactate dehydrogenase in polyacrylamide gel is discussed below.

Lactate dehydrogenase catalyses the oxidation of lactate into pyruvate. An intermediate electron carrier, phenazine methosulphate (PMS) transfers electrons from NADH to reduce the yellow tetrazolium salt. The tetrazolium salt on reduction becomes intense blue. Hence, the gel containing lactate dehydrogenase, when incubated in a solution containing lactate, NAD, PMS and nitro blue tetrazolium chloride (NBT), develops an intense blue colour spot. The main advantage of this technique is speed with which it detects. However, there are some disadvantages:

1. *Non-specific staining* Any protein which donates electrons to PMS will give positive result, which could be avoided by running appropriate substrate control.
2. *Impermanence of the stained gels* Permanent recording either by photography or densitometry tracing should be done immediately to overcome this problem.

Immunological Methods

Immunological methods are quite specific, safe and sensitive. Antibodies react specifically with the antigen. This reaction forms the basis for all the immunological methods for analysing proteins. Each antibody is made of four polypeptide chains, two light chains and two heavy chains. The part of the antibody, which reacts with antigen specifically, is known as **antigen-binding site**. Each antibody carries two such antigen-binding sites (bivalent) (Figure 10.31a). The antibody-binding site of an antigen varies according to the type of an antigen. Antigen has one or many antibody binding sites (Figure 10.31b). When antigen–antibody reaction is allowed in a solution or in a gel matrix, a visible precipitation occurs. The immunoprecipitate is formed only when antigen and antibody are present at a certain ratio called **equivalence point**. If there is excess of antibodies or antigens in the reaction medium, the resulting antigen–antibody complexes are soluble so there will not be any visible precipitation. So it is important to apply only the required concentration of antibodies to detect an antigen in immunological methods. The concentration of antibodies in antiserum is expressed as the titre of an antibody. The amount of antigen (mg) required to form an immunoprecipitate with a known volume of antibody is known as the titre of an antibody.

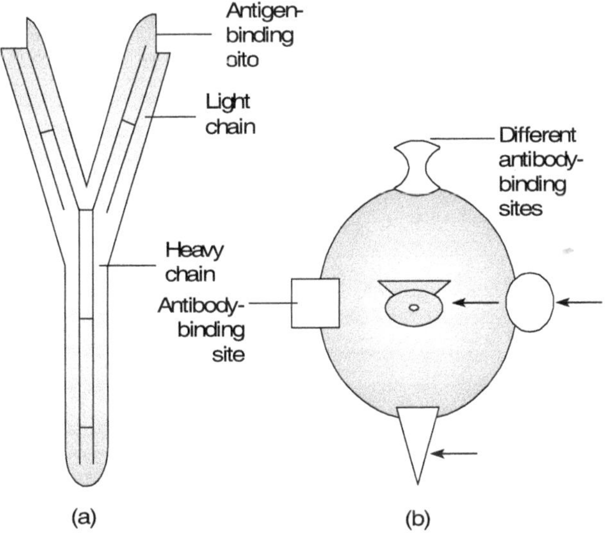

Figure 10.31 (a) Antibody (b) Antigen

Antigen–antibody reaction in solution shows increase in turbidity with the increase in the concentration of immunoprecipitate. By monitoring the turbidity in a nephelometer one can determine the concentration of antigen/antibody present in a solution. However, this method is not suitable for samples containing unknown antigens or antibodies.

Ouchterlony's technique Ouchterlony in 1948 developed a technique which evaluates antigen–antibody reaction. It is a double diffusion technique, in which antigen and antibody are allowed to diffuse towards each other in a gel. The diffusing antigen and antibodies form a white precipitin line at a place where their concentration reaches equivalence (Figure 10.32 a,b,c,d and e).

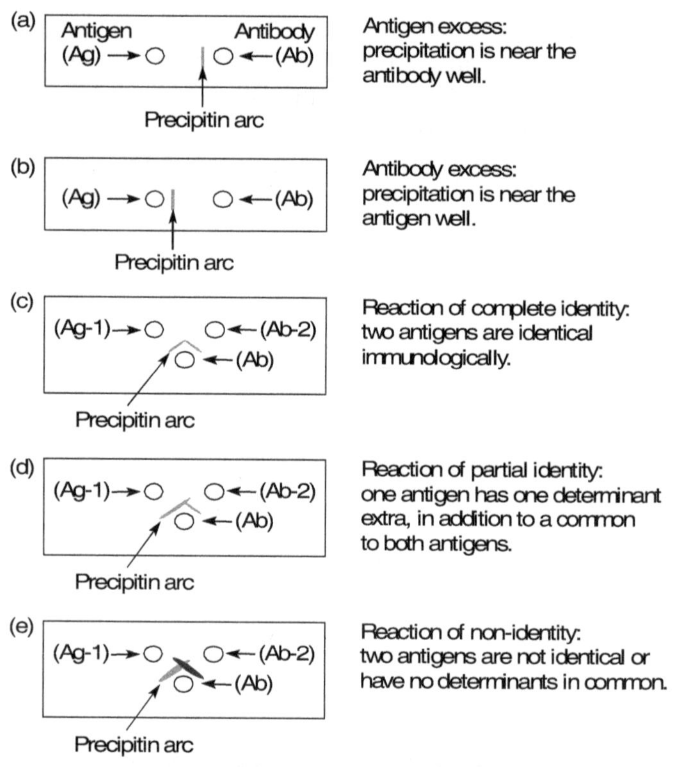

Figure 10.32 Ouchterlony's technique

This technique is also useful to determine the optimum concentration of antigen or antibody for precipitation reactions. But this technique is not employed for quantitative estimations.

Graber–William's technique Graber–William's technique is a classic immunoelectrophoresis technique devised by Graber and William in 1953. This method is widely used for detection of several antigens in complex biological fluids. The method involves the following steps:

1. The separation of antigens into different components in agar and agarose gels is carried out under the influence of electrical current (Figure 10.33a).
2. A rectangular trough is made parallel to the direction of antigen separation. Antiserum is applied to the rectangular trough and antigens–antibodies are allowed to diffuse (Figure 10.33b).
3. Precipitin arcs are formed when antigens bind with specific antibodies at their equivalence zone. By employing known antigen it is possible to identify the corresponding proteins in complex biological fluids such as blood serum. But quantitative analysis is not possible (Figure 10.33c).

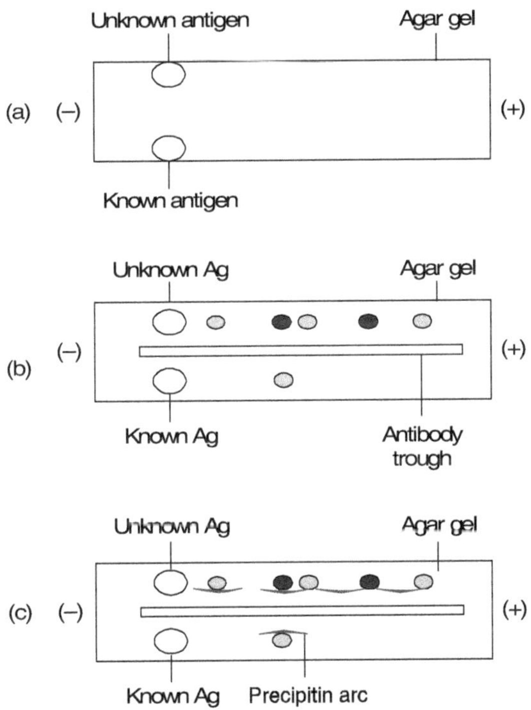

Figure 10.33 Graber–William's technique

Mancini technique Mancini technique is a simple quantitative method devised in 1965. This is the simplest of all immunotechniques for quantitative determination of antigens. In this technique, 2% agar or agarose gel containing specific antibodies is prepared on a petri dish or on glass slides. Wells are cut (Figure 10.34a) and test samples in different dilutions are loaded (2 μl) using a Hamilton syringe. Gel plates are incubated in a moist chamber for 24 to 48 hours (Figure 10.34b). Diameter of the precipitin rings is measured (Figure 10.34c). If the precipitin ring is faint, the gel can be stained with amido black after washing the slides thoroughly with saline. A standard curve is drawn with a known antigen (Figure 10.34d). By using the standard curve, concentration of the antigen in unknown sample can be determined. Mancini technique is very simple at the same time it is a quantitative method. No special equipment is needed. But specific antibodies are required for estimating antigen. It is time-consuming as visible precipitation takes a day or two.

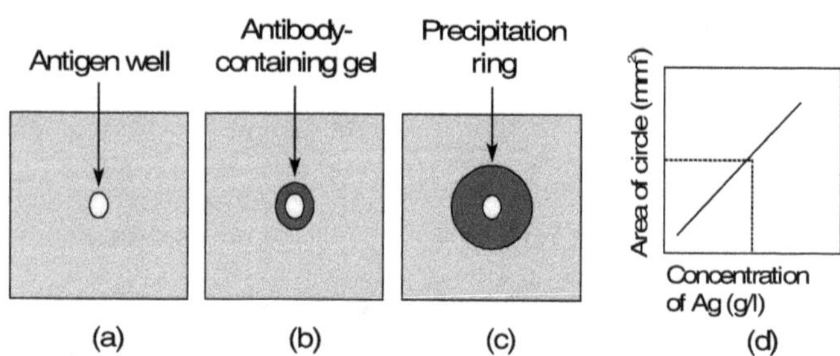

Figure 10.34 Mancini single immunodiffusion technique

Rocket electrophoresis Laurell in (1965) described a method to determine the concentration of a single antigen using electrical force to facilitate antigen migration in antibody gel. When antigen samples are placed in a gel containing monospecific antibodies, and an electrical field is applied, antigens tend to migrate towards anode. During this process, antigen molecules meet the corresponding antibodies and form a rocket-like precipitin peak (Figure 10.35a). The height of the rocket varies depending on the concentration of antigen. A standard graph can be drawn, i.e., concentration of antigen versus height of the peak (Figure 10.35b). Using this curve, concentration of unknown antigen can be determined.

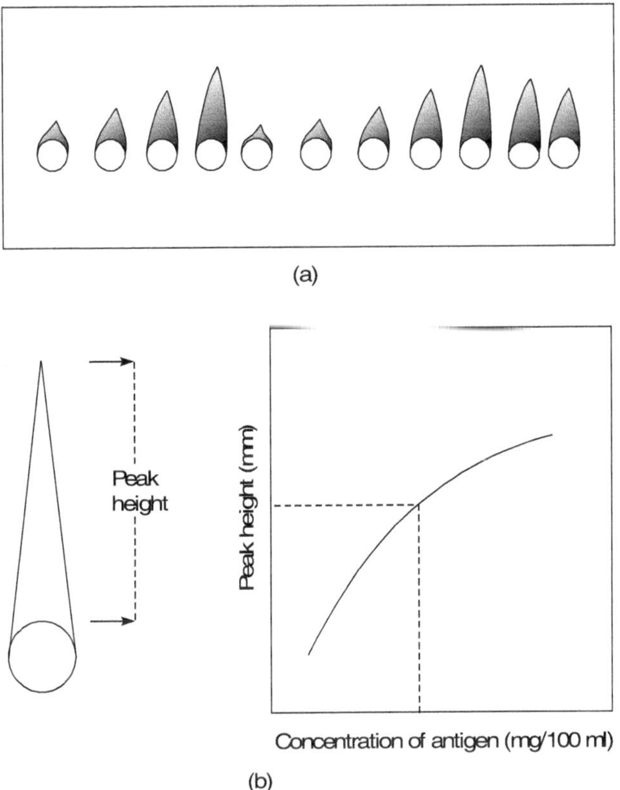

Figure 10.35 (a) Rocket immunoelectrophoresis (b) A standard curve for determination of antigens by rocket immunoelectrophoresis

BLOTTING TECHNIQUE

Southern (1975) described a novel method called blotting technique. In blotting technique, the molecules separated in a slab gel are eluted through the broad face of the gel on to a membrane filter that binds with the molecule as they emerge. As proteins or nucleic acids stay predominantly on the surface of the membrane, they become concentrated and easily accessible for detection procedure. There are different methods of blotting.

Southern blot This was the first blotting technique described by Southern (1975) for the transfer of DNA from agarose gel on to nitrocellulose membrane. Hence, this technique is named after him as Southern blot.

The method involves placing the agarose gel on filter paper wick in a tray as shown in the Figure 10.36. Nitrocellulose membrane is then carefully layered on to the gel. Dry blotter papers are stacked on the membrane. A suitable weight is kept on it. Transfer buffer is added to the tray so that the paper wick is in contact with it. When this arrangement is left undisturbed, the dry blotter absorbs buffer from the gel and there is an upward capillary flow of buffer from the tray to the blotter paper through the membrane. During this process, macromolecules also move along with the buffer. As they cannot pass through the tiny holes of the membrane, the membrane blocks them. The DNA in the gel is ultimately blotted on to the membrane. Since the membrane is in firm contact with the gel and blocks DNA in the same pattern as on the original gel, the result is a true copy of the original. The membrane is then taken for analysis. Southern blot is driven by capillary force. Furthermore, upward capillary flow is a slow movement. Hence, this technique is a very slow method requiring 2 to 24 hours for transfer to complete.

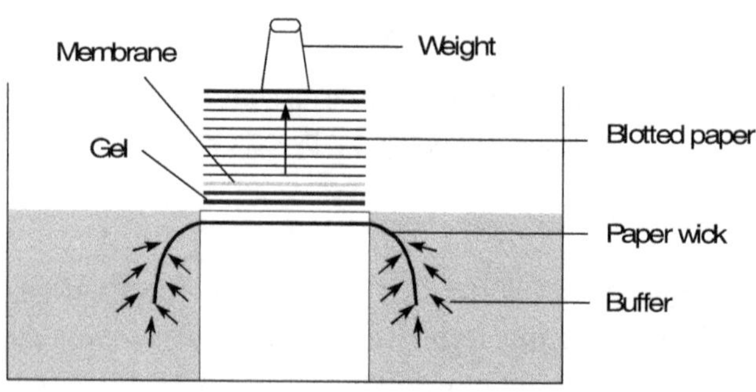

Figure 10.36 Southern blot transfer

A similar transfer technique for RNA is known as 'northern blotting. Any blot transfer to proteins is then called as western blotting.

Turbo blotter Turbo blotter is a downward capillary transfer system. It is simply the reverse process of Southern blotting system. In this system an arrangement is made with the paper wick so that the buffer flows in a downward direction (Figure 10.37). Naturally, the downward flow is faster due to gravitational force and the entire transfer is completed in 3 hours time.

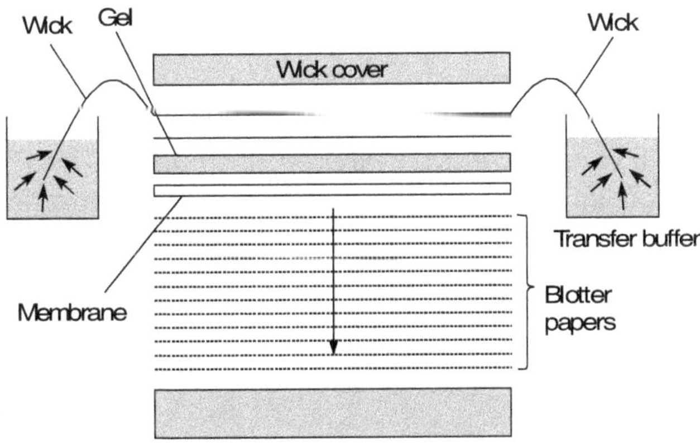

Figure 10.37 Turboblotter

Vacuum blotter Vacuum blotter quickly and efficiently transfers nucleic acids. Transfer of nucleic acids from agarose gels on to nylon membranes is facilitated quickly by applying vacuum.

A vacuum blotter is made of durable moulded plastic which is resistant to the reagents commonly used in transfers, such as HCl and NaOH. The gel-membrane assembly is placed on the perforated stage and it is sealed (Figure 10.38). When vacuum is applied, the transfer buffer along with the molecules is drawn down from the gel. During the downward flow they bind on to the membrane immediately. Vacuum blotting generally requires about 90 minutes for transfer of genomic DNA and 30 minutes for transfer of small fragments ($\leq$ 1500 bp).

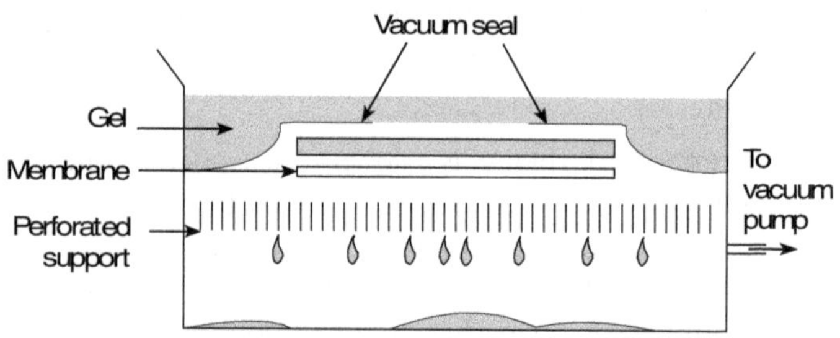

Figure 10.38 Vacuum blotter

Electroblotting Electrophoretic transfer was found faster than capillary action and takes about 30 min. to 2 hours. The method of transferring molecules from gel to membrane by electrical force is known as **electroblotting** (Figure 10.39). It became popular when Towbin *et al.* (1979) reported it for the transfer of proteins from polyacrylamide gels. Since then numerous publications have appeared discussing the various aspects of electroblotting of proteins, DNA and RNA (Gershoni and Palade, 1983; Symington, 1983; Sasse and Gallagher, 1991).

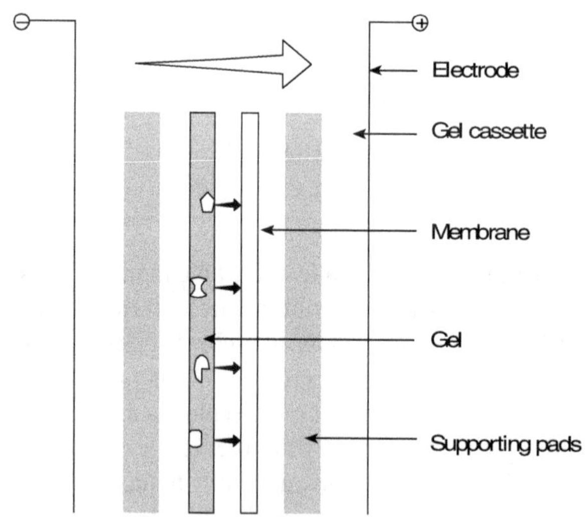

Figure 10.39 Electroblotting

Advantages of Electroblotting

1. It is a rapid process and any type of gel (agarose, polyacrylamide, gradient, urea gel, O'Farrel 2D gel) can be transferred in 30 minutes to few hours.
2. It is a quantitative method and recommended for preparative or for analytical applications involving low levels of sample or low sensitive probes.
3. As the extent of elution is high, the sensitivity is also high when compared to capillary transfer.
4. Since the transfer process is faster under controlled electrical conditions, transfer is exact. Resolution of the bands is not altered after transfer.
5. Membranes with high binding capacity such as supported nitrocellulose and Zeta-probe are available, so the technique becomes more efficient.

REVIEW YOUR LEARNING

1. Define electrophoretic mobility.
2. Explain the principle and applications of electrophoresis.
3. Discuss the principle, method and applications of SDS-PAGE.
4. Discuss the methodology of immunoelectrophoresis and its applications.
5. Explain the principle, methodology and applications of capillary electrophoresis.
6. Write descriptive notes on the following.
   i. Moving boundary electrophoresis
   ii. Paper electrophoresis
   iii. Starch gel electrophoresis
   iv. Agar gel electrophoresis
   v. Agarose gel electrophoresis
   vi. Submarine electrophoresis

vii. Preparative electrophoresis
viii. Pulse field gel electrophoresis
ix. Isoelectric focusing
x. Continuous flow electrophoresis
xi. Counter immunoelectrophoresis
xii. Crossed immunoelectrophoresis
xiii. Densitometer
xiv. Gel documentation system
xv. Autoradiography
xvi. Ouchterlony's technique
xvii. Graber–William's technique
xviii. Mancini technique
xix. Rocket electrophoresis
xx. Southern blot
xxi. Turbo blotter
xxii. Vacuum blotter
xxiii. Electroblotting

SPECTROSCOPY

Spectroscopy is the branch of science dealing with the study of interaction of **electromagnetic radiation** with matter. The most important consequence of such interaction is that energy is absorbed or emitted by matter in discrete amounts called **quanta**. The absorption or emission processes are known throughout the electromagnetic spectrum ranging from the gamma region (nuclear resonance absorption or the Mössbauer effect) to the radio region (nuclear magnetic resonance). The ways in which the measurements of radiation frequency (emitted or absorbed) are experimentally made and the energy levels from these are deduced comprise the practice of spectroscopy.

ELECTROMAGNETIC RADIATION

Light or electromagnetic radiation is a form of energy that is transmitted through space at a constant velocity of $3 \times 10^8\ ms^{-1}$. These radiations are said to have dual nature exhibiting both wave and particle characteristics. The dual character is indeed useful for understanding the interactions of radiations with matter.

Wave Theory of Electromagnetic Radiation

According to this theory, the electromagnetic radiations travel in the form of waves. This wave motion consists of oscillating electric and magnetic fields directed perpendicular to each other and perpendicular to the direction of propagation of the wave as shown in the Figure 11.1.

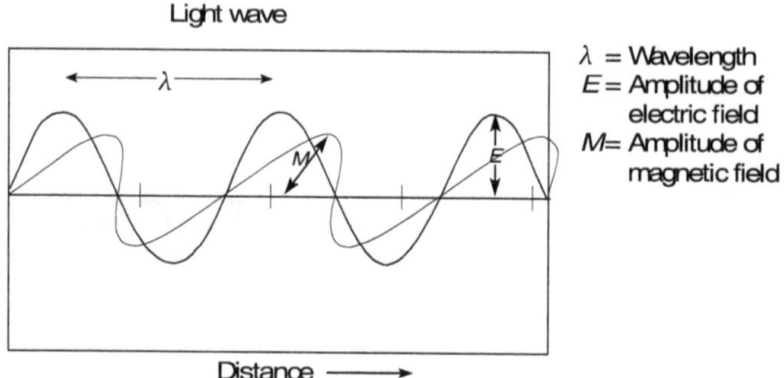

Figure 11.1 The electric and magnetic components of electromagnetic radiation

In spectroscopic studies, the effects associated with the electrical component of the electromagnetic wave are important. The propagation of vibrations in the electrical field only is shown in the Figure 11.2.

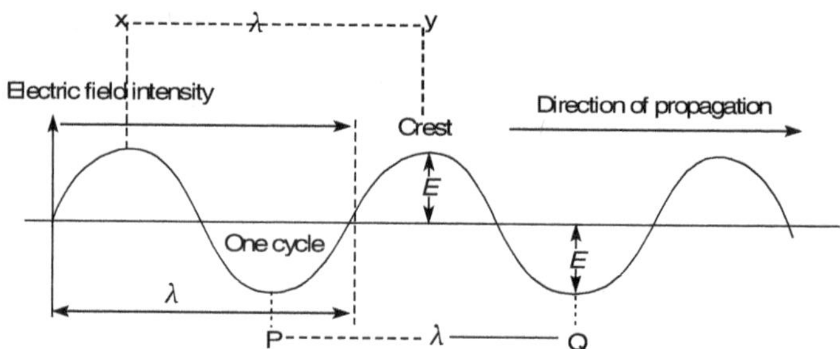

Figure 11.2 Propagation of vibration in the electric field of electromagnetic radiation

The points X, Y, P and Q on the wave represent the maximum disturbances in the electric field. The distance from the mean position is known as the amplitude of the wave. The distance from the crest X to crest Y (or from valley P to valley Q) is the wavelength λ. The number of complete wavelength units passing through a given point per second is called frequency ν. These two quantities are related to each other, and given

$$\nu = \frac{c}{\lambda} \tag{1}$$

where c is the velocity of the electromagnetic wave. Since c is constant (3×10^8 ms^{-1} in vacuum) for all types of electromagnetic radiations, the above relation may be expressed as

$$\nu \alpha \frac{1}{\lambda} \qquad (2)$$

Reciprocal of wavelength, i.e., $\frac{1}{\lambda}$ is called wave number, $\bar{\nu}$. Hence equation 1 may be written as,

$$\nu = c\bar{\nu} \qquad (3)$$

The **wavelength** is expressed in terms of centimetre (cm), metre (m), micron (μ) or micrometer (μm) or angstrom (Å) units. The other commonly used unit is nanometer (nm) where 1 nm $= 10^{-9}$.

Frequency is measured as cycles per second (cps) called Hertz (Hz) or kilocycles per second (kHz) or megacycles per second (MHz).

1 kHz = 10^3 Hz

1 MHz = 10^6 Hz or cps.

The **wave number** is the number of waves per unit distance and is expressed in the units of cm^{-1} called **Kaysers** (K). Sometimes kilokayser (kK) is also used.

1kK = 1000 K = 1000 cm^{-1}

Quantum Theory of Electromagnetic Radiation

The quantum theory describes electromagnetic radiation as one consisting of a stream of energy packets called **photons** or **quanta**, which travel in the direction of propagation of the beam with the velocity of light. Orbital electrons accept only those radiations which have the exact amount of energy to push it into the permitted next higher energy level. In other words, it can absorb radiations of only a particular wavelength, which provides exact requirement of energy. Hence, the absorption spectrum is always specific for a substance. During the emission or absorption of light by chemical substances, the energy changes take place directly, always as integral multiples of small units of energy, i.e., photon. The energy

E of photon is proportional to the frequency of radiation and is given by the equation,

$$E = hv \qquad (4)$$

where h is the Planck's constant (6.625×10^{-27} erg/second; 6.625×10^{-34} Joule/second). The energy of photon is called quantum of energy.

By substituting $\frac{c}{\lambda}$ for v as expressed in equation 1, equation 4 can be alternatively expressed as

$$E = hv = \frac{hc}{\lambda}$$

By substituting $c\bar{v}$ for v as expressed in equation 3, equation 4 can be alternatively expressed as,

$$E = hc\bar{v}$$

Electromagnetic Spectrum

The electromagnetic spectrum for most spectroscopic purposes is considered to be consisting of the region of radiant energy ranging from wavelengths of 10 metres to 1×10^{-12} centimetres. When a molecule absorbs electromagnetic radiation, it can undergo various types of excitation. This excitation may be electronic excitation, rotational excitation, excitation leading to a change in nuclear spin, excitation resulting in bond deformation and so on. Vacuum UV, UV, visible and near infrared ranges of spectrum are produced due to transitions which occur at the valence electron level. Far infrared range of spectrum is produced due to molecular vibration and rotation. If the energy available approaches the ionization potential of the molecule, an electron may be ejected and ionization may occur. Since each mode of excitation requires a specific quantity of energy, different absorptions appear in different regions of the electromagnetic spectrum. The various regions of electromagnetic spectrum are shown in the Table 11.1.

Table 11.1 Different regions of the electromagnetic spectrum

Type of radiation	Wavelength	Wave number	Type of molecular spectrum
Radio frequency	> 100 mm	$< 3 \times 10^9$ Hz	NMR (spin orientations)
Microwave	1 to 100 mm	10 to 0.1 cm^{-1}	Rotational
Far-IR	50 μm to 1 mm	200 to 10 cm^{-1}	Vibrational fundamentals or rotational
Mid-IR	2.5 to 50 μm	4000 to 667 cm^{-1}	Vibrational fundamentals
Near-IR	780 nm to 2.5 μm	$(13 \text{ to } 4) \times 10^3$ cm^{-1}	Vibrational (overtones)
Visible	380 to 780 nm	$(2.6 \text{ to } 1.3) \times 10^4$ cm^{-1}	Electronic (valence orbitals)
Near UV	200 to 380 nm	$(5 \text{ to } 2.6) \times 10^4$ cm^{-1}	Electronic (valence orbitals)
Vacuum UV	10 to 200 nm	10^6 to 5×10^4 cm^{-1}	Electronic (valence orbitals)
X-rays	10 pm to 10 nm	10^9 to 10^6 cm^{-1}	Electronic (core orbitals)
Gamma rays	10^{-10} cm	10^{10} cm^{-1}	Mossbauer (Nuclear transitions) (excited states of nuclei)
Cosmic rays	10^{-12} cm	10^{12} cm^{-1}	Mossbauer (Nuclear transitions) (excited states of nuclei)

The major characteristics of various spectrum regions are outlined as follows:

1. γ-ray region The γ-rays are short waves emitted by atomic nuclei involving energy changes of 10^9 to 10^{11} Joules/gram atom.

2. X-ray region X-rays emitted or absorbed by movement of electrons close to the nuclei of relatively heavy atoms, involve energy changes of the order of 10000 kilo Joules.

3. Visible and ultraviolet region This region further consists of vacuum ultraviolet, ultraviolet and visible regions. The distinction between vacuum ultraviolet and ultraviolet is made because air and quartz start absorbing below 180 nm. Radiations, which we call light, form the visible region of the electromagnetic spectrum.

4. *Infrared region* This has been further divided into near infrared, mid infrared and far infrared regions. All the three subregions of the infrared part of the electromagnetic spectrum are associated with the changes in the vibration of molecules and distinction between them is due to difference in instrumentation.

5. *Microwave region* This region corresponds to changes in the rotation of molecules.

6. *Radio frequency region* The energy change involved in this region arises due to the reversal of a spin of nucleus or electron.

When a molecule absorbs radiation, its energy increases in proportion to the energy of the photon. The lowest state of energy of an atom or molecule is called the ground state. By absorbing one quantum of energy, $h\nu$, the molecule is raised to the next higher level and is said to be in the excited state. Absorption of more energy in integral multiples of $h\nu$, will result in further excitation to higher energy levels.

Study of absorbed radiations from a continuous source that are utilized in raising the internal energy of a molecule constitutes **absorption spectroscopy**. After the absorption of energy, the excited species returns to the ground state by emitting this energy as radiations. The study of this emitted radiation constitutes **emission spectroscopy**.

The internal energy of a molecule may be regarded as the sum of the **translational, rotational, vibrational** and **electronic energies**. The kinetic energy component due to free motion of molecules through space is called translational energy, which is not of concern in molecular spectroscopy. Rotational energy is associated with the rotational motion of a molecule as a whole. The energy component associated with the vibration of the constituent atoms in the molecule is called vibrational energy. Electronic energy is associated with the motion of electrons.

TYPES OF SPECTROSCOPY

Absorption of photons by the molecules may change its internal energy (electronic, vibrational or rotational energy) or may cause transitions between different spin orientations of nuclei in the magnetic field. It is thus possible to affect a change in a particular type of molecular energy using appropriate frequency wavelength or wave number of the incident radiation. However, the spectra may result from transitions in which more than one type of molecular energy changes. For example, electronic spectra arise from transitions between electronic energy levels accompanied by changes in both vibrational and rotational states. Although almost all parts of the electromagnetic spectra are used for studying matter, the following types of spectroscopy are in common use:

1. γ-ray spectroscopy
2. Mössbauer spectroscopy
3. X-ray spectroscopy
4. Colorimetry
5. Ultraviolet and Visible spectroscopy (UV-VIS)
6. Electronic micromotility meter
7. Flame photometry
8. Nephelometry and turbidimetry
9. Fluorimeter and phosphorimeter
10. ORD and CD spectroscopy
11. Infrared spectroscopy (IR)
12. Raman spectroscopy
13. Electron Spin Resonance spectroscopy (ESR)
14. Nuclear Magnetic Resonance spectroscopy (NMR)
15. Atomic spectroscopy
16. Mass spectrometry (MS)

γ-RAY SPECTROSCOPY

γ-rays are of nuclear origin, but they are also part of electromagnetic spectrum. Due to their considerable penetrating power, their main applications in biology are in imaging and radiotherapy. An important application of the use of γ-ray emission spectroscopy is the use of the element technicium (Tc), which is not a natural element but produced in nuclear industry. This element is used in medical studies, because if it is complexed to a compound that is concentrated in biological tissues like bone, liver or brain, its location can be determined by its emission spectrum. The emitted radiation is detected using a device known as γ-ray emission spectroscopy, in which the γ-camera enables the shape and structure of the tissue under study to be investigated.

MÖSSBAUER SPECTROSCOPY

Principle

The γ-ray energy from a radioactive nucleus may be modulated by giving a Doppler velocity to the source. The Doppler effect is recognized as the apparent change in frequency that occurs when the source is moving relative to the detector. The change in frequency is proportional to the source velocity and any velocity may be chosen to give the required frequency. γ-rays of discrete energy can be absorbed resonantly by appropriate nuclei. The source used is usually ^{57}Co which emits a range of γ-rays with different energies, an appropriate one of which may be selected. The selected ray is then modulated by the imposed Doppler phenomenon.

Instrumentation and Applications

Figure 11.3 shows a simplified diagram of the arrangement required to perform Mössbauer spectroscopy. The Doppler velocity is imposed by rapidly vibrating the ^{57}Co source. γ-rays are absorbed by the sample and finally detected by the detector.

The major application of this technique is in the study of the coordination of metal atoms by ligands of an appropriate complexing agent. This has biological importance. For example, in sickle cell anaemia the iron atom is distorted out of the plane of the heme moiety, when compared with the normal haemoglobin. This can be identified by using Mössbauer spectroscopy.

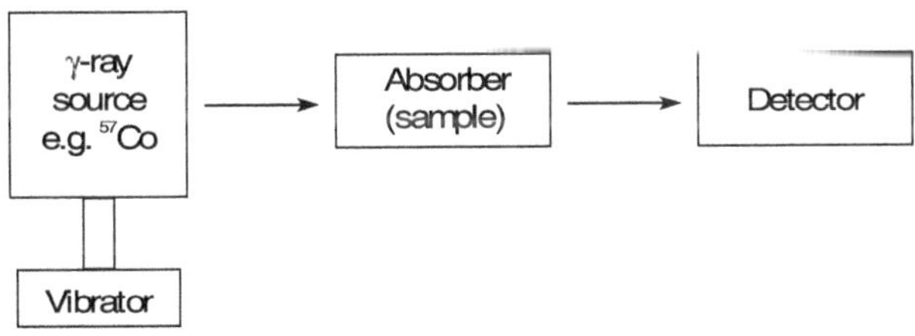

Figure 11.3 Schematic diagram of Mössbauer spectrometer

X-RAY SPECTROSCOPY

The X-ray region of the electromagnetic spectrum consists of wavelengths in the region of about 0.1 to 100Å. The X-rays can interact with matter in three ways—absorption, scattering and diffraction.

Absorption If a beam of X-ray is allowed to pass through matter, it loses energy partly by scattering and partly by true absorption. The electrons of the atoms constituting the matter absorb energy from X-rays and get excited. Then, these excited electrons emit secondary (secondary fluorescent) radiations characteristic of those atoms.

Scattering and diffraction When a beam of X-radiation is incident upon a substance, the electrons constituting the atoms of the substance become small oscillators. These, on oscillating at the same frequency as that of incident X-radiation, emit electromagnetic radiations in all directions at the same frequency as the incident X-radiation. These scattered waves are coming from electrons, which are arranged in a regular manner in a crystal lattice and then travelling in certain directions. If these waves undergo constructive interference, they are said to be diffracted by the crystal plane. Every crystalline substance scatters the X-rays in its own unique diffraction pattern producing a fingerprint of its atomic and molecular structure.

X-ray techniques and methods can be classified into three main categories:

1. X-ray absorption
2. X-ray fluorescence and
3. X-ray diffraction methods

X-ray Absorption Methods These are analogous to absorption methods in the other regions of the electromagnetic spectrum. In these methods, a beam of X-rays is allowed to pass through the sample, and the attenuation or fraction of X-ray photons absorbed is considered to be a measure of the concentration of the absorbing substance. X-ray absorption methods are only helpful in certain cases like elemental analysis and thickness measurements. As compared with other X-ray methods, these are undoubtedly the least used.

X-ray Fluorescence Methods X-rays arise from displacement of inner, extranuclear electrons. Matter can absorb X-rays and this gives rise to X-ray absorption spectra. X-rays with shorter wavelength emit X-rays of a frequency different from that of the incident ray. The phenomenon is called X-ray fluorescence and gives rise to X-ray fluorescence analysis (XRFA).

Thus in this method, X-rays are generated within the sample and by measuring the wavelength and intensity of the generated X-rays, one can perform qualitative and quantitative analyses. X-ray fluorescence method is non-destructive and frequently requires very little sample preparation before the analysis can be carried out.

X-ray Diffraction Methods These methods are based on the scattering of X-rays by crystals. By these methods, one can identify the crystal structures of various solid compounds.

Instrumentation

X-ray absorption, X-ray fluorescence and X-ray diffraction are the three main fields of X-ray spectroscopy. Only the optical system varies in each case although the component parts of the instrument are the same. The main components are described below.

X-ray tube X-rays are generated when high velocity electrons impinge on a metal target. Many types of X-ray tubes are available which are used for producing X-rays. A typical X-ray tube is shown in Figure 11.4. The cathode which is a filament of tungsten metal is heated by a battery to emit the thermionic electrons. This beam of electrons constitutes the cathode ray stream. If a positive voltage in the form of an anode having a target is kept near these electrons, the electrons are accelerated towards the target. On striking the target, the electrons transfer their energy to its metallic surface which then gives off X-ray radiation.

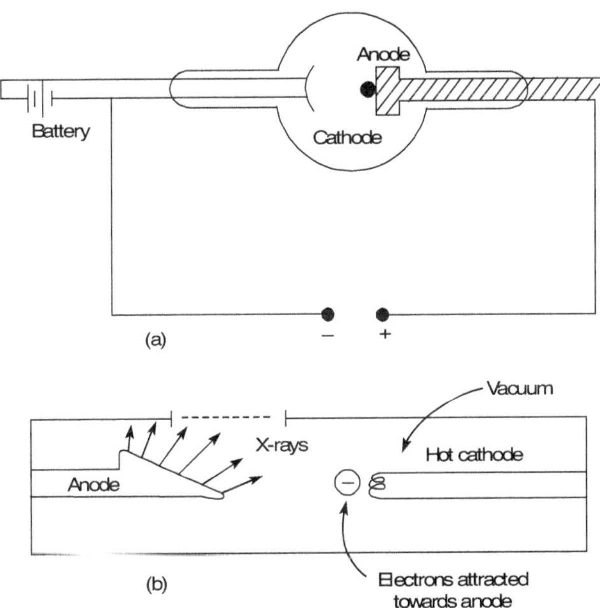

Figure 11.4 (a) X-ray tube for production of X-rays, (b) Schematic representation of an X-ray source

Collimator The X-rays produced by the target material are randomly directed. They form a hemisphere with a target at the centre. In order to get a narrow beam of X-rays, the X-rays generated by the target material are allowed to pass through a collimator which consists of two sets of closely packed metal plates separated by a small gap. The collimator absorbs all the X-rays except the narrow beam that passes between the gap.

Monochromator Monochromatization of X-rays is done either by using the filter or a crystal.

1. **Filter** The X-ray beam may be partly monochromatized by the insertion of a suitable filter. A filter is a window of material that absorbs undesirable radiations but allows only particular radiation of required wavelength to pass through.
2. **Crystal monochromator** The beam is split up by the crystalline material into the component wavelengths in the same way as a prism splits up the white light into a rainbow. Such a crystalline substance is called as an analysing crystal.

Detectors The X-ray intensities can be measured and recorded either by photographic or counter methods. Both these methods depend upon the ability of the X-rays to ionize matter and differ only in the subsequent fate of the electrons produced by the ionizing process.

1. **Photographic methods** In order to record the position and intensity of the X-ray beam, a plane or cylindrical film is used. The film, after exposing to X-rays, is developed. The photographic method is mainly used in diffraction studies since it reveals the entire diffraction pattern on a single film. But this method is time-consuming and uses exposures of several hours. At the same time, the photographic method is rarely used for a quantitative measurement because it involves the use of a densitometer, the operation of which can be both time-consuming and subject to considerable error.

2. **Counter methods** Geiger–Müller tube counter and scintillation counters are used to measure the X-rays. Counter methods are explained in chapter 12.

X-RAY FLUORESCENCE SPECTROMETER

A suitable X-ray source is required that can be focused into the specimen chamber where the substance under test is excited by the incident beam. Monochromators disperse the fluorescent (emitted) radiation and are finally detected by the detector and processed by data processor (Figure 11.5).

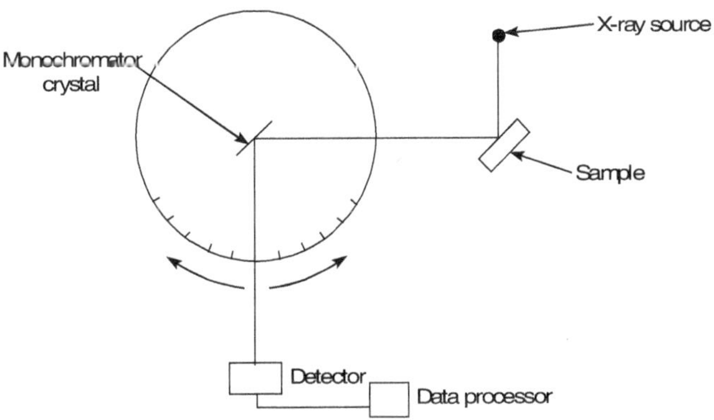

Figure 11.5 X-ray fluorescence analysis. Dispersion of fluorescent X-rays may be detected at various angles.

Applications The technique has wide applications in forensic science and environmental pollution studies, because it enables many elements to be detected and their concentrations to be measured. A clinical application for performing bone densitometry measurements involves either single-photon or dual energy X-ray absorptiometry (DEXA). These studies are useful for monitoring hormone replacement therapy (HRT) in female patients. X-ray spectrometers obviously require a more rigorous approach to the incorporation of safety features.

X-Ray Diffraction

A crystal is a solid with a regular repeating internal three-dimensional arrangement of atoms. This periodic arrangement can be exploited to determine molecular identity and structure when the crystal is exposed to X-rays. X-rays are used because their wavelengths correspond to inter-atomic distances. When X-rays are beamed at the crystal, electrons diffract the X-rays, which causes a diffraction pattern. Using the mathematical Fourier transform, these patterns can be converted into electron density maps. These maps show contour lines of electron density. Since electrons more or less surround atoms uniformly, it is possible to determine where atoms are located. Since hydrogen has only one electron, it is difficult to map hydrogen. To get a three-dimensional picture, the crystal is rotated while a computerized detector produces two-dimensional electron density maps for each angle of rotation. The third dimension comes from comparing the rotation of the crystal with the series of images. Computer programs use this method to come up with three-dimensional spatial. Probably the most famous X-ray diffraction image is the photograph of the B-form of DNA (Figure 11.6) taken by **Rosalind Franklin** in May 1952. With

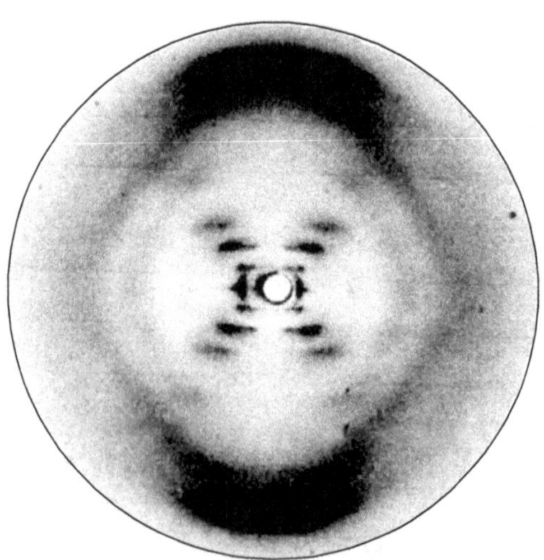

Figure 11.6 X-ray diffraction image of B-form DNA

the help of the Oxford crystallographer, **Dorothy Hodgkin**, **Franklin** described the helical backbone and correct crystallographic space group for DNA. Her work was instrumental in Watson's and Crick's correct modelling of DNA, for which they received the Nobel Prize.

X-Ray Crystallography

It is a technique in crystallography, in which the pattern produced by the diffraction of X-rays through the closely spaced lattice of atoms in a crystal is recorded and then analysed to reveal the nature of that lattice. This generally leads to an understanding of the material and molecular structure of a substance. The spacings in the crystal lattice can be determined. The electrons that surround the atoms, rather than the atomic nuclei themselves, are the entities which physically interact with the incoming X-ray photons. This technique is widely used in chemistry and biochemistry to determine the structures of immense variety of molecules, including inorganic compounds, DNA and proteins. X-ray diffraction is commonly carried out using single crystals of a material, but if these are not available, microcrystalline powdered sample may also be used, although this requires different equipment.

COLORIMETRY

Colorimetry is a form of photometry, which deals with measurement of light absorption by coloured substances in solutions. The instrument which measures the intensity of the colour is known as colorimeter.

Principle

The colorimeter is one of the most widely used instruments in biological research. It is based on the principle that when a beam of incident light passes through a coloured solution, the coloured substances in the solution absorb a part of light and hence, the intensity of the transmitted light is always less than that of the incident light. As the number of light-absorbing molecules increases, the intensity of light coming out of the medium decreases exponentially and vice-versa. The difference in intensities between the incident and transmitted light, in turn reflects the number of absorbing molecules or in other words, the concentration of the absorbing molecules in that solution.

Beer–Lambert's law

Two basic laws proposed by Beer and Lambert governs absorption of light by a substance in solution.

Beer's law states that, "when a parallel beam of monochromatic light passes through an isotropic, light-absorbing medium, the amount of light that is absorbed is directly proportional to the number of light-absorbing molecules in that medium or in other words, the concentration of the substance in that medium."

i.e.,

$$A \alpha C$$

where,

A is the absorbance or optical density and
C is the concentration of the light-absorbing substance in the medium.

Lambert's law states that, "when a parallel beam of monochromatic light passes through an isotopic, light-absorbing medium, the amount of light that is absorbed is directly proportional to the length of the medium through which the light passes."

i.e.,

$$A \alpha L,$$

where,

A is the absorbance and
L is the length of the medium.

Since the measurement of light absorption depends on both the laws, it is popularly known as Beer–Lambert's Law.

Thus,

$$A \alpha C \times L$$

$$A = eCL$$

where e is known as the extinction coefficient.

$$A = \log\frac{I_i}{I_t}$$
$$= \log\frac{100}{T}$$
$$= \log 100 - \log T$$
$$= 2 - \log T$$

where,

I_i is the intensity of the incident light,
I_t is the intensity of the transmitted light and
T is transmittance.

Molar extinction coefficient In the above equation, when concentration is 1M, i.e., one mole per litre and path length of light is 1 cm, then e is known as **molar extinction coefficient** (Σ_M). Σ_M is constant for a particular compound at particular wavelength and has a maximal value when the compound is in its purest state.

Specific extinction coefficient When the molecular weight of compounds such as proteins, nucleic acids, etc. are not known, then the term specific extinction coefficient can be conveniently used. This is the extinction of a 1% (w/v) solution of the compound when the light path is 1 cm. For example, the specific extinction coefficient of BSA = 0.667, IgG = 1.35 and IgM = 1.2 at 280 nm.

Wavelength Electromagnetic radiation is composed of an electric vector and magnetic vector, which oscillate in planes at right angles to each other and both at right angles to the direction of propagation. The distance along the direction of propagation for one complete cycle is known as wavelength (λ).

Absorption spectrum The pattern of energy absorption by a solution of any substance when light of different wavelengths passes through it is a characteristic of that substance. This pattern is known as absorption spectrum. The absorption spectrum of a substance is established by measuring either optical density or transmittance of a particular concentration of the substance at different wavelengths.

Emission spectrum When a solid is heated to a high temperature, light is emitted. On examining this light by a spectroscope, a spectrum is found to be produced which is known as emission spectrum. The emission spectra of sodium and potassium are simple, consisting of only a few wavelengths. For other elements such as iron and uranium, thousands of distinct reproducible wavelengths are present.

Visible spectrum Light from the sun contains the entire visible spectrum (Table 11.2). Daylight or white light is a combination of seven colours (i.e., it is polychromatic). "VIBGYOR" is a code to indicate the component colours in order. The wavelength of light, which the human eye can perceive, ranges from 400 to 700 nanometres (nm). This is called visible spectrum. Radiation of shorter wavelength, less than 400 is called **ultraviolet** (UV) light and radiation of longer wavelength, i.e., more than 700 is called **infrared** (IR) light.

Table 11.2 Visible spectrum

Colour of the solution	Range of wavelength	Complementary or subtraction colour
Violet	400–465	Greenish yellow
Blue	465–482	Yellow
Green	498–530	Red purple (magenta)
Yellow	576–580	Blue
Orange	587–610	Greenish blue
Red	617–660	Bluish green
Purple red	670–720	Green

Colorimeter is a device that measures the intensity of light before and after it has passed through a coloured solution (Figure 11.7). A colorimeter consists of mainly six parts.

1. A light source
2. A condensing lens to render the light rays parallel
3. A filter to generate monochromatic light
4. A sample holder

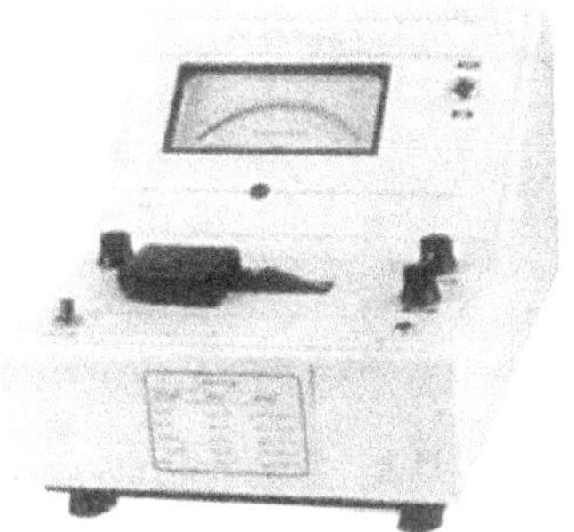

Figure 11.7 Colorimeter

5. A photocell to convert light energy (photons) into electrical energy
6. A galvanometer to measure the electrical energy (current) which is thus generated

The light from a tungsten lamp (composed of wavelengths between 400 and 900 nm) passes through a slit, a condensing lens, and a filter and finally emerges as a parallel beam of monochromatic light (Figure 11.8). Filters consist of selected glass (or sometimes dyed gelatin), which is capable of transmitting light over a limited portion of the spectrum only. The instrument is provided with a set of replaceable filters marked "V, B, G, Y or R" violet, blue, green, yellow or red filters. Instead of the above mark a number may be written on each, which indicates the wavelength of light that the filter transmits. For example, a filter of number '54' absorbs all light except that of wavelengths around 540 nm, which pass through. Other filters bear the following numbers: 42, 49, 59 and 65. In some types of instruments, the filters are not removable and they are fixed onto a disc, which can only be rotated to bring the appropriate filter in the light path. Filters are of limited specificity, and one that is designed to transmit 540 nm may actually transmit light between 520 and 560 nm with a peak transmittance at 540 nm.

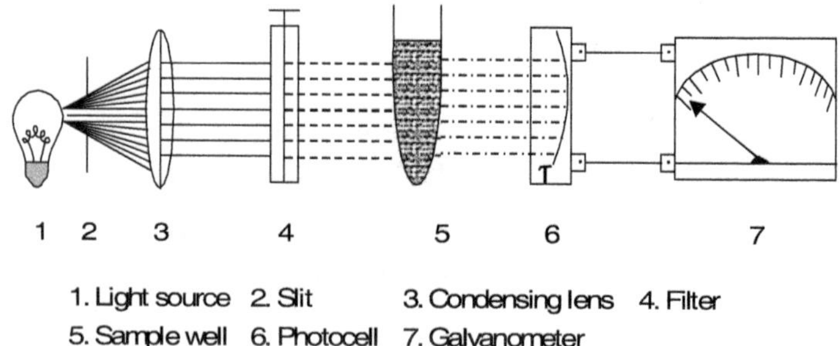

Figure 11.8 Schematic diagram of colorimeter

The monochromatic light passes through the sample solution and the transmitted light falls on the photocell. The photocell converts the transmitted light energy into electrical energy, which is amplified and measured by the galvanometer. The galvanometer is calibrated to read the absorbency/transmittance directly. It is necessary to standardize the instrument or set it at zero using the blank, change cuvettes and read the absorbance. The best analytical procedure requires the zero to be reset between each measurement as colorimeters and some filters are influenced by temperature changes.

Applications

Colorimetry has perhaps the widest application in biological sciences. The concentration of any unknown substance can be determined using a colorimeter. If the substance by itself is colourless, it can be chemically converted to a coloured substance stoichiometrically by adding a chromophoric group, and the concentration measured. The usual procedure is to prepare a set of standards and produce a concentration versus absorbance calibration curve. The absorbance of an unknown substance is measured and its concentration is determined using the standard curve.

SPECTROPHOTOMETRY

Spectrophotometer is an instrument which measures light absorption as a function of wavelength in the UV as well as visible regions (Figure 11.9). It also follows essentially the laws of light absorption, viz. the Beer–Lambert's law. Unlike colorimeters, in spectrophotometers the compounds can be measured at precise wavelengths. Spectrophotometry has become a powerful tool for qualitative and quantitative measurements.

A spectrophotometer consists essentially of six parts.

1. Light source
2. Condensing lens
3. Monochromator
4. Sample holder(s)
5. Detector(s) and
6. Recorder

Figure 11.9 Spectrophotometer

A spectrophotometer has two light sources, an UV light (for measuring light absorption from ~200 to ~400 nm) and white light (for measuring light absorption from ~400 to 900 nm). With the help of a shutter, only one of the lights is allowed to fall on a silvered mirror (SM). The reflected light from the mirror passes through an entrance slit and a condensing lens. The lens renders the light rays into parallel beams and the parallel beams of light now fall on a monochromator (grating). The monochromator disperses the light into its component wavelengths. Using the wavelength selector, the desired wavelength is selected. Now the selected beam of monochromatic light passes again through a lens to a light-tight compartment where the sample is kept in a cuvette. After passing through the sample, the transmitted light falls on a photomultiplier tube (PMT). The PMT converts the light energy into electrical energy, which is amplified, measured and recorded on the analog/digital read-out (Figure 11.10).

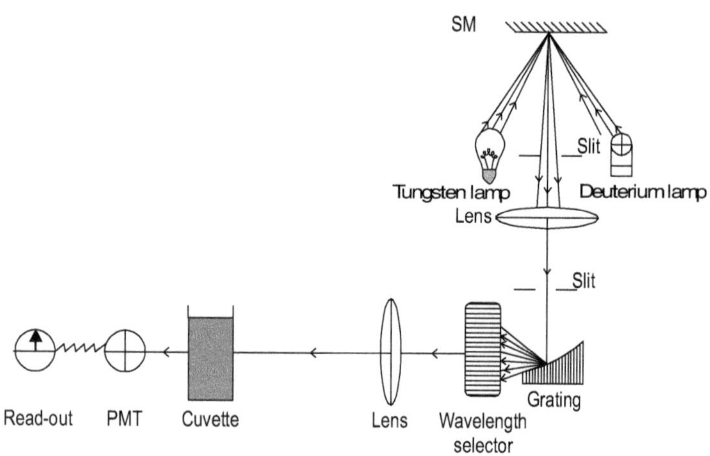

Figure 11.10 Schematic diagram of a single-beam spectrophotometer

In double beam spectrophotometers, the monochromatic light coming out from the lens is split into two halves by placing a half-silvered mirror (HSM) on its path. Now 50% of the light passes directly through the mirror and falls on the reference cuvette and 50% of the light reflected onto a second silvered mirror and then allowed to fall on the sample cuvette. At any given time, the intensities of the transmitted light from the reference and sample cuvettes are measured, amplified, the difference in intensities computed and sent to the read-out (Figure 11.11).

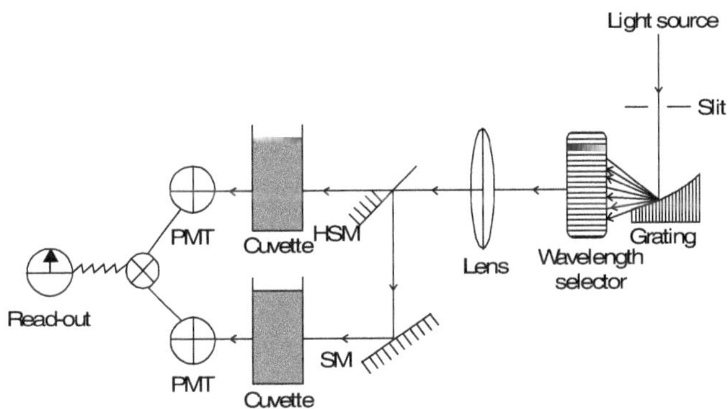

Figure 11.11 Schematic diagram of a double-beam spectrophotometer

ULTRAVIOLET AND VISIBLE SPECTROSCOPY (UV-VIS)

Ultraviolet and visible regions of the electromagnetic spectrum and their associated techniques are most widely used in analytical and biological research.

Parts of a UV-VIS Spectrophotometer

Light sources A UV-VIS spectrophotometer has two light sources, a tungsten lamp for visible light, and a deuterium or a hydrogen lamp for UV light respectively (deuterium lamp gives wider and more intense light in UV region than a hydrogen lamp). The light from the light source is composed of a wide range of wavelengths. This light is called polychromatic or heterochromatic. The polychromatic light reflected back using a plane mirror, passes through an entrance slit and through a condensing lens and falls onto a monochromator. The monochromator disperses the light and the desired wavelength is focused on the exit slit using the wavelength selector.

Monochromators The monochromators which produce radiations of single wavelength are based either upon refraction by a prism or diffraction by a grating. Prisms are made of glass for visible region and of quartz or silica for UV region. A grating consists of ruled lines (as many as 2000 lines per millimetre) on a transparent or reflecting base. The resolving power of a grating is directly proportional to the closeness of these lines. Gratings are superior to prisms as they yield linear resolution of the spectrum for the entire range of wavelengths. The efficiency of monochromation is enhanced by using double monochromators in which a selected part of the spectrum from the first grating is further resolved by a second grating, resulting in a bandwidth of as low as 0.1 mm.

Cuvettes The optically transparent cells (cuvettes) are made up of glass, plastic, silica or quartz. Glass and plastic absorb UV light below 310 nm. Hence, they cannot be used for light measurements in UV region. Silica and quartz do not absorb UV light and hence they are used for both UV and visible light measurements. Since quartz absorbs light below 190 nm, cuvettes of lithium fluoride can be used which transmit radiation down to 110 nm. Oxygen also absorbs light at wavelengths less than 200 nm. Therefore, if spectra are required in this region the apparatus must be evacuated. The standard cuvettes are made up of quartz, have an optical path of 1 cm, and hold a volume of 1–3 ml. Microcuvettes (0.3–0.5 ml) are used for measurement of expensive chemicals.

Photocell and photomultiplier tubes A photocell (Figure 11.12) is a photoelectric device, which converts light energy into electrical

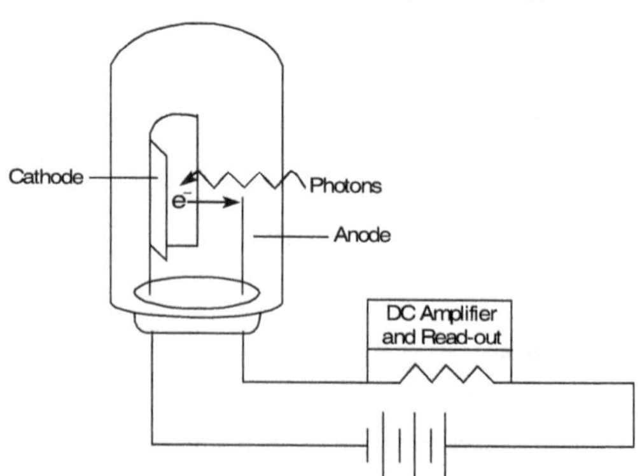

Figure 11.12 Schematic diagram of a photocell

energy, which is then amplified, detected and recorded. In photocells, the photons strike a semicylindrical photoemissive cathode in vacuum. This causes emission of electrons, which is proportional to the intensity of radiation. When a potential difference is applied across the electrodes, the emitted electrons flow to the anode wire generating a photocurrent. This current is amplified electronically and measured.

A **photomultiplier tube** (Figure 11.13), has a cathode with photoemissive surface (a selenium layer) and a wire anode. In addition to the photoemissive cathode, it also contains a circular array of nine additional cathodes called dynodes.

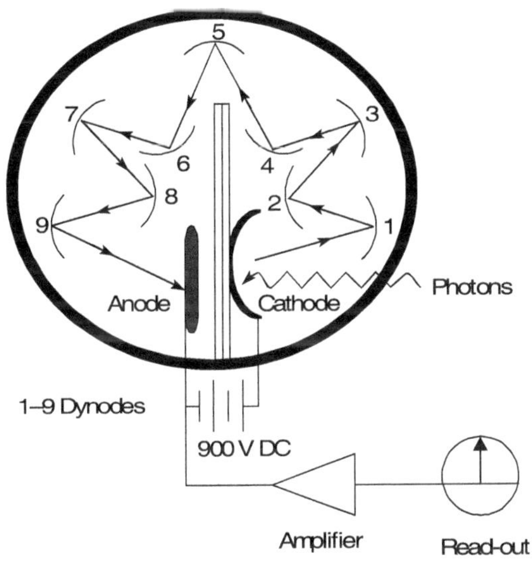

Figure 11.13 Cross-section of a photomultiplier tube

The electrons emitted from the photoemissive cathode strike dynode 1, which emit several additional electrons. The electrons are accelerated towards dynode 2, which again emit several electrons. The amplified electrons flow to the anode generating a much larger photoelectric current than in a photocell.

Applications

UV-VIS spectrophotometer is a more refined instrument and it gives a far better precision and resolution than a colorimeter. It has a wide range of applications in biological research.

1. It is used to estimate the concentration of both coloured as well as colourless solutions, which could absorb light.
2. Because of its higher sensitivity, it is used to estimate extremely small quantities of substances in a matter of a few minutes.
3. It usually does not degrade or modify the materials studied (unless a photochemical reaction occurs) and hence the materials can be recovered and reused.
4. It is also used to find out the absorption maxima of compounds with a wide range of wavelengths.
5. It offers selectivity in that each component in a solution or reaction mixture can be singled out and estimated.
6. It also enables one to follow details of fast reactions and fast enzyme kinetics.
7. It is also used to measure the growth of bacteria and yeasts and to determine the number of cells in a culture.
8. Small volumes (as small as 0.3 ml) can be used for estimation of precious samples.

ELECTRONIC MICROMOTILITY METER

Electronic micromotility meter (EMM) is highly sensitive and gives accurate quantitative measures of the motility of bacteria, larval and adult trematodes and nematodes. This instrument can also be used to record the motility of sperms. A new application has been developed for UV bio-spectrophotometer and the instrument is renamed as electronic micromotility meter (EMM) (Figure 11.14), which can be used to monitor the motor activity of microorganisms (Veerakumari, 2003).

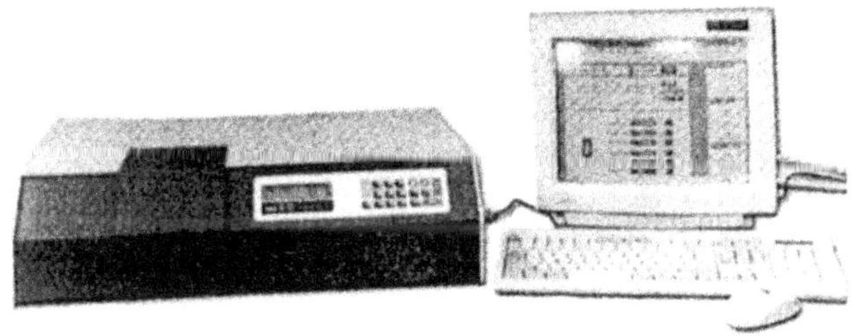

Figure 11.14 Electronic micromotility meter

Principle This follows the principle of spectroscopy. This is governed by Beer–Lambert's law.

Instrumentation The light source (LS) most commonly used is the tungsten filament lamp for visible region and deuterium lamp for the UV region. The monochromator (M) selects a particular wavelength from the continuous spectrum produced by the lamp. The monochromator beam is chopped by a light chopper (CH). The light transmitted from the sample and reference material is sensed by the detector (D) and a voltage is produced across the detector.

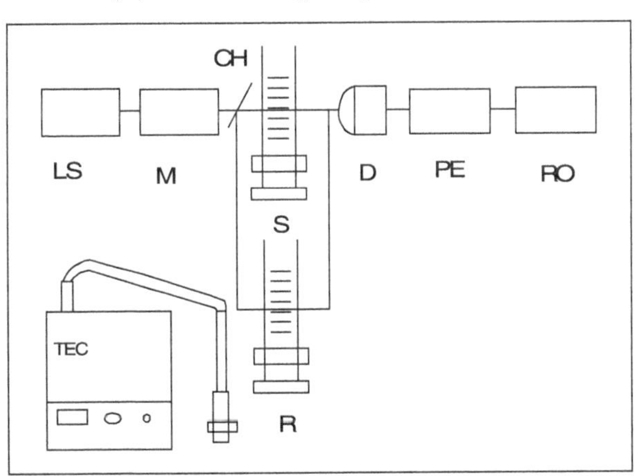

LS—Light source M—Monochromator CH—Light chopper S—Sample
R—Reference D—Detector PE—Processing electronics
RO—Read-out TEC—Temperature controller

Figure 11.15 Schematic diagram of electronic micromotility meter

The signal corresponding to the difference between the transmitted radiation of a reference material and that of a sample at selected wavelengths are measured (PE) and the optical density of the sample is read. Temperature controller (TEC) maintains the temperature required by the organisms. The schematic diagram of an electronic micromotility meter is shown in the Figure 11.15.

Methodology

The motor activity of microorganisms can be recorded with the help of electronic micromotility meter (EMM). When microorganisms are placed in the cuvette, the movement of the microorganisms perturbs the light passing through the cuvette. This perturbation of light produces a voltage signal, which will vary from the reference cuvette. This difference is processed and is represented as optical density. The wavelength can be selected by scanning the particular medium, which is used for the maintenance of the microorganisms in living condition.

FLAME PHOTOMETRY

Flame photometry is based on the measurement of intensity of light emitted when a metal is introduced into a flame. The wavelength of the colour indicates the nature of the element and the intensity of the colour indicates the quantity of the element present.

Principle

When a liquid sample containing a metallic salt solution is introduced into a flame, the processes involved in flame photometry are complex, but a simplified version of the events is listed below:

1. The solvent is vaporized, leaving particles of the solid salt.
2. The salt is vaporized or converted into the gaseous state.
3. A part or all of the gaseous molecules are progressively dissociated to give free neutral atoms or radicals. These neutral atoms are excited by the thermal energy of the flame.

The excited atoms which are unstable quickly emit photons and return to lower energy state, eventually reaching the unexcited state. The measurement of the emitted photons, i.e., radiation forms the basis of flame photometry.

Instrumentation

A schematic diagram of the equipment used in flame photometry is shown in Figure 11.16. The instrument possesses the same basic components as that of a spectroscopic apparatus. The flame photometer also includes a burner, which is utilized for burning the sample solution and exciting the atoms produced in the flame after burning.

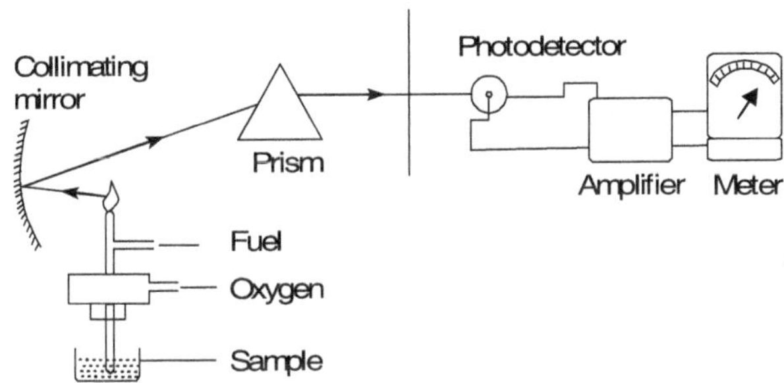

Figure 11.16 Schematic diagram of a flame spectrophotometer

The various components of the instrument are as follows:

Burner The flame used in the flame photometer must possess the following functions:

1. The flame should possess the stability to evaporate the liquid droplets from the sample solution, resulting in the formation of solid residue.
2. The flame should decompose the compounds in the solid residue formed in the step (1), resulting in the formation of atoms.
3. The flame must have the capability to excite the atom formed in step (2) and cause them to emit radiant energy for analytical

purposes; it becomes essential that emission intensity should be steady over reasonable periods of time (1–2 min.)

The temperature of the flame, which is primarily responsible for the occurrence of the above-mentioned processes, is controlled by several factors, which are summarized as follows:

1. Type of fuel and oxidant and fuel-to-oxidant ratio
2. Type of solvent for preparing the sample solution
3. Amount of solvent which is entering into the flame
4. Type of burner employed in flame photometer
5. The particular region in the flame, which is to be focused into the entrance slit of the spectral isolation unit

Mirrors The radiation from the flame is emitted in all directions in space. Thus, a major portion of emitted radiation will not be reaching the detector, and, will thereby yield poor results. In order to increase the amount of radiation reaching the detector, a concave mirror is used which is often set behind the burner to reflect radiation that would have otherwise lost. The focal point of the mirror lies at the entrance of the monochromator.

Slits In most of the good quality flame photometers, entrance and exit slits are generally employed before and after the monochromator. The entrance slit is kept between the flame and optical system. The entrance slit will not allow the entrance of radiation from the surroundings and only permits the radiation from the flame and the mirrored reflection of the flame to enter the optical system.

The exit slit is kept between the monochromator and detector. The purpose of the exit slit is to allow only a selected wavelength range from the radiation travelling from the monochromator to the detector. The exit slit also prevents the entry of interfering lines.

Monochromator In simple flame photometers, the monochromator is the prism. But in expensive models, grating monochromators are used. Quartz is the material most commonly used for making prisms even though its dispersive power is less than that of glass. The reason for this is that quartz is transparent over the entire region.

The grating monochromator employs a grating, which is essentially a series of parallel straight lines cut into a plane surface.

Filters In some elements, the emission spectrum contains a few lines. In such cases wide wavelength ranges will be allowed to enter the detector without causing any serious error. In such a situation, an optical filter may be used in the place of the slit and monochromator system. The filter is made from such a material, which is transparent over a narrow spectral range. When a filter is kept between the flame and detector, the radiation of the desired wavelength from the flame will be entering the detector, which can be measured. The remaining undesired wavelength will be absorbed by the filter and not measured.

The flame photometers, which use filter monochromators, are very convenient for simple repetitive analyses. However, such instruments can be used for a small number of elements. The reason for this is that a large number of filters are employed.

Detectors The radiation coming from the optical system is allowed to fall on the detector which measures the intensity of radiation falling on it. The detector should be sensitive to radiation of all wavelengths that may be examined.

In good flame photometers, photomultiplier detectors are employed which produce an electrical signal from the radiation falling on them.

Applications

Flame photometry is used to detect and quantify the elements sodium, potassium, lithium, magnesium, calcium, strontium and barium.

NEPHELOMETRY AND TURBIDIMETRY

Nephelometry and turbidimetry are techniques of analysis that are closely allied to colorimetry. Both nephelometry and turbidimetry are based on the scattering of light by non-transparent particles suspended in a solution. However, the two techniques differ only in the manner of measuring the scattered radiation.

When light is allowed to pass through a suspension, a part of the incident radiant energy is dissipated by absorption, reflection, and refraction while the remaining is transmitted. Measurement of the intensity of transmitted light as a function of the concentration of the suspended particles forms the basis of turbidimetric analysis. This is illustrated in Figure 11.17.

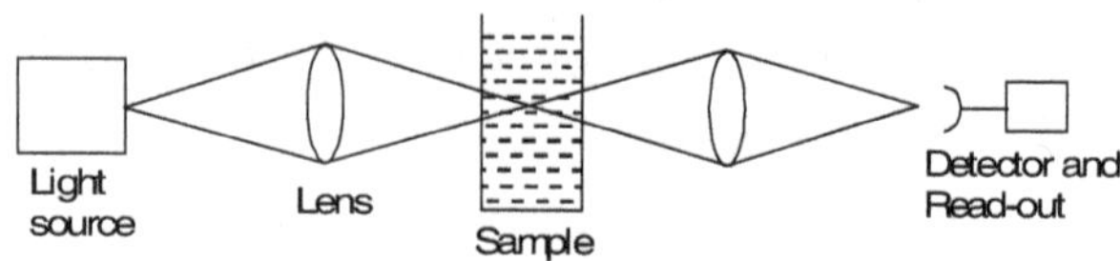

Figure 11.17 Schematic diagram of turbidimeter

In nephelometry, the light is also allowed to pass directly through the sample solution having suspended particles. The amount of radiation scattered by the particles is measured at an angle (usually 90°) to the incident beam. The measurement of the intensity of the scattered light as a function of the concentration of the dispersed phase forms the basis of nephelometric analysis (Figure 11.18a,b).

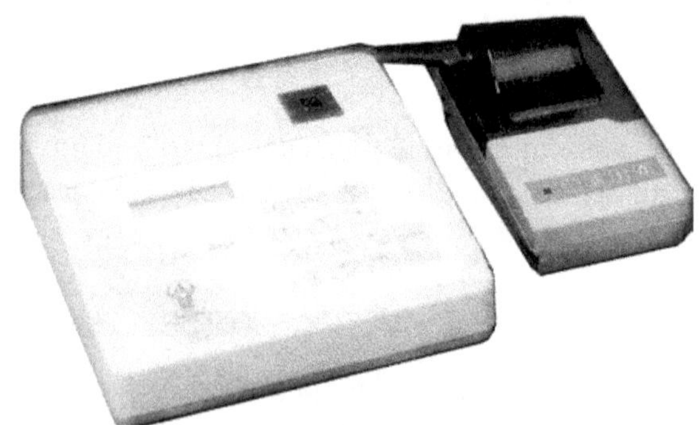

Figure 11.18 (a) Nephelometer

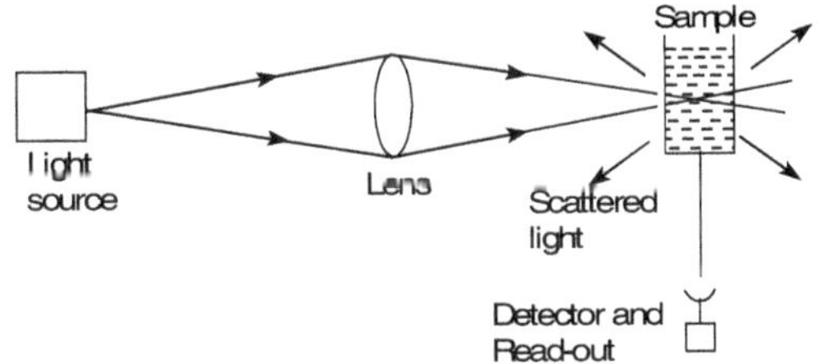

Figure 11.18 (b) Schematic diagram of nephelometer

The choice between a nephelometric and turbidimetric analysis depends upon the amount of light scattered by suspended particles present in the solution. Turbidimetry is useful for determining relatively high concentrations of suspended particles because the scattering is quite extensive due to the presence of many particles. On the other hand, nephelometry is most suited when the suspension is less dense and decrease in power of the incident beam is small. In such a case, more accurate results are obtained due to the small amount of scattered light which would be measured against a black background. In such an instance, turbidimetry should not be used since a comparison would have to be made of two large quantities of nearly equal values.

Theory

Both reflection and scattering phenomena are very important in turbidimetry and nephelometry. If light is allowed to pass through a solution having suspended particles, reflection will take place if the dimensions of suspended particles are larger than the wavelength of incident light. On the other hand, scattering will take place if the dimensions of suspended particles are of the same order of magnitude or smaller than the incident wavelength. This distinction plays an important role in nephelometry and turbidimetry because it affects the sensitivity of the measurement as well as how the measurement is made.

In nephelometry, suspended particles should neither be too large nor too small otherwise the scattering efficiency falls off. For measurements to be made in the ultraviolet and visible regions of spectrum, the optimum particle size should be in the range of about 0.1 to 1 μm.

In turbidimetric measurements, particles larger than the wavelength of light do not pose much problem because measurement depends on the total radiation removed from the primary beam irrespective of the mechanism by which it is removed or the angle through which it undergoes deviation. But with larger particles, another problem arises, i.e., the relationship between absorbance and concentration does not remain linear. Thus, in such cases measurements may not be very accurate.

Instrumentation

Much of the instrumentation used in nephelometry and turbidimetry is very similar to that of spectrophotometric devices.

Light sources One may use white light in nephelometers but it is advantageous to use monochromatic radiation. Similarly, monochromatic radiation is used in turbidimeters to minimize absorption. In either case it is necessary to use sources providing high intensity monochromatic radiation and wherever possible short wavelengths are used to increase the efficiency of Rayleigh scattering. A mercury arc or a laser, with appropriate filter combinations for isolating one of its emission lines, is undoubtedly the most convenient source. However, if one has to determine the concentration of a particular material, a polychromatic source such as a tungsten lamp may be used. Even in such a case best results can be obtained if we use a blue spectral region; a filter may be employed to block other wavelengths.

Cells Although we can use cylindrical cells, they must have flat faces where the entering and exiting beams are to be passed. This is to minimize reflections and multiple scatterings from the cell walls. In general, a cell with a rectangular cross-section is preferred. In places where measurements are to be made at angles other than 90°, semioctagonal cells are used (Figure 11.19). The octagonal faces will allow measurements to be made at 0°, 45°, 90° or 135° to the primary beam. Generally, walls through which light beams are not to pass are painted a dull black to absorb unwanted radiation and minimize stray radiation. In experimental cells, a blackened curved horn is frequently affixed to the wall directly opposite the entering beam to trap the entire beam, which is not scattered. Alternatively, one can put a light trap for this purpose in the cell of the chamber in which the cell is located.

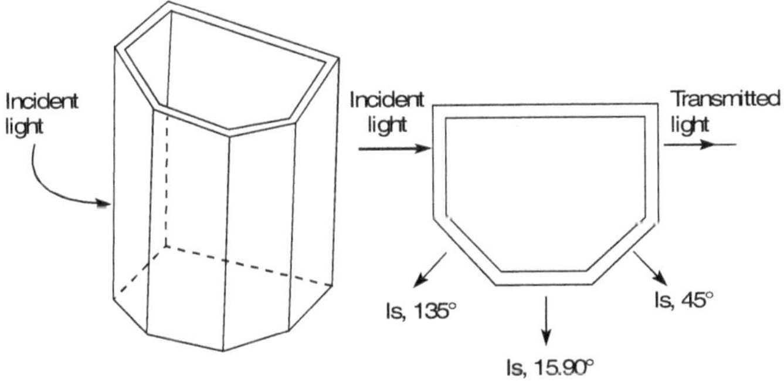

Figure 11.19 Semioctogonal cell

Detectors In nephelometers, photomultiplier tubes should be used as detectors because the intensity of scattered radiation is usually very small. In most of the nephelometers, the detector is generally fixed at 90° to the primary beam but for maximum versatility and sensitivity, it is desirable to vary the detector angle, which is generally close to the primary beam. In some nephelometers, the detector is mounted on a circular disc which allows measurement at many angles, i.e., at 0° and from 30° to 135°. The outer edge of the disc is usually graduated in degrees and is readable from the outside.

In turbidimeters, ordinary detectors such as phototubes may be used.

Applications

1. **Inorganic analysis** The important use of nephelometry and turbidimetry are the determination of sulphates such as $BaSO_4$,

carbonates such as $BaCO_3$, chlorides such as AgCl, fluorides such as CaF_2, cyanides such as silver cyanide, calcium as oxalate or oleate and zinc as ferrocyanide. Out of all these, sulphate determination is of particular importance and serves for the routine determination of total sulphur in coke, coal, coils, rubbers, plastics and other organic substances.

Another important application of nephelometry and turbidimetry is in the determination of carbon dioxide. The method involves bubbling of the gas through an alkaline solution of a barium salt and then analysing the barium carbonate suspension with nephelometry or turbidimetry.

2. **Organic analysis** In food and beverages, turbidimeter is used for the analysis of turbidity in sugar products, and clarity of citrus juices. Another interesting application is in the determination of benzene in alcohol by dilution with water to make an immiscible suspension.
3. **Biochemical analysis** An important application of turbidimetry is to measure the amount of growth of a test bacterium in a liquid nutrient medium. It is also used to find out the amount of amino acids, vitamins and antibiotics. Nephelometry has been used for the determination of protein and the determination of yeast, glycogen and beta and gamma globulin in blood serum and plasma.
4. **Air and water pollution** Turbidimetry and nephelometry are used for the continuous monitoring of air and water pollution. In air, dust and smoke are monitored whereas in water, turbidity is monitored.
5. **Turbidimetric titration** This titration may be carried out in a manner analogous to photometric titration. In this titration, the absorbance is to be plotted against the volume of titrant added. With the increase in the volume of the titrant, the concentration of precipitate increases and hence the absorbance increases. When all the substance gets precipitated, the absorbance becomes constant. Thus, an abrupt change in the slope indicates the endpoint.

6. **Determination of molecular weight of high polymers** The measurement of the intensity of light scattered by polymer solutions constitutes an important method for determining molecular weights of macromolecules. The turbidity of a solution of macromolecules is related to its molecular weight by the following relation,

$$C \to 0 \frac{HC}{Turbidity} = \frac{1}{M}$$

where,

H is a constant for a given polymer and given dispersion medium,

C is the concentration of the solution in grams per ml and

Turbidity is the fraction of incident light scattered per cm length of the solution through which it passes.

In order to determine molecular weight of a polymer, turbidity is measured at different concentrations of its solution in a suitable solvent. The plot of (*HC/turbidity*) against concentrations is then extrapolated as shown in Figure 11.20. The intercept gives the value of 1/*M*.

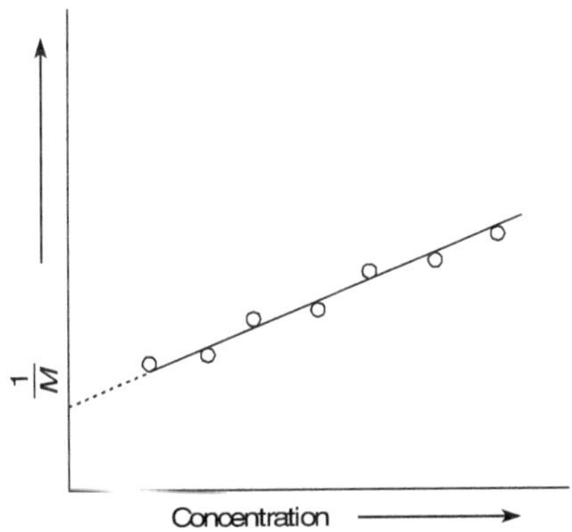

Figure 11.20 Determination of molecular weight

FLUORIMETRY AND PHOSPHORIMETRY

A large number of substances are known which can absorb ultraviolet or visible light energy. But these substances lose excess energy as heat through collisions with neighbouring atoms or molecules. However, a number of important substances are also known which lose only part of this excess energy as heat and emit the remaining energy as electromagnetic radiation of a wavelength longer than that absorbed. This process of emitting radiation is collectively known as luminescence.

In luminescence, light is produced at low temperatures. Therefore, light produced by this process is regarded as "light without heat" or "cold light".

Luminescence is of two types.

1. **Fluorescence** When a beam of light is incident on certain substances, they emit visible light or radiations. This phenomenon is known as fluorescence and the substances showing this phenomenon are known as fluorescent substances. The phenomenon of fluorescence is instantaneous and starts immediately after the absorption of light and stops as soon as the incident light is cut off.
2. **Phosphorescence** When light radiation is incident on certain substances, they emit light continuously even after the incident light is cut off. This type of delayed fluorescence is called phosphorescence and the substances are called phosphorescent substances.

Materials exhibiting fluorescence generally re-emit excess radiation within 10^{-6} to 10^{-4} second of absorption. On the other hand, materials exhibiting phosphorescence re-emit excess radiation within 10^{-4} to 20 seconds or longer. Thus, the lifetime of phosphorescence is much longer than fluorescence.

FLUORIMETERS

The basic arrangement for a single-beam 90° filter fluorimeter is as follows:

Fluorimeter employs a mercury vapour lamp, a condensing lens, a primary filter, a sample container, a secondary filter and a receiving photocell (Figure 11.21). Generally, the primary filter is used to select ultraviolet but not visible radiation whereas the secondary filter is used to transmit visible fluorescent radiation and to absorb incident ultraviolet radiation.

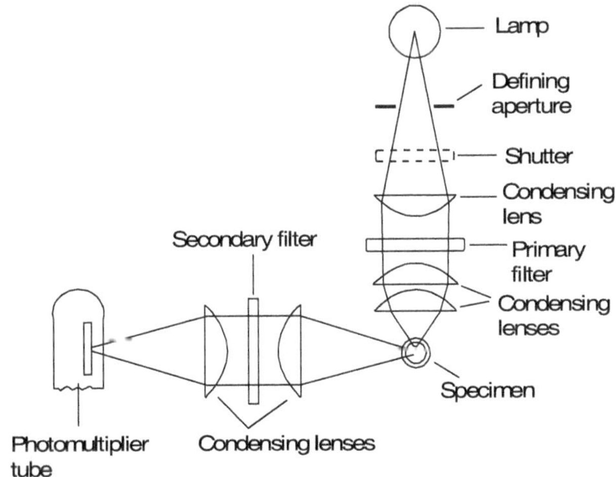

Figure 11.21 Schematic diagram of single-beam 90° filter fluorimeter

The light from a mercury vapour lamp is allowed to pass through a condensing lens followed by its passage through a primary filter. The primary filter selects only ultraviolet radiation but absorbs visible radiation. The ultraviolet radiation from the primary filter is passed through a sample container. From the sample, ultraviolet and fluorescent radiations are obtained which are passed through a secondary filter that absorbs the primary radiant energy but transmits the fluorescent radiation. This is received by a photocell placed in a position at right angles to the incident beam. The output of the photocell is measured by a sensitive galvanometer or other device.

The light source must be stable because fluorescence intensity is proportional to the intensity of irradiation. In most of the fluorimeters, fluctuations in the intensity of irradiation are not corrected. In such instruments, two photocells are used and the readings are recorded on a potentiometer in balancing the photocells against each other.

A double-beam filter fluorimeter (Figure 11.22) is an instrument in which a specially designed mercury vapour lamp is employed. This has two anodes on the opposite sides of a central structure. This lamp is made to work by alternating current. This makes two anodes to receive the discharge and hence produce radiation on alternate half-cycles of the exciting voltage. This makes the sample and standard to receive identical radiation. The two beams from the lamp are allowed to pass through two primary filters, followed by their passage through the sample and reference simultaneously. The fluorescent radiations from the sample and reference are converged through a common secondary filter into a single photomultiplier tube which records the ratio of two fluorescence signals (one from the sample and the other from the reference). This system minimizes the effect of temperature changes and variations in the source and detector.

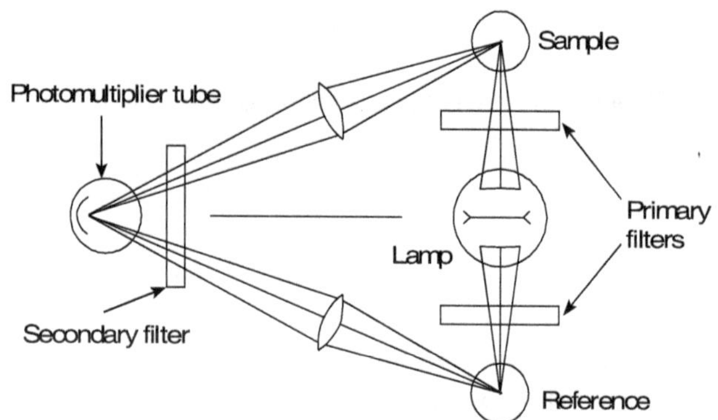

Figure 11.22 Schematic diagram of a double-beam filter fluorimeter

SPECTROFLUORIMETERS

A spectrofluorimeter designed by Aminco–Bowman is a single-beam instrument incorporating two Czerny–Turning grating monochromators with a 90° geometry. Gratings ensure a higher throughput of light than prisms. As absorption and fluorescence peaks are broad in most of the cases, the resolution and spectral slit width of monochromator should not be as fine as for atomic or gas-phase spectrometric studies.

In Aminco–Bowman spectrofluorimeter, the source is generally a high-pressure xenon arc, fed by a stable direct current power supply. The light radiation emitted by the lamp is passed onto the excitation monochromator through a slit. Light from the excitation monochromator enters the sample through a series of slits. The fluorescence radiation emitted by the sample leaves the cell compartment at right angles to the excitation source and received by the emission monochromator. The fluorescence radiation from the emission monochromator is passed onto the photomultiplier detector through a slit. The output of the photomultiplier is further amplified and then displayed on a meter, oscilloscope or recorder.

The spectrofluorimeter shown in Figure 11.23 is mainly used for the detection and determination of small concentrations of organic substances. This is equally suitable for carrying out quantitative work.

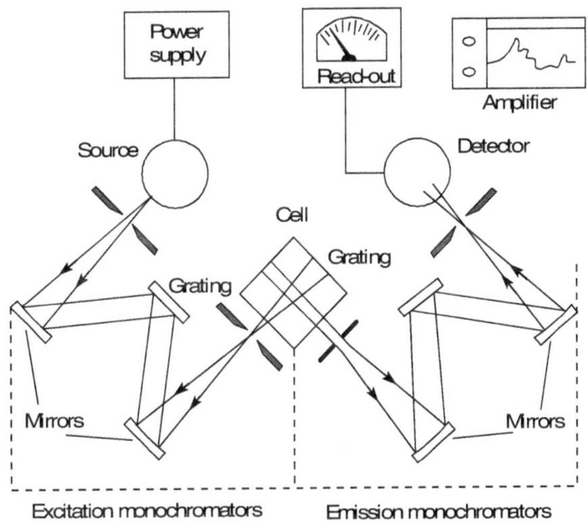

Figure 11.23 Schematic diagram of spectrofluorimeter

SPECTROPHOSPHORIMETER

Compounds with phosphorescence are likely to fluoresce as well. Therefore, a phosphorimeter must be capable of making the distinction between the two. This has been achieved by means of a rotating shutter, which introduces a delay between the times during which the sample gets irradiated and observed. A single-beam phosphorimeter involving a rotating shutter is shown in the Figure 11.24. Filters may be located in either the primary or secondary beam or both.

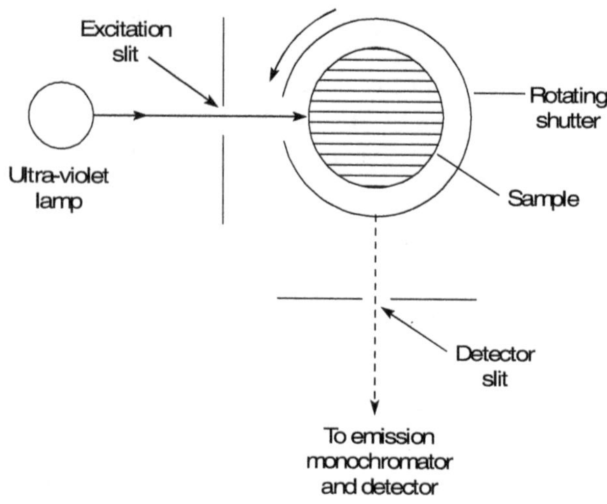

Figure 11.24 Rotating shutter in single-beam phosphorimeter

A spectrophosphorimeter is similar to a spectrofluorimeter except that the former must be fitted with (i) a rotating shutter device commonly called a phosphoroscope and (ii) a sample system which is maintained at liquid nitrogen temperature.

Phosphoroscopes are of two types that are as follows:

The rotating-can phosphoroscope It consists of a hollow cylinder having one or more slits which are equally spaced in the circumference. This is rotated by a variable-speed motor. When the rotating-can is rotated by a motor, the sample is first illuminated and then darkened. In the dark, phosphorescent radiation passes onto the emission monochromator and can be measured. A rotating-can phosphoroscope is shown in Figure 11.25.

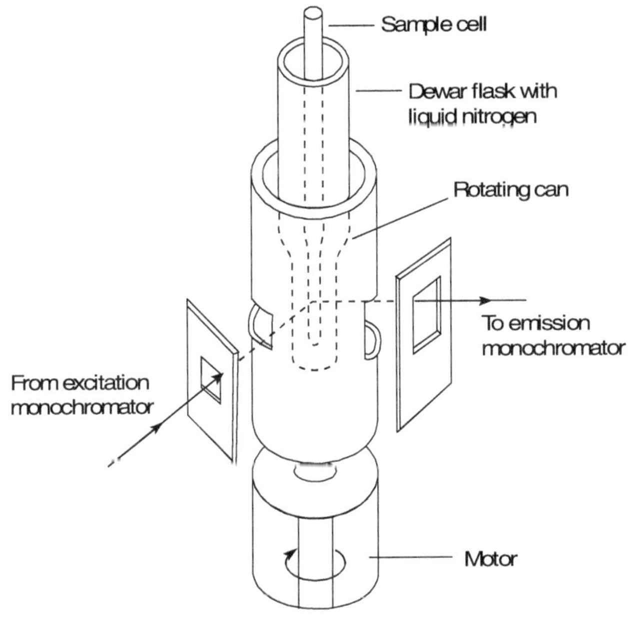

Figure 11.25 Rotating-can phosphoroscope

The Becquerel or rotating disc phosphoroscope It has two discs which are mounted on a common axis turned by a variable-speed motor. Both the discs are having openings equally spaced in their circumference. On moving Becquerel disc phosphoroscope, the sample is first illuminated and then darkened; whenever it is dark, phosphorescence radiation passes onto the emission monochromator and can be measured. One can measure shorter decay times by Becquerel rotating-disc phosphoroscope than rotating-can phosphoroscope because one can cut a greater number of openings in the discs than the can. A Becquerel rotating-disc phosphoroscope is shown in Figure 11.26.

In spectrophosphorimeters, quartz tubes are used for samples. These are placed in liquid nitrogen (–196°C at 1 atm.) and held in quartz Dewar flask. For phosphorimetric studies, the common solvent used is EPA (EPA is a mixture of ethyl ether, isopentane and ethanol in the volume ratio of 5:5:2). This on freezing yields a clear rigid glass. But some snow formation is also possible with the solvent. This problem has been overcome by employing a rotating sample. This also increases the precision in phosphorimetric measurements by a factor of 10 or more.

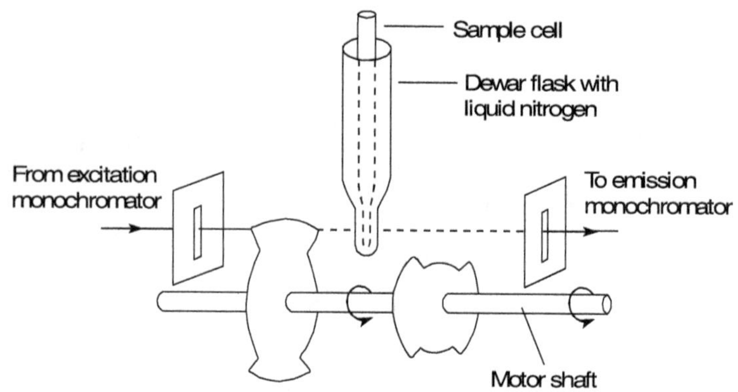

Figure 11.26 Rotating-disc phosphoroscope

Applications of Fluorimetry

Fluorimetry is a well established analytical tool. A large number of applications are known. Fluorimeter is used in the analysis of elements, determination of vitamin B_1 and B_2, organic analysis, and also in the analyses of food products, pharmaceutical, clinical samples and natural products.

1. *Analysis of elements* Fluorimeter is used in the determination of uranium in salts. This is used extensively in the field of nuclear research. Further fluorimeter is used in the determination of ruthenium ion in the presence of other platinum metals, determination of aluminium in alloys, estimation of traces of boron in steel by means of the complex formed with benzoin, estimation of cadmium by precipitating it with 2-(2-hydroxyphenyl)-benzoxazole in the presence of tartrate and estimation of calcium with calcein solution.

2. *Determination of vitamin B_1* Vitamin B_1 (thiamine) is non-fluorescent, whereas its oxidation product, thiochrome, fluoresces with blue colour. This property is used for the determination of vitamin B_1 in food samples like meat, cereal, etc.
3. *Determination of vitamin B_2 (Riboflavin)* Determination of vitamin B_2 is done by a fluorescence method because the fluorescent power depends upon the exact conditions and upon the nature and amount of impurities.
4. *Organic analysis* Fluorimetry has been used to carry out qualitative as well as quantitative analysis for a great many aromatic compounds present in cigarette smoke, air-pollutant concentrates and automobile exhausts. A specific example is the determination of benzopyrene in the nanogram range.
5. The most important applications of fluorimetry are to be found in the analyses of food products, pharmaceuticals, clinical samples and natural products. The sensitivity of the method makes it particularly valuable in these fields.

Applications of Phosphorimetry

Phosphorimetry finds applications in biology and medicine.

1. Concentration of aspirin (acetylsalicylic acid) in blood serum is determined by phosphorimetry at liquid nitrogen temperatures.
2. Low concentrations of procaine, cocaine, phenobarbital and chlorpromazine in blood serum have been determined by employing phosphorimetry in combination with extraction procedures.
3. Cocaine and atropine in urine have been determined by employing phosphorimetry in combination with extraction procedures.

OPTICAL ROTARY DISPERSION (ORD) AND CIRCULAR DICHROISM (CD) SPECTROSCOPY

The optical rotatory dispersion (ORD) and the circular dichroism (CD) are special variations of absorption spectroscopy in the UV and visible region of the spectrum. The basic principle of the two methods is the interaction of polarized light with optically active substances. If a linearly polarized light wave passes through an optically active substance, the direction of polarization will change. This change is wavelength-dependent. This phenomenon is called **optical rotatory dispersion** (ORD). Linearly polarized light waves can be described as a superposition of two circularly polarized light waves. If a substance absorbs these two circularly polarized components to a different extent, the difference is described as **circular dichroism** (CD). In other words, circular dichroism is the difference in absorption between right and left hand circularly polarized (RCP and LCP) light in chiral molecules. A **chiral molecule** is one with a low degree of symmetry, which can exist in two mirror-image isomers. An example of circular dichroism in glucose, a simple sugar, is shown in the Figure 11.27. When a molecule exhibits a combination of ORD and CD in the region of absorbance, then the transmitted light is said to be elliptically polarized.

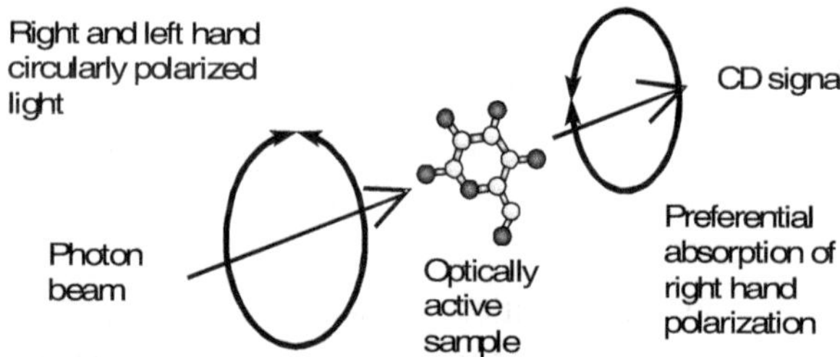

Figure 11.27 Circular dichroism in glucose molecule

A curve that shows wavelength dependence of optical rotation is called as **optical rotatory dispersion spectrum** . If the material is optically inactive, the absorption of RCP and LCP components is equal. However, an optically active medium has unequal molar absorption coefficient for RCP and LCP components. The difference in the molecular extinction coefficient of the RCP and LCP rays ($\Delta\varepsilon$) is called **differential dichroic absorption.**

i.e., $$\varepsilon_L - \varepsilon_R = \Delta\varepsilon \neq 0$$

where $\Delta\varepsilon$ is called the circular dichroism or CD. It is positive if $\varepsilon_L - \varepsilon_R > 0$ negative if $\varepsilon_L - \varepsilon_R < 0$. The ellipticity ($\theta$) is related to $\Delta\varepsilon$ by the equation:

$$\theta = 3300\Delta\varepsilon$$

A curve showing the dependence of θ on wavelength is called a **CD curve or CD spectrum.**

Cotton effect If the RCP and LCP components of a beam of plane polarized light are transmitted with equal speed through an optically inactive medium, no rotation of the plane of polarization will be observed. However, if a beam of plane polarized light passes through an optically active medium, it rotates the original plane of polarization and hence the emerging RCP and LCP components are no longer in phase and the resultant vector has been rotated. This property of the medium to rotate the original plane of polarization is called **circular birefringence.** The combination of CD and circular birefringence known as **Cotton effect** may be studied by plotting the change of optical rotation with wavelength. Such plots may now be obtained quite readily down to about 220 nm by means of photoelectric spectropolarimeters.

Advantages of CD over ORD The principal advantage of CD analysis is its greater ability to resolve bands due to different optically active transitions. A flat baseline is never reached in an ORD curve but in a CD curve, there is a defined zero baseline outside the absorption band. Also, CD bands are narrow and allow good resolution of nearby bands, while ORD spectra totally overlap and weak bands are almost undetectable.

Instrumentation

A CD spectropolarimeter (Figure 11.28a) is a hybrid instrument consisting of a variable wavelength polarimeter and absorption spectrophotometer. Although the technique was 'invented' by Cotton in 1896, it was only in 1960 that it came to the forefront of analytical research.

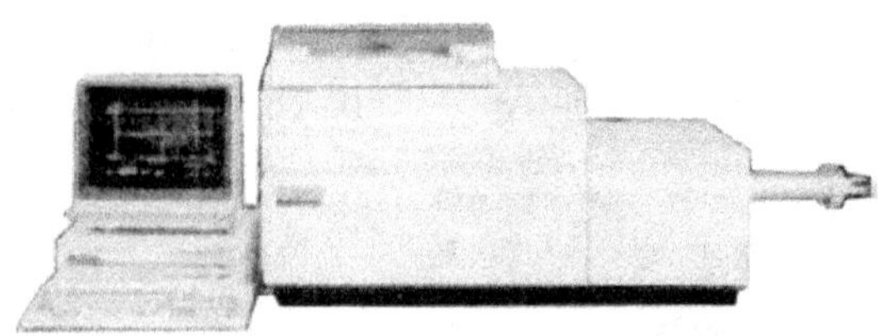

Figure 11.28 (a) CD spectropolarimeter

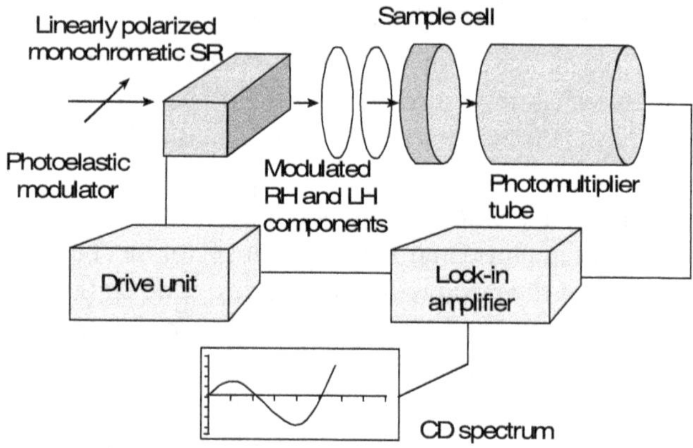

Figure 11.28 (b) Schematic diagram of CD apparatus

Many biological macromolecules exhibit chirality. Proteins and enzymes, for example, are composed of many chiral subunits and can have large CD signals. An important region for CD in proteins is the UV and in particular below 200 nm. At these short wavelengths, commercial CD spectrometers suffer from a lack of flux. Synchrotron radiation sources on the other hand maintain their high flux levels throughout the UV and visible making them ideal for such studies. The CD apparatus is shown schematically in the Figure 11.28b.

Solutions are generally used for ORD and CD measurements, but films and solids are also used. The solution is contained in a cell. The basic instrumentation requires light, a system for measuring the polarization after the light has passed through the cell and a detector by which the amount of light can be measured.

Application of ORD and CD

In general, this phenomenon will be exhibited in absorption bands of any optically active molecule. As a consequence, circular dichroism is exhibited by biological molecules, because of the dextrorotary (e.g. some sugars) and laevorotary (e.g. some amino acids) molecules they contain.

CD is closely related to the ORD technique, and is generally considered to be more advanced. CD is measured in or near the absorption bands of the molecule of interest, while ORD can be measured far from these bands. In principle, these two spectral measurements can be interconverted through an integral transform, if all the absorptions are included in the measurements.

The ultraviolet CD spectrum of proteins can predict important characteristics of the secondary structure of their proteins. CD spectra can be readily used to estimate the fraction of a molecule that is in the alpha-helix conformation, the beta-sheet conformation, the beta-turn conformation, or some other (random) conformation. CD cannot, in general, say where the alpha helices that are detected are located within the molecule or even completely predict how many there are. Despite this, CD is a valuable tool, especially for showing changes in conformation. It can, for instance, be used to study how the secondary structure of a molecule changes as a function of temperature or of the concentration of denaturing agents. Thus, it can reveal important thermodynamic information about the molecule that cannot otherwise be easily obtained.

CD/ORD instruments are also used extensively in a number of other applications given below:

- Enzyme kinetics
- Organic stereochemistry studies
- Purity testing of optically active substances
- Quantitative analysis of pharmaceuticals

- Natural organic chemistry
- Biochemistry and macromolecules
- Metal complex chemistry
- Polymer chemistry
- Medical science
- Agrochemistry
- Physical chemistry
- Rapid scanning (time-resolved) experiments

INFRARED SPECTROPHOTOMETRY

The technique is based upon the simple fact that a chemical substance shows marked selective absorption in the infrared region. After absorption of IR radiations, the molecules of a chemical substance vibrate at many rates of vibration, giving rise to close-packed absorption bands, called an IR absorption spectrum, which may extend over a wide wavelength range. Various bands will be present in IR spectrum that will correspond to the characteristic functional groups and bonds present in a chemical substance. Thus, an IR spectrum of a substance is a fingerprint for its identification.

SINGLE-BEAM INFRARED SPECTROPHOTOMETER

A diagram of the optical system of a single-beam infrared spectrophotometer is shown in Figure 11.29.

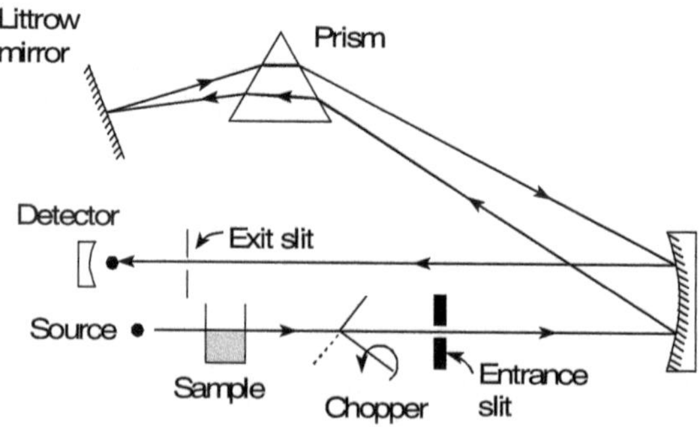

Figure 11.29 Schematic diagram of single-beam infrared spectrophotometer

In the single-beam system, the radiation is emitted by the source through the sample and then through a fixed prism and a rotating Littrow mirror. Both prism and Littrow mirror select the desired wavelength and then allow it to pass on to the detector. The detector measures the intensity of radiation after it passes through the sample. Knowing the original intensity of radiation, one can measure how much radiation has been absorbed. By measuring the degree of absorption of wavelengths, the absorption spectrum of the sample can be obtained.

Disadvantages

The various disadvantages of a single-beam spectrophotometer are as follows:

1. This type of instrument has a basic disadvantage in that the intensity of the emission of the radiation source varies from point to point in IR absorption spectrum; therefore, the resulting spectrum is considerably deformed. The necessary correction by the continuous variation of slit is cumbersome.
2. When the sample is analysed in solution, the bands of solvent appear in the spectrum. In this case, the spectrum of the sample is obtained by subtracting the spectrum of the solvent from the resultant spectrum, the former must be recorded under identical conditions (thickness of layer, etc.)

In order to overcome the above mentioned difficulties, a double-beam spectrophotometer is used.

DOUBLE-BEAM INFRARED SPECTROMETER

The essential features of double-beam infrared spectrometer (Figure 11.30a, b) are described briefly.

Radiation source The source of radiation in a typical infrared spectrometer is a small ceramic rod, heated electrically in the range 1100–1800°C and is made of either silicon carbide

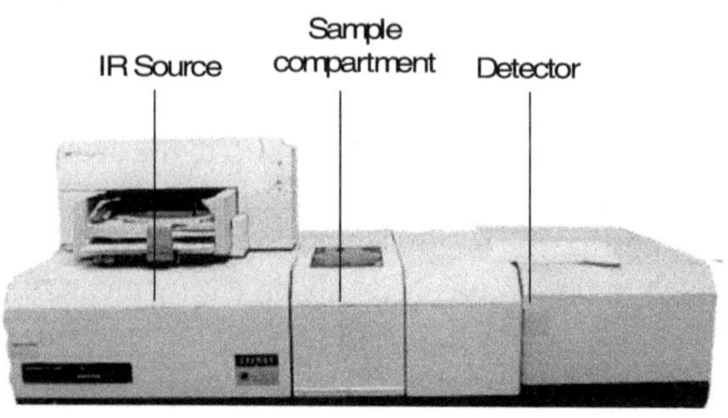

Figure 11.30a Infrared spectroscopy

(Glowbar) or Nernst filament (a high-resistance, brittle element composed of mixture of a sintered oxides of zirconium, thorium and cerium held together by a binding material). The radiation is divided into two beams, one of which passes through the sample while the other functions as a reference beam. The reference and the sample beams are then passed alternately into a monochromator at very short intervals by means of a rotating mirror.

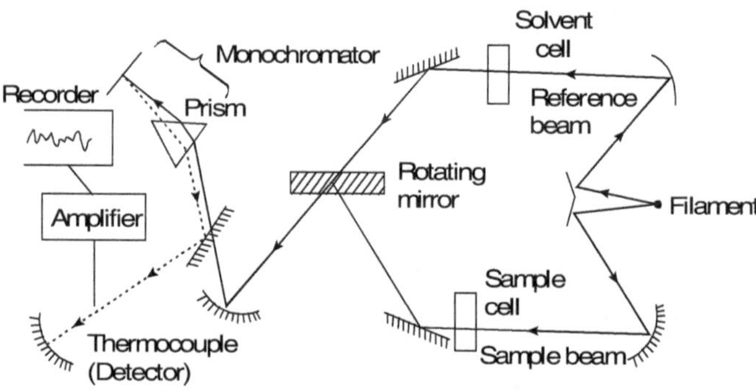

Figure 11.30b Schematic diagram of double-beam infrared spectrophotometer

1. *Absorption cells and sample preparation* The cells generally employed are made up of rock salt or potassium bromide (glass and quartz cells are unsuitable as these materials themselves absorb IR radiation). A compound (0.55 mg)

may be examined in solution in a suitable minimum absorbing solvent (CCl_4, $CHCl_3$, etc.) as a liquid film. In the case of solids which are not soluble in these solvents the compound can be examined as a mull in nujol or as a pellet obtained by pressing the sample in a hydraulic press.

2. *Monochromator* The pulse beam enters the monochromator through an entrance slit and is dispersed by a grating or by a Littrowmount prism. In the monochromator, the emergent beams are sorted out into individual wavelengths by means of a sodium chloride prism, which is transparent to infrared radiation throughout the range 4000–650 cm^{-1}. Lithium fluoride has more favourable dispersion properties than sodium chloride at high wave numbers but is not transparent below 1000 cm^{-1}. In order to extend the operating range below 650 cm^{-1} prisms made up of potassium bromide (transparent up to 400 cm^{-1}) and caesium iodide (transparent up to 200 cm^{-1}) are also used.

In high resolution rating instruments, a filter is used to reject radiation of unwanted orders.

Detector, Amplifier and Recorder The pulsating single beam now emerging through the slit is a narrow band consisting of only a very few frequencies. After dispersion, the beams are focused alternately at each particular wavelength throughout the spectral range, by means of a mirror system on to the detector, usually a sensitive fast thermocouple. The signals from this are amplified electronically. The spectrum, which is really a measure of the difference in intensities of the reference and sample beams throughout the wavelength range, is recorded on a special graph paper mounted on a rotating drum.

Applications

Infrared spectroscopy is an ideal and rapid method for measuring certain contaminants in foodstuff. It is also widely used in biochemical research to identify metabolic intermediates, drug metabolites and environmental pollutants. Further, infrared spectroscopy is used in the study of photosynthesis and respiration in plants. Infrared spectroscopy is one of the valuable methods for characterizing both qualitatively and quantitatively the multitude of inorganic compounds encountered in research as well as in industry.

FOURIER TRANSFORM INFRARED SPECTROSCOPY

Fourier transform infrared spectroscopy is a technique for obtaining high quality infrared spectra by mathematical conversion of an interference pattern into a spectrum.

Principle

Infrared radiations consisting of all wavelengths (e.g. 5000–400 cm^{-1}) is split into two beams, which recombine after a path difference has been introduced (one beam is of fixed length and the other is of variant path length). A condition is, therefore, set up under which interference between the beams can occur. When the difference in the corresponding wavelengths is an integral multiple of the invariant beam, the interference is constructive; destructive interference occurs when the difference is an odd integer multiple of one quarter of the wavelength. The result of a complete variation of wavelengths is an oscillatory series of destructive and constructive combinations. This is called **interferogram**. Fourier transformation converts this interferogram from the time domain into one spectral point on the more familiar form of the frequency domain. Similar transform at successive points throughout this variation gives rise to the complete IR spectrum.

Instrumentation

A simple form of FTIR spectrometer including Michelson interferometer is shown in the Figure 11.31.

It consists of two mutually perpendicular plane mirrors, one of which is fixed (M_1) while the other (M_2) can move along an axis that is perpendicular to its plane. The movable mirror is either moved at a constant velocity or is held at equally spaced points for fixed, short time periods and rapidly stepped between these points. A beam of radiation from the source S, is focused on a beam splitter, placed between the fixed and movable mirrors. The beam splitter is constructed of suitable material with the necessary optical properties (KBr-coated germanium is used for beam splitting in mid-infrared and polyethylene terphthalate in far infrared region) such that half the beam is transmitted to the moving mirror which reflects the beam back to the beam splitter which reflects part of this beam through a sample to a detector. The other half of the beam from the source is reflected from the beam splitter to a fixed mirror, which reflects the beam through the beam splitter to detector, via the sample. The interferogram contains all the information associated

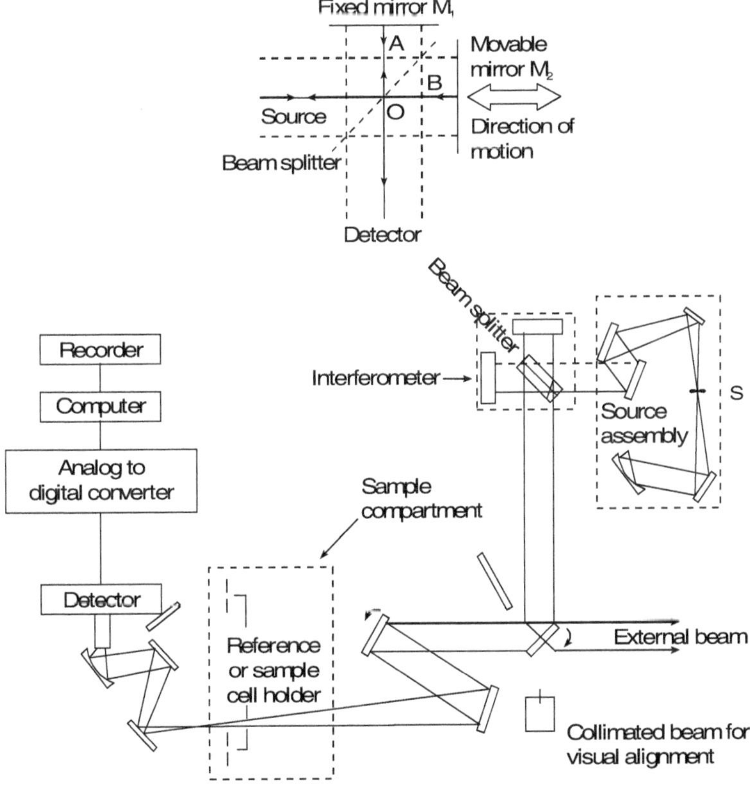

Figure 11.31 Optical layout of an FTIR spectrometer

with a conventional spectrum. All the spectral information can be extracted by performing the appropriate inverse Fourier transformation carried out with the assistance of a digital computer.

RAMAN SPECTROSCOPY

In 1923, Smekel predicted theoretically that if a substance in the gaseous, liquid or solid state is irradiated with monochromatic light, the scattered light should contain radiations with different frequencies than the frequency of incident light. In 1928, **Sir C.V. Raman** discovered that when a beam of monochromatic light was allowed to pass through a substance in the solid, liquid or gaseous state, the scattered light contains some additional frequencies over and above that of the incident frequency. This is known as Raman effect and is a beautiful confirmation of the Smekel's prediction. The difference between Raman spectra and infrared spectra are listed in Table 11.3.

Table 11.3 Differences between Raman spectra and Infrared spectra

S.No	Raman Spectra	Infrared Spectra
1.	It is due to the scattering of light by the vibrating molecules.	It is the result of absorption of light by vibrating molecules.
2.	Polarizability of the molecule will decide whether the Raman spectra will be observed or not.	The presence of a permanent dipole moment in a molecule may be regarded as a criterion of infrared spectra.
3.	It can be recorded only in one exposure.	It requires at least two separate runs with different prisms to cover the whole region of infrared.
4.	Water can be used as a solvent.	Water cannot be used as a solvent because it is opaque to infrared radiation.

(Contd.)

Table 11.3 (Continued)

S.No	Raman Spectra	Infrared Spectra
5.	The method is very accurate, but is not very sensitive.	The method is accurate and very sensitive.
6.	Optical systems are made of glass or quartz.	Optical systems are made up of special crystals such as CaF_2, NaBr, etc.
7.	Sometimes photochemical reactions take place in the regions of Raman lines and thus create difficulties.	Photochemical reactions do not take place.
8.	Substances under investigation must be pure and colourless.	This condition is not rigid.
9.	In Raman effect, vibrational frequencies of large molecules can be measured.	In this, the vibrational frequencies of very large molecules cannot be measured.
10.	As Raman lines are weaker in intensity, concentrated solutions must be utilized to increase the intensity of Raman lines.	Generally, dilute solutions are preferred.
11.	Homonuclear diatomic molecules arc often found to be active.	Homonuclear diatomic molecules are not found to be active.

Instrumentation

The basic instrumentation for obtaining Raman spectrum is shown in the Figure 11.32.

C is a Raman tube which acts as a container for the liquid to be investigated. It is made of glass and is 1–2 cm in diameter and 10–25 cm long. The one end of the tube is drawn like a horn and blackened outside to provide a suitable background. The other end of the tube is closed with an optically plane glass plate. The scattered light emerges through the window W. The Raman tube C is surrounded by water jacket J in which cold water is circulated in order to prevent the overheating of liquid because of the proximity of the heated arc.

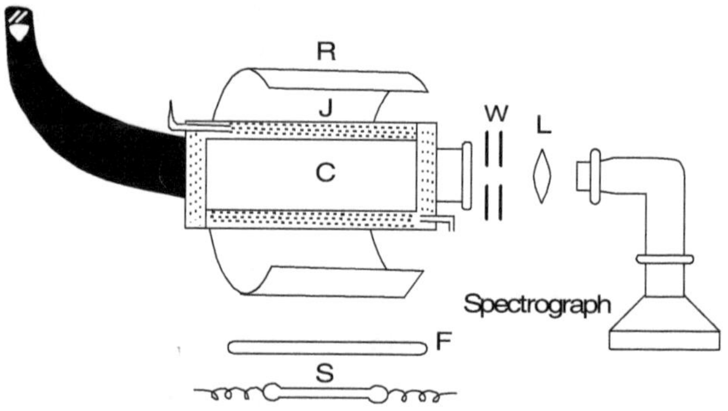

Figure 11.32 Raman spectroscope

S is a helium discharge tube which acts as a source of light and is filtered by nickel oxide filter F to get a monochromatic light. In general practice, mercury arc is used as a source of light because it is difficult to construct and obtain monochromatic light from helium discharge tube. The mercury arc is placed quite near to the Raman tube and a semi-cylindrical aluminium reflector enhances the intensity of light. A lens L in front of the plane window directs the scattered radiations upon the slit of the spectrograph, and the Raman lines are obtained on the photographic plate.

There are at present several manufacturers of Raman spectrographs. A Raman spectrograph consists of the following components:

Source of light The Raman effect is relatively weak. Therefore, it is essential to have a source of high intensity. The mercury arc is the most useful source of excitation, which is used nowadays. It is used after obtaining a single wavelength by the use of suitable filters. The line (obtained from mercury arc) corresponding to 4358 Å is the most commonly used radiation in Raman spectroscopy.

Nowadays, Higler lamp is also used. It consists of four low-pressure mercury discharge tubes completely enclosed in a hollow jacket whitened internally to act as a reflector to increase the intensity.

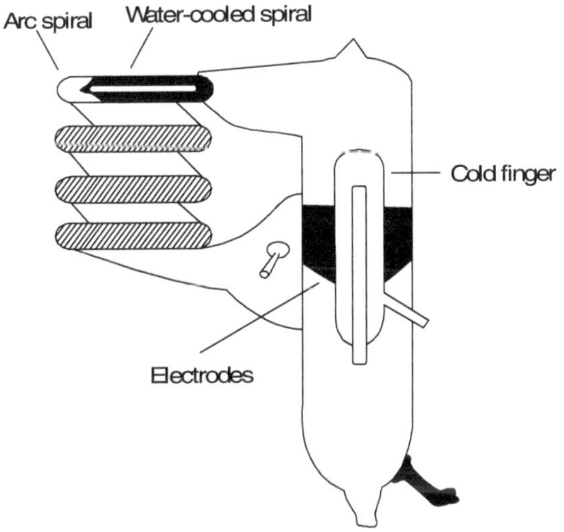

Figure 11.33 Toronto lamp

Welsh and Crafford devised a lamp known as Toronto lamp (Figure 11.33) which consists of a tube water-cooled low pressure mercury lamp. In this, the sample tube is kept in the central axis of the spiral.

Before the development of the laser as an excitation source, Raman spectroscopy suffered from the following disadvantages:

1. Samples had to be restricted to clear, colourless, non-fluorescent liquids.
2. The low intensity of the Raman effect required relatively concentrated solutions.

3. Much larger volumes of sample solutions were needed than for infrared spectrophotometer.

These three disadvantages were the major factors in limiting the use of Raman spectroscopy; but the helium–neon laser source now available has gone a long way in overcoming these difficulties. A schematic illustration of a complete helium–neon laser is shown in Figure 11.34. The tube is filled with a 7: 1 mixture of helium and neon gas for optimum output of 6328 Å laser line. Although radio frequency excitation was used in the early models, high voltage excitation is preferred nowadays. To start the laser, 5 to 10 kV DC is used but thereafter the beam can be maintained on the lower operating voltage.

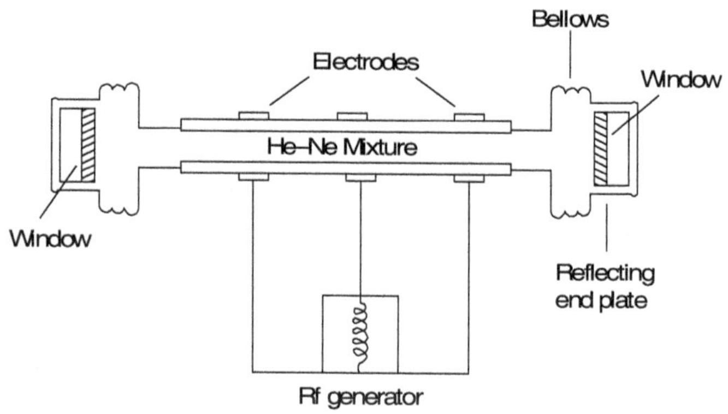

Figure 11.34 Simplified schematic diagram of helium–neon laser

The scope of Raman spectroscopy is greatly widened with the use of laser as the exciting source. The advantages of using laser are:

1. A single, intense frequency source replaces the multiple lined mercury lamp. Thus no filtering is necessary.
2. The line width of a laser line is smaller than the mercury excited line, therefore the resolution will be better.
3. There is greater ease in focusing and collimating the radiation because of highly coherent character of laser light.

4. A large number of exciting frequencies are available and thus it is possible to study coloured solutions without causing any electronic transition. This is of utmost practical importance in the study of solutions of inorganic salts because many of them are coloured compounds.

Filters In the case of non-monochromatic incident light, there will be overlapping of Raman shifts which will make the interpretation of the Raman spectrum difficult. It is therefore essential to have monochromatic radiations. For getting monochromatic radiation, filters are used. They may be made of nickel oxide, glass, or quartz. Sometimes, a suitable coloured solution such as an aqueous solution of ferricyanide or iodine in carbon tetrachloride may be used as a monochromator.

Sample holder For the study of Raman effect, the type of sample holder to be used depends upon the intensity of source, and the nature and the availability of the sample. The study of Raman spectra of gases requires sample holders (Figure 11.35), which are generally bigger in size than those for liquids.

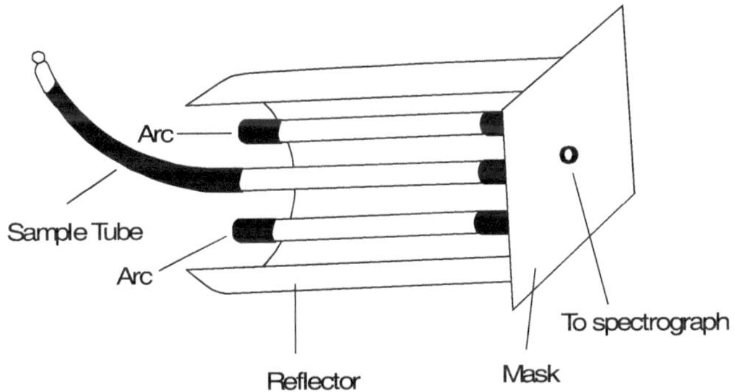

Figure 11.35 Sample holder

Solids are dissolved before subjecting to Raman spectrograph. Any solvent which is suitable for the ultraviolet spectra, can be used for the study of Raman spectra. Water is regarded as a good solvent for the study of inorganic compounds in Raman spectroscopy. For taking spectrum of a small amount of the sample, a four-traversal tube is used (Figure 11.36). In this tube, M_1 and M_2

are two concave mirrors both having focal lengths equal to half the distance between them. M_1 is only half-mirrored and the remaining half allows the light to enter. A lens L makes a virtual image of M_2 at infinity. Light is allowed to pass through the clear half of mirror M_1 and then passed on to the mirror M_2. After this, the light undergoes a series of two reflections, i.e., four-traversals, i.e., it emerges out through the same aperture.

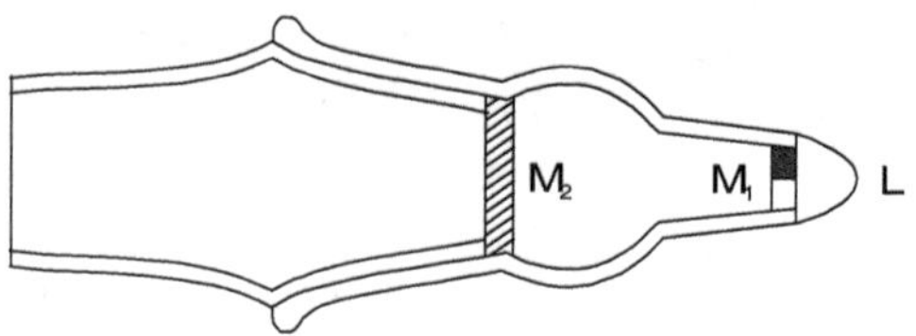

Figure 11.36 Four-traversal tube

As the intensities of Raman lines of gases are very small, very intense light, extremely sensitive detectors and multiple traversal tubes should be employed for recording their Raman spectra.

Spectrograph The spectrograph used for the study of Raman spectrum should possess the following characteristics:

i. It should have large gathering power.
ii. Special prisms of high resolving power should be employed.
iii. A short-focus camera should be employed.

A lens in front of the plane window directs the scattered radiation upon the slit of the spectrograph and the Raman lines can be obtained on a photographic plate.

Raman spectrographs have either photographic or automatic recording devices. Photographic emulsion or photomultiplier tubes are therefore employed as receptors or detectors for the study of Raman spectra.

Due to the weak nature of the Raman lines, photographic method is preferred because it is more sensitive.

In order to study the Raman effect in solids and gases, certain modifications in the experimental arrangements are to be made. In the case of solids like gypsum and quartz, a container is not used to keep the sample. But in such cases, the light from the arc is directly focused on the material with the help of a lens. Raman effect for the powdered solids can be studied by reflecting the light from the crystal surface.

The intensity of light scattered by gases is very small. However, it can be increased by the intense illumination of the gas under high pressure and using the spectrographs of large light-gathering power.

Rasetti developed the technique of investigation of Raman spectra under high pressure. Nowadays, different types of tubes are being used for the study of Raman spectra of gases under high pressure. A tube made of transparent silica enclosed in a steel tube is commonly used. This can be used up to a pressure of 50 atm.

A schematic diagram of Perkin–Eimer Raman spectrometer is shown in the Figure 11.37. Light from the helium–neon laser beam is allowed to enter the sample compartment horizontally. Then the Raman scattering from the sample cell is focused on the monochromator entrance slit. If depolarization measurements are to be made, the Raman emission is first allowed to pass through an analyser prism before entering a monochromator, which is a double-pass Littrow-mounted grating type. A 13-Hz chopper is used between the first and second passes and the detector is made to respond only to this signal. This decreases interference from stray radiation. The wavelength is scanned automatically.

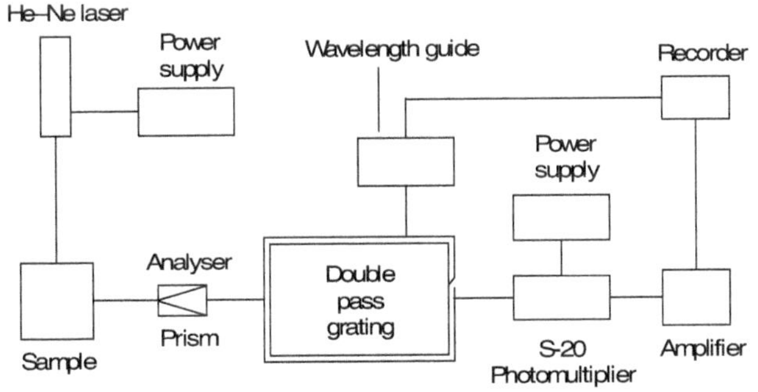

Figure 11.37 Schematic diagram of Perkin–Elimer Raman spectrometer

Intensity of Raman Peaks

The intensity or power of a Raman peak has been found to depend upon the following factors:

i. Polarizability of the molecule
ii. Intensity of the source and
iii. Concentration of the active group

In the absence of absorption, the power of Raman emission increases with the frequency of the source. Raman intensities are directly proportional to the concentration–intensity relationship, where the concentration-intensity relationship is logarithmic.

Applications of Raman Spectroscopy

Spectra of crystal powders, single crystals, polymers and coloured substances may be recorded. Furthermore, the Raman spectra of aqueous solutions and of air-sensitive and corrosive substances can be obtained more easily than infrared spectra. All substances to be analysed by Raman spectroscopy should be free from impurities likely to be absorbed in the range of the Raman spectrum. Gas chromatography or vacuum distillation can be used for purification of liquids. Crystalline substances can be examined in solution, as a melt, in powdered form or as single crystals. Solvents are purified by distillation in grease-free apparatus or by filtration through alumina.

The Raman effect is very important because of its large number of applications. In fact, it is a useful tool for solving the intricate research problems concerning the constitution of compounds.

Raman spectrum is a molecular spectrum and therefore it helps in revealing the molecular structure of organic compounds. It has been observed that each group will have its own characteristic frequency. A Raman spectrum provides information about the following facts:

1. Presence or absence of specific linkages in a molecule
2. Structure of simple compounds
3. Study of isomers

4. Presence of impurities in dyes
5. Classification of the compounds

Raman spectroscopy can be successfully used for rapid, easy and accurate analysis of mixtures that are troublesome with any other method. Raman spectroscopy has two main advantages over infrared spectrophotometry in quantitative analysis. Firstly, the height of Raman peaks varies linearly with concentration, whereas with infrared spectrophotometry there is a logarithmic relationship between concentration and the transmitted light. Secondly, Raman spectroscopy gives simple spectra as compared with the infrared spectroscopy. In addition, Raman instrumentation is not subject to attack by moisture and small amounts of water in a sample do not interfere. Despite these advantages, Raman spectroscopy has not been widely exploited for quantitative analysis.

ELECTRON SPIN RESONANCE SPECTROMETRY (ESR)

When the molecules of a solid exhibit paramagnetism as a result of unpaired electron spins, transitions can be induced between spin states by applying a magnetic field and then supplying electromagnetic energy, usually in the microwave range of frequencies. The resulting absorption spectra are described as electron spin resonance (ESR) or electron paramagnetic resonance (EPR). The technique may be used for detecting transition metal ions and their complexes, free radicals and excited states.

Principle

An electron possesses both spin and charge and therefore they behave like magnets. In the presence of an external magnetic field, they can exist in two states— parallel to the field in a low-energy state or antiparallel in a high-energy state. For an electron to change from the low-energy state to the high-energy state, it must absorb the appropriate quantum of energy. The quantum of energy ($h\nu$) required to cause the resonance and the transition between energy states in an ESR experiment may be quantified as follows:

$$h\nu = g\beta H$$

where,

g is the spectroscopic splitting factor (a constant),
β is the magnetic moment of the electron and
H is strength of the applied external field.

Peak absorption of the microwave frequency in an electron spin resonance spectrum corresponds to a paramagnetic spectrum. The area under the peak is a measure of the concentration of that spectrum which may be quantitatively determined if a standard containing a known concentration of unpaired electrons is available.

Instrumentation

The basic components of an electron spin resonance spectrometer are as shown in the Figure 11.38. The monochromatic microwave radiation is produced in Klystron oscillator, the wavelength being of the order of 3×10^{-2} m (9000 Mhz). Samples must be in solid state, therefore biological samples are usually frozen in liquid nitrogen. The microwave radiations are passed towards the sample and the reflected microwave radiation is detected by a crystal

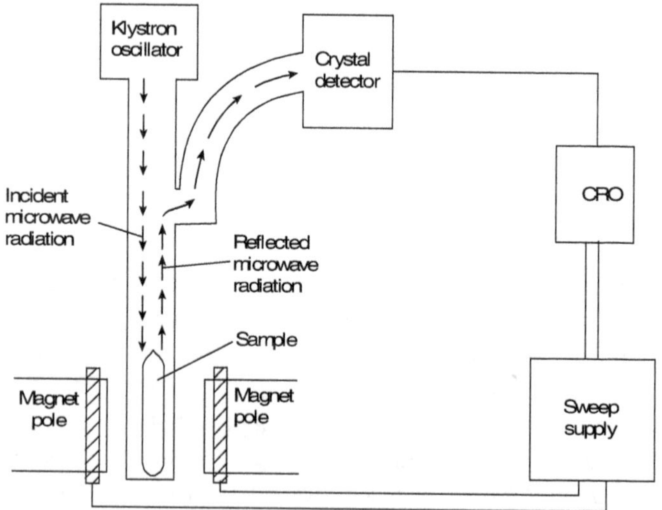

Figure 11.38 Schematic diagram of electron spin resonance spectrometer

detector, which converts the microwave into electrical current. This is amplified by the amplifier and the signal from the amplifier is recorded by cathode ray oscilloscope (CRO).

Applications

Electron spin resonance has been used as an investigative tool for the study of radicals formed in solid materials, since the radicals typically produce an unpaired spin on the molecule from which an electron is removed. Study of the radicals produced by the ionizing radiation gives information about the location and mechanism of radiation damage.

ESR is used to detect the free radicals. Free radicals are found in metabolic pathways and as degradation products of drugs and toxins. In general, free radicals are found in low concentrations in tumours than in normal tissues.

Electron spin resonance is one of the main methods used to study metalloproteins, particularly those containing molybdenum (xanthine oxidase), copper (cytochrome oxidase and copper blue enzymes) and iron (cytochrome, ferredoxin, etc.).

Both copper and non-heme iron which do not absorb radiation in the ultraviolet or visible regions possess ESR absorption peaks in one of their oxidation states. Hence, the appearance and disappearance of their electron spin resonance signals are used to monitor their activity in the multienzyme systems of intact mitochondria, chloroplasts, as well as in the isolated enzymes.

In metalloproteins, the metal atom has a characteristic number of ligands coordinated to it in a definite geometrical arrangement. These ligands are frequently amino acid residues of the protein. ESR studies reveal the distortion in structural geometry and the distortion may be related to biological function. Intramolecular motions and lateral diffusion of lipid through the membrane may be observed and measured. This study is achieved by either concentrating the spin-labelled lipids into one region of the bilayer or randomly incorporated, usually into model membranes.

NUCLEAR MAGNETIC RESONANCE (NMR)

This is a technique for detecting atoms, which have nuclei that possess a magnetic moment. The phenomenon of nuclear magnetic resonance was first reported independently in 1946 by two groups of physicists—**Block**, **Hansen** and **Packard** at Stanford University detected a signal from the protons of water, and **Purcell**, **Torrey** and **Pound** at Harvard University observed a signal from the protons in paraffin wax. **Block** and **Purcell** were jointly awarded the Nobel Prize for physics in 1952 for this discovery. Since that time, the advances in the NMR techniques leading to widespread applications in chemistry, clinical, solid state and biophysical sciences. In 1991, **Richard R. Ernst** was awarded Nobel prize for his contributions to the development of methodology of high resolution nuclear magnetic resonance (NMR) spectroscopy.

Principle

NMR spectroscopy involves transition of a nucleus from one spin state to another with the resultant absorption of electromagnetic radiation by spin active nuclei (having nuclear spin not equal to zero) when they are placed in a magnetic field.

Many nuclei have spin and all nuclei are electrically charged. When an external magnetic field is applied, an energy transfer is possible between the base energy to a higher energy level (generally a single energy gap). The energy transfer takes place at a wavelength that corresponds to radio frequencies (RF) and when the spin returns to its base level, energy is emitted at the same frequency. The signal that matches this transfer is measured and processed in order to yield an NMR spectrum for the nucleus concerned.

Instrumentation

Two types of NMR spectrometers are commonly encountered. They are: a) Continuous Wave (CW) NMR Spectrometer b) Fourier Transform (FT) NMR Spectrometer.

Continuous Wave (CW) NMR Spectrometer The CW-NMR spectrometer detects the resonance frequencies of nuclei in a sample placed in a magnetic field by sweeping the frequency of RF radiation through

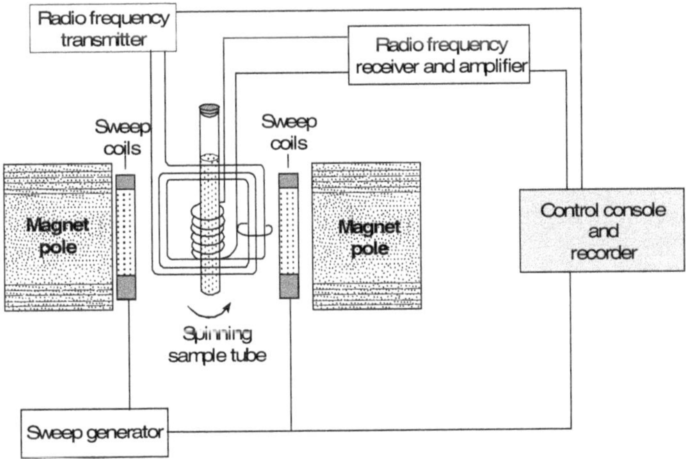

Figure 11.39 NMR Spectrometer

a given range and directly recording the intensity of absorption as a function of frequency. The spectrum is usually recorded and plotted simultaneously with a recorder synchronized to the frequency of the RF source.

The basic components of NMR spectroscopy are shown in Figure 11.39.

A **radio frequency transmitter** generates an appropriate RF to irradiate the sample. Radiofrequencies are generated by electronic multiplication of the natural frequency of a quartz crystal contained in a thermostat block. Different crystal sources and transmitters are uscd for different resonance frequencies. The radiofrequency oscillator coil is installed perpendicular to the magnetic field and transmits radiowaves of fixed frequency such as 60, 100, 200 or 300 MHz to a small coil that encircles the sample in the probe. The RF source is associated with a power-controlling device so that the level of RF power can be varied and adjusted empirically to get the best response.

The sample must be dissolved in a solvent, which lacks protons. The sample probe is a device which holds the sample contained in the NMR tube in between the poles of the magnet, the strongest and most homogeneous part of the magnetic field.

The magnets used in NMR spectrometers are large permanent or electromagnets, capable of producing strong, stable and homogeneous magnetic fields (up to 2.5 tesla or 25000 gauss) over the area occupied by the sample. A very small air gap between the poles (about 3–5 inches) and also the spinning of the sample between them serve as a source of homogeneous magnetic field. To minimize variation in the magnetic field, the sample kept in a tube of high-precision diameter is rotated at high speed by an air turbine.

The magnet is further annexed by a sweep generator in order to change the magnetic field precisely and continuously over a small range. The sweep generator possesses a pair of Helmholtz coils which are fixed in pole spaces of the magnets. These coils produce a magnetic field which can be changed by varying current flow.

The radiofrequency transmitter transmits radiofrequency radiation through the coil of the oscillator wound around the sample holder. In the magnetized sample, the radiofrequency undergoes absorption and dispersion. The absorption signal is detected by a radio receiver. The sample in the NMR probe yields information or data, which has been detected as an analogue signal or voltage. Detectors used in NMR spectrometers are to be very sensitive as the signal levels are very small (less than 1 millivolt). Therefore, multiple amplification of the signals is required before they are fed to the output devices. The signal is amplified and sent to the recorder.

In routine analysis, the signal is usually sent directly to the recorder or oscilloscope. The oscilloscope is useful for the fast scanning of the spectrum and allows preliminary adjustments to be made, e.g. setting optimum sensitivity or gain and electronic filtering of the signal.

The recorder records the spectrum as a plot of the resonance signal on the y-axis (vertical) versus the strength of the magnetic field on the x-axis (horizontal). The strength of the resonance signal is directly proportional to the number of nuclei resonating at that field strength.

Computer enhancement known as computer averaging technique (CAT) is very useful for studying very weakly absorbing biological samples. In this technique, the read-out from $10^1 - 10^3$ spectra of the sample are superimposed and back-ground noise minimized.

Uses of NMR spectroscopy

Nuclear Magnetic Resonance (NMR) spectroscopy is a technique in analytical chemistry used for determining the content and purity of a sample as well as its molecular structure. For example, NMR can quantitatively analyse mixtures containing known compounds. For unknown compounds, NMR can either be used to match against spectral libraries or to infer the basic structure directly. Once the basic structure is known, NMR can be used to determine the molecular conformation in solution as well as to study physical properties at the molecular level such as conformational exchange, phase changes, solubility, and diffusion.

FOURIER TRANSFORM NMR SPECTROSCOPY

In FT-NMR spectroscopy, the sample is subjected to a high-power, short-duration pulse of radio frequency (RF) radiation. This pulse of radiation contains a broad band of frequencies and causes all the spin-active nuclei to resonate all at once at their Larmor frequencies (transition frequency). Immediately following the pulse, the sample radiates a signal called **free induction decay (FID)**, which is modulated by all the frequencies of the nuclei excited by the pulse. The signal that is detected as the nuclei return to equilibrium (intensity as a function of time) is recorded, digitized and stored as an array of numbers in a computer. Fourier transformation of the data affords a conventional intensity as a function of frequency representation of the spectrum.

ATOMIC SPECTROSCOPY

The science of atomic spectroscopy has yielded three techniques for analytical use namely atomic emission, atomic absorption, and atomic fluorescence. In atomic spectroscopy atoms have to be volatilized either in a flame or electrothermally in an oven. In this state, the elements will readily emit or absorb monochromatic radiation at the appropriate wavelength. Usually nebulizers (atomizers) will be used to spray the standard or the test solution into the flame through which light is passed. Alternatively the light beam is passed in an oven through a cavity containing the vaporized material. The wavelengths emitted from excited atoms may be identified using a spectroscope, spectrograph, or a direct reading spectrophotometer that use the human eye, a photographic plate or a photoelectric cell, respectively as detectors.

ATOMIC EMISSION SPECTROSCOPY

In emission spectroscopy, a sample is excited by absorbing thermal or electric energy and the radiation emitted by the excited sample is studied for both qualitative and quantitative analysis. Most of the spectroscopic techniques are related to molecules but emission spectroscopy is related to atoms. Therefore, this technique is used as a method of elemental metal analysis. With it, all metallic elements can be identified and quantitatively determined in very low concentrations, as can the metalloids such as arsenic, silicon and selenium. Solids, liquids or gases can be analysed quite easily by this method. But most of the samples are solids or evaporated solutions. Liquid samples are occasionally analysed whereas gaseous samples are rarely analysed. If proper precautions are taken, the method can be used for quantitative analysis of about seventy elements at concentration level as low as 1 ppm.

Principle

If an atom is in the ground state, its electrons are present in the lowest permitted energy levels. When electric or thermal methods excite the atom, its electrons move from the inner orbital to the outer orbital. The excited electrons rapidly emit a photon of energy and occupy the orbital with the lowest energy or ground state. The emitted radiation from the excited atoms in the form of discrete spectral lines forms the basis of emission spectroscopy.

Instrumentation

In emission spectroscopy, the sample is excited by thermal or electric methods and thus becomes the source of radiation. Then, the radiation is passed through a monochromator to select the desired wavelength and, finally, passed on to the detector, which measures the radiation for qualitative analysis. The various individual components of the equipment are described below:

Excitation sources The excitation sources in emission spectroscopy are the flame, direct or indirect current arc and alternating current spark. Any excitation source used for emission spectroscopy must accomplish the following process

1. The sample must be vaporized.
2. It must be dissociated into atoms.
3. The electrons in the atoms must be excited to higher energy levels above the ground state.
4. It should be capable of exciting atoms of all the elements of interest.
5. It should provide sufficient line intensity so as to detect these lines within detection limit.
6. It should provide reproducible excitation conditions from sample to sample.

Flame is used for those molecules which do not require very high temperature for excitation and dissociation into atoms. Different flames having different temperatures used in emission spectroscopy are shown in the Table 11.4.

Table 11.4 Temperatures of commonly used fuels and oxidants in flames

Fuel	Oxidant	Temperature °C
Natural gas	Air	1700
Natural gas	Oxygen	2700
Acetylene	Air	2200
Acetylene	Oxygen	3200
Hydrogen	Oxygen	2800
Acetylene	Nitrous oxide	3400

Nebulizer The nebulizer passes a stream of air over a capillary tube whose other end dips into the solution under test. In indirect injection systems, large droplets will not remain in the hottest part of the flame for long time, and hence are allowed to settle in a cloud chamber. Combustion of air and natural gas gives a temperature of 1500°C which is adequate for sodium determination. Calcium is better analysed at 2000–2500°C, magnesium and iron require 2500°C obtained from air/acetylene gas mixture.

Monochromators In emission spectroscopy both prism and grating monochromators are used. The function of the monochromator is to separate the various lines of a sample's emission spectrum.

Detectors Two types of detectors are widely used—photomultipliers and photographic plates or films. For qualitative analytical studies, photographic plates or films are used. Photomultipliers are used for quantitative analysis.

Emission spectra may be examined by means of an optical arrangement that will identify component frequencies and their intensities. This optical arrangement is known by several names. For example, it is termed as a spectroscope if it involves a visual device; it is termed as a spectrograph if it records many wavelengths simultaneously on a photographic plate or film; it is termed as a spectrometer if it scans a spectrum. Except spectroscope, both spectrograph and spectrometer determine intensities. The basic layout of atomic emission spectroscopy is given in the Figure 11.40. Multichannel polychromators allow the emission of up to six elements at a time to be measured.

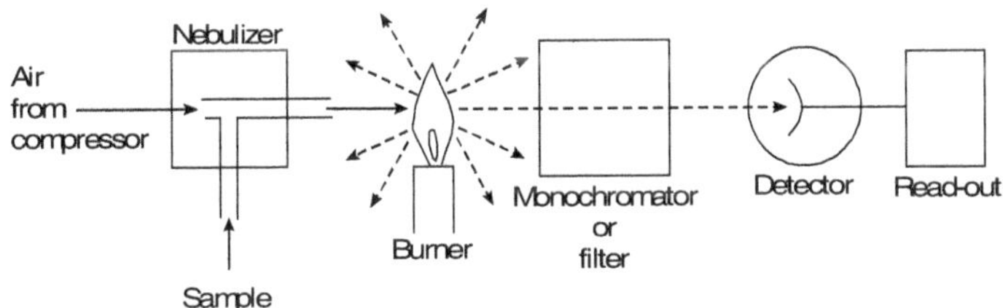

Figure 11.40 Components of atomic emission spectrophotometer

Applications of Emission Spectroscopy

1. Emission spectroscopy is used for elemental qualitative and quantitative analysis. But this provides very little direct information about the molecular form of the sample. In order to perform qualitative analysis, the emission spectrum of the sample is recorded on a photographic plate or film. The emission lines are then identified by comparing them with lines from the known samples.
2. Emission spectroscopy has been used in many fields to provide useful information. The iron and steel as well as the aluminium industries are based on emission spectroscopic analysis in all steps of their processes. This is done to control the composition of the molten metals before further processing. For example, the presence of nickel, chromium, silicon, manganese, molybdenum, copper, aluminium, arsenic, tin, cobalt, vanadium, lead, etc., in iron and steel has been determined on a routine basis by using automatic emission spectrograph. All these elements have been estimated in concentration ranges varying from 0.001% in iron to 30% in steels. Numerous elements can be determined simultaneously by this method with very little sample preparation.
3. It has also been used for the analysis of aluminium alloys, magnesium alloys, zinc alloys, lead alloys, and tin alloys. Some

of the trace elements such as, vanadium, copper, nickel and iron have been estimated by employing automatic emission spectrograph.

4. It has been used for the analysis of many elements including sodium, potassium, zinc, copper, calcium, magnesium, nickel and iron present in tissues of humans and animals. Changes in trace metal concentrations have been studied as related to the ageing process.
5. Emission spectroscopy has been used to detect elements in plants and soils. This has been done to diagnose deficiency problems in plants.

ATOMIC ABSORPTION SPECTROSCOPY

Since its introduction by Alan Walsh in the middle of the 1950s atomic absorption spectroscopy has been improved to become the most powerful technique for the determination of trace metals in liquids. This method provides the total metal content of a sample and is almost independent of the molecular form of the metal in the liquid.

Principle

The absorption of energy by ground state atoms in the gaseous state forms the basis of atomic absorption spectroscopy. When a solution containing metallic species is introduced into a flame, the vapour of the metallic species will be obtained. Some of the metal atoms may be raised to an energy level sufficiently high to emit the characteristic radiation of the metal—a phenomenon that is utilized in the familiar technique of emission flame photometry. But a large percentage of the metal atoms will remain in the non-emitting ground state. These ground-state atoms of particular elements are receptive of light radiation of their own specific resonance wavelength. Thus, when a light of this wavelength is allowed to pass through a flame having atoms of the metallic species, part of that light will be absorbed and the absorption will be proportional to the density of the atoms in the flame. Thus, in atomic absorption spectroscopy, the amount of light absorbed can be determined. Once this value of absorption is known, the concentration of metallic element can be determined because the absorption is proportional to the density of the atoms in the flame.

Instrumentation

Atomic absorption spectrophotometer and a schematic diagram of the atomic absorption spectrophotometer are shown in Figure 11.41a, b. The principle of the instrumentation is similar to other spectroscopic absorption methods. Light of a certain wavelength (produced by a special kind of lamp), which is able to emit the spectral lines corresponding to the energy required for an electronic transition from the ground state to an exciting state, is allowed to pass through the flame. Meanwhile, the sample solution is aspirated into the flame. Before it enters the flame, the solution gets dispersed into a mist of very small droplets, which evaporate in the flame to give first the dry salt, and then the vapour of the salt. At least a part of this vapour will be dissociated into atoms of the element to be measured. Thus, the flame possesses free unexcited atoms which are capable of absorbing radiation from an external

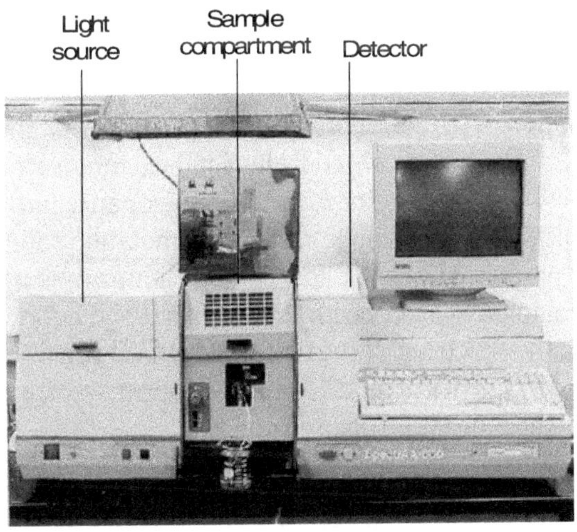

Figure 11.41 (a) Atomic absorption spectrophotometer

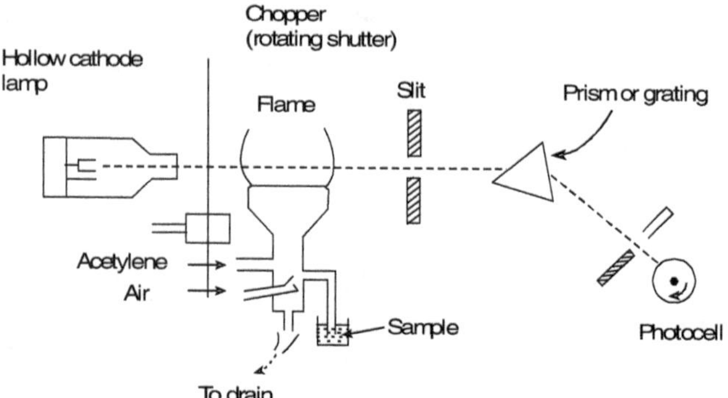

Figure 11.41 (b) Schematic diagram of atomic absorption spectrophotometer

source, when the radiation corresponds exactly to the energy required for a transition of an element from the ground electronic state to an upper excited level. Then the unabsorbed radiation from the flame is allowed to pass through a monochromator, which isolates the exciting spectral lines of light source. From the monochromator, the unabsorbed radiation is sent to the photo detector, which is then recorded, the output of which is amplified and measured on a recorder. Absorption is measured by the difference in the transmitted signal in the presence and absence of the test element.

For all types of atomic absorption spectrometers the following components are required.

Radiation source The radiation source for absorption spectrophotometer should emit stable, intense radiation of the element to be determined, usually a resonance line of the element. A hollow cathode lamp is used as the radiation source. The spectral lines produced by the hollow cathode lamp are so narrow that they are completely absorbed by the atoms. Each hollow cathode lamp emits the spectrum of that metal which is used in the cathode. For example, copper cathode emits the copper spectrum, zinc cathode emits the zinc spectrum and so on.

In atomic absorption spectrophotometer, gaseous discharge lamps are also used. These are called as arc lamps. Gaseous discharge lamps contain an inert gas at low pressure and a metal or metal salt. These lamps are useful for the alkali metals, zinc, cadmium and mercury.

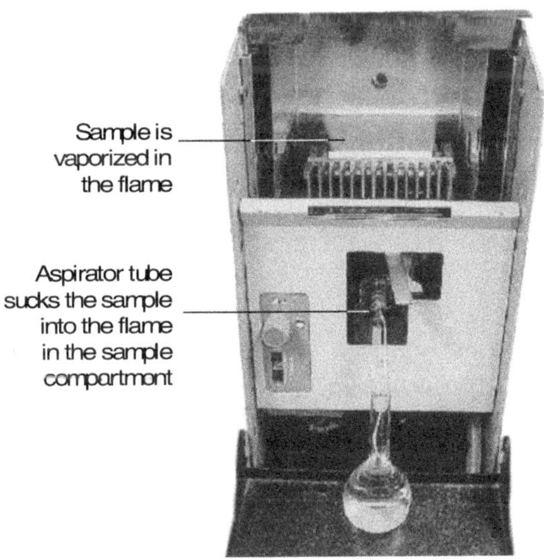

Figure 11.42 Vaporization of sample in atomic absorption spectrophotometer

Chopper A rotating wheel is interposed between the hollow cathode lamp and the flame. This rotating wheel is known as chopper and is interposed to break the steady light from the lamp into an intermittent or pulsating light. This gives a pulsating current in the photocell. There is also steady current caused by light, which is emitted by a flame. But only the pulsating (or alternating) current is amplified and recorded and, thus, the absorption of light will be measured without interference from the light emitted by the flame itself.

Production of the atomic vapour Before the liquid sample enters the burner, it is first converted into small droplets. This method of formation of small droplets from the liquid sample is called **nebulization**. A common method of nebulization is by the use of gas moving at high velocity, called pneumatic nebulization. The Beckman total consumption burner is commonly used in atomic absorption measurements. The fuel and oxidizing gases are passed through separate passages to meet at the opening of the base of the flame. The flame breaks up the liquid sample into droplets which are then evaporated or burnt, leaving the residue which is reduced to atoms (Figure 11.42). Total consumption burners do use oxygen, with hydrogen or acetylene, and give very hot flames. A typical total consumption burner is shown in the Figure 11.43. However, total consumption burner is noisy and hard to use. The efficiency of this burner is not good. Hence a pre-mixed burner is used. A typical pre-mixed burner is illustrated in Figure 11.44. In the pre-mixed burner, a mixture of the sample (liquid) and pre-mixed gases

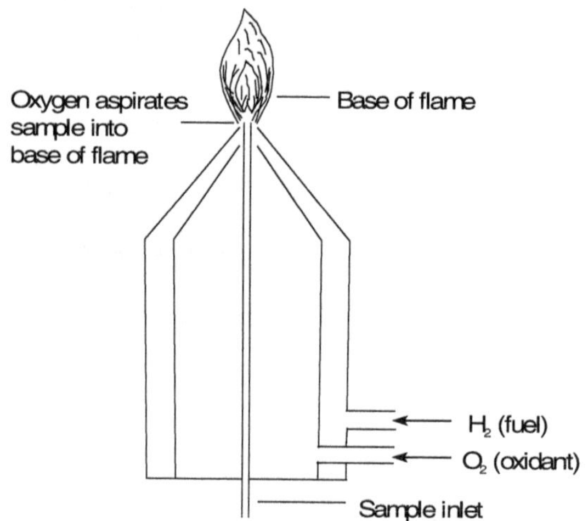

Figure 11.43 Total consumption burner

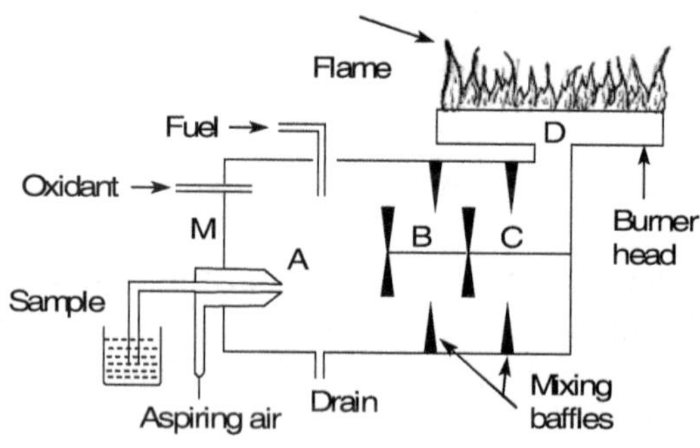

Figure 11.44 Premix or laminar-flow burner

($C_2H_2+O_2$) is allowed to enter the base, M and then to the region A. From the region A, the unburnt hydrocarbon gaseous mixture and liquid droplets enter the region B. In the region B, the liquid is evaporated leaving a residue. This residue is burnt in the regions C and D to produce atoms.

Monochromators The function of a monochromator is to select a given absorption line from spectral lines emitted from the hollow cathode. Prism or gratings are most commonly used as monochromators to select a particular wavelength of light. When the cathode in the hollow cathode lamp is made up of transition metals, the emission spectrum from the hollow cathode is so complicated that high dispersion is essential. For such cases, large dispersion and high resolving monochromators are advantageous for resolving spectra.

Detectors For atomic absorption spectroscopy, the photomultiplier tube is most suitable. In the photomultiplier tube, there is an evacuated envelope, which contains a photocathode, a series of electrodes called dynodes and an anode. The photocathode is fixed to the terminal of the power supply. As soon as a photon strikes the photocathode, an electron is dislodged and the photon is accelerated to dynode 1, resulting in the liberation of two or more electrons. Thus, the current multiplied at each dynode and the resultant electron current is received by the anode to produce much larger photoelectric current, which goes to the external amplifier and read-out system.

Amplifier The electric current from the photomultiplier detector is fed to the amplifier, which amplifies the electric current many times.

Read-out device In most of the atomic absorption measurements, chart recorders are used as read-out devices. In some atomic absorption measurements, digital read-out devices are also used.

Applications

Atomic absorption spectroscopy finds valid application in every branch of science including, analytical chemistry, ceramics, mineralogy, biochemistry, physiology, pharmacology, water supplies, metallurgy and soil analysis. The techniques are widely used in clinical laboratories, for the determination of metals in body fluids. In physiological and pharmacological research sodium, potassium, calcium, magnesium, cadmium and zinc may be measured directly, but copper, lead, iron and mercury require prior extraction from the biological source. The methods are also widely used in element determination in soil and plant materials. Further, atomic absorption spectroscopy may be used for the analysis of metals in tissues after adopting suitable ashing procedures.

ATOMIC FLUORESCENCE SPECTROPHOTOMETRY

Excitation of atoms by electromagnetic radiation rather than by thermal energy is required for analysis in atomic fluorescence spectrophotometry. Further, analysis of solids or solutions are not possible with atomic fluorescence spectrophotometry and hence atoms should be vaporized. The source beam must be intense but less spectrally pure than that required for atomic absorption spectrophotometry, as only the resonant wavelengths will be absorbed and lead to fluorescence. Emission can be modulated by detector, amplifier and can be recorded. This technique is highly sensitive and can detect few metals like zinc and cadmium at levels as low as 1 and 2 parts per 10^{10} respectively.

MASS SPECTROMETRY (MS)

In the techniques of mass spectrometry, the compound under investigation is bombarded with a beam of electrons, which produce an ionic molecule, or ionic fragments of the original species. The resulting assortment of charged particles is then separated according to their masses. The spectrum produced, known as **mass spectrum**, is a record of information regarding various masses produced and their relative abundances. A detailed interpretation of the mass spectrum frequently makes it possible to place functional groups into certain areas of the molecule and to see how they are connected to one another.

Mass spectrum is an analytical technique, which can provide information concerning the molecular structure of organic and inorganic compounds. It can be used to determine directly molecular weights as high as 4000. It is one of the few methods that can be used as a qualitative analytical tool to characterize different organic substances. With it, one can do analysis of mixtures (gases, or liquids, and in some cases solids) quantitatively. A mass spectrometer is also useful to investigate reaction mixtures and in tracer work. It is also used in understanding kinetics and mechanisms of unimolecular decomposition reactions.

MASS SPECTROMETER

The mass spectrometer has long been an indispensable tool in science. The mass spectrometers used for the investigation of any compound may vary in their types depending on the compound investigated but generally all contain the following components.

i. Inlet system (or sample handling system)
ii. Ion source (or ionization chamber)
iii. Electrostatic accelerating system
iv. Ion separator

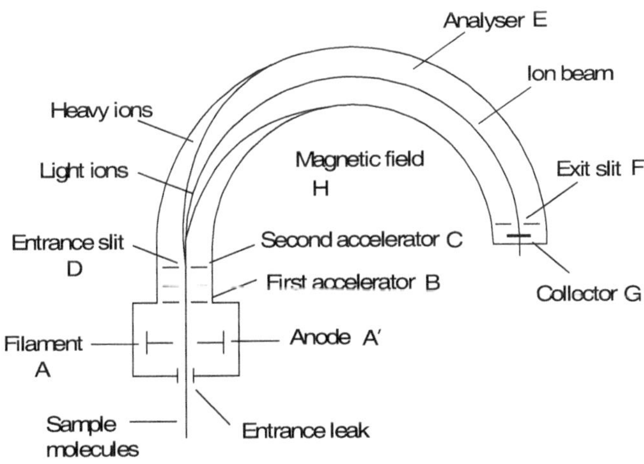

Figure 11.45 Schematic diagram or single deflection 180° mass spectrometer

v. Ion collector (detector and read-out system)

vi. Vacuum system

A typical arrangement for an 180° mass spectrophotometer is shown in the Figure 11.45. A beam of positive ions of various masses is produced in the ionization chamber. This beam is then withdrawn by electrode B and accelerated in the electric field between B and C. Then, the ions emerge through the slit D, and pass through the evacuated tube E that is placed in a strong magnetic field H, perpendicular to the plane of the diagram. Only those ions with the proper mass-to-charge (m/e) ratio will be allowed to pass through the exit slit F and strike collector plate G. Finally, the mass spectrum is obtained by varying the accelerating voltage on C.

Kinds of MS (Mass Spectrometry)

1. Gas chromatography–MS (GC–MS) A common form of mass spectrometry is gas chromatography–mass spectrometry (GC/MS or GC–MS) (Figure 11.46). In this technique, a gas chromatograph is used to separate compounds. This stream of separated compounds is fed on-line into the ion source, a metallic filament to which voltage is applied. This filament emits electrons, which ionize the compounds. The ions can then further fragment, yielding predictable patterns. Intact ions and fragments pass into the mass spectrometer's analyser and are eventually detected.

Figure 11.46 Gas chromatography–Mass spectrometry (GC–MS)

2. *Liquid chromatography–MS (LC–MS)* Similar to gas chromatography–MS (GC–MS), liquid chromatography–mass spectrometry (LC–MS) (Figure 11.47) separates compounds chromatographically before they are introduccd to the ion source and mass spectrometer. It differs from GC–MS in that the mobile phase is liquid, usually a combination of water and organic solvents, instead of gas. Most commonly, an electrospray ionization source is used in LC–MS.

Figure 11.47 Liquid chromatography

3. *Ion mobility spectrometry–mass spectrometry (IMS–MS)* Ion mobility spectrometry/mass spectrometry is a technique where ions are first separated by drift time through some pressure of neutral gas and given an electrical potential gradient before being introduced into a mass spectrometer. The drift time is a measure of the radius relative to the charge of the ion. The duty cycle of IMS (time over which the experiment takes place) is longer than most mass spectrometers such that the mass spectrometer can sample along the course of the IMS separation. This produces data about the IMS separation and the charge-to-mass ratio of the ions in a manner similar to liquid chromatography–mass spectrometry (LC–MS). However, the duty cycle of IMS is short relative to liquid chromatography or gas chromatography separations and can thus be coupled to such techniques producing triply hyphenated techniques such as LC–IMS–MS.

Tandem MS (MS–MS) Tandem mass spectrometry involves multiple steps of mass selection or analysis, usually separated by some form of fragmentation. A tandem mass spectrometer is one capable of multiple rounds of mass spectrometry. For example, one mass analyser can isolate one peptide from many entering a mass spectrometer. A second mass analyser then stabilizes the peptide ions while they collide with a gas, causing them to fragment by collision-induced dissociation (CID). A third mass analyser then catalogs the fragments produced from the peptides. Tandem MS can also be done in a single mass analyser over time as in a quadrupole ion trap. There are various methods for fragmenting molecules for tandem MS, including collision-induced dissociation (CID), electron capture dissociation (ECD), infrared multiphoton dissociation (IRMPD) and blackbody infrared radioactive dissociation (BIRD).

Isotope ratio MS (IR–MS) Mass spectrometry is also used to determine the isotopic composition of elements within a sample. Differences in mass among isotopes of an element are very small, and the less abundant isotopes of an element are typically very rare, so a very sensitive instrument is required. These instruments are called isotope ratio mass spectrometers (IR-MS) and usually use a single magnet to bend a beam of ionized particles towards a series of Faraday cups [A Faraday cup is a metal (conductive) cup meant to recatch secondary particles] which convert particle impacts to electric current.

Fourier Transform Mass Spectrometry (FTMS) Instead of measuring the deflection of ions with a detector such as an electron multiplier, the ions are injected into a pennin trap (a static electric/magnetic ion trap) where they effectively form part of a circuit. Detectors at fixed positions in space measure the electrical signal of ions which pass near them over time producing cyclical signal. Since the frequency of the ions cycling is determined by its charge-to-mass ratio, this can be deconvoluted by performing a Fourier transform on the signal. FTMS has the advantage of improved sensitivity (since each ion is "counted" more than once) as well as much higher resolution and thus precision.

Mass spectrometry of proteins Mass spectrometry is an important emerging method for the characterization of proteins. The two primary methods for ionization of whole proteins are electrospray ionization and matrix-assisted laser desorption ionization (MALDI). In keeping with the performance and mass range of available mass spectrometers, two approaches are used for characterizing proteins. In the first, intact proteins are ionized by either of the two techniques described above, and then introduced to a mass analyser. In the second, proteins are enzymatically digested into smaller peptides using an agent such as trypsin or pepsin. Other proteolytic digest agents are also used. The collection of peptide products are then introduced into the mass analyser. This is often referred to as the "bottom-up" approach of protein analysis.

Whole protein mass analysis is primarily conducted using either time-of-flight (ToF) MS (Figure 11.48) or Fourier transform ion cyclotron resonance (FT-ICR). These two types of instruments are preferable because of their wide mass range, and in the case of FT-ICR, its high mass accuracy. Mass analysis of proteolytic peptides is a much more popular method of protein characterization, as cheaper instrument designs can be used for characterization. Additionally, sample preparation is easier once whole proteins have been digested into smaller peptide fragments. The most widely used instrument for peptide mass analysis is the quadrupole ion trap. Multiple stage quadrupole-time-of-flight and MALDI time-of-flight instruments also find use in this application.

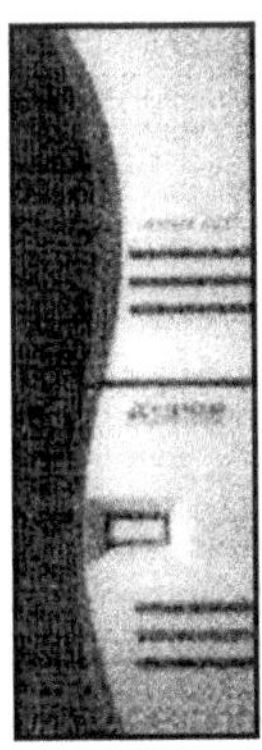

Figure 11.48 Time-of-flight (ToF) MS

Protein and peptide fractionation coupled with mass spectrometry Proteins of interest to biological researchers are usually part of a very complex mixture of other proteins and molecules that co-exist in the biological medium. This presents two significant problems. First, the two ionization techniques used for large molecules only work well when the mixture contains roughly equal amounts of constituents, while in biological samples, different proteins tend to be present in widely differing amounts. If such a mixture is ionized using electrospray or MALDI, the more abundant species have a tendency to "drown" signals from less abundant ones. The second problem is that the mass spectrum from a complex mixture is very difficult to interpret due to the overwhelming number of mixture components. This is exacerbated by the fact that enzymatic digestion of a protein gives rise to a large number of peptide products.

To contend with this problem, two methods are widely used to fractionate proteins, or their peptide products from an enzymatic digestion. The first method fractionates whole proteins and is called two-dimensional gel electrophoresis. The second method, high performance liquid chromatography is used to fractionate peptides after enzymatic digestion. In some situations, it may be necessary to combine both of these techniques.

Gel spots identified on a 2D gel are usually attributable to one protein. If the identity of the protein is desired, the gel spot can be excised, and digested proteolytically. The peptide masses resulting from the digestion can be determined by mass spectrometry using peptide mass fingerprinting. If this information does not allow unequivocal identification of the protein, its peptides can be subject to tandem mass spectrometry.

Characterization of protein mixtures using HPLC/MS is also called ''shotgun proteomics'' and ''mudpit''. A peptide mixture that results from digestion of a protein mixture is fractionated by one or two steps of liquid chromatography. The eluent from the chromatography stage can be either directly introduced to the mass spectrometer through electrospray ionization, or laid down on a series of small spots for later mass analysis using MALDI.

Applications

The various applications of mass spectrometry are as follows:

1. **Molecular mass determination** Mass spectrometry is one of the best methods to determine the molecular mass accurately. When a substance is bombarded with moving electrons and the mass spectrum is recorded, the mass of the peak at the highest m/e reveals the molecular mass accurately. This method is only accurate when no ions heavier than the parent ion are formed.
2. **Isotopic abundance** The isotopic abundance of easily vaporizable elements can be determined using a mass spectrometer.
3. **Quantitative analysis of mixtures** The compounds present in the mixture are separated by gas chromatography and then their amounts are determined by mass spectrometry by allowing various compounds to enter a mass spectrometer one by one.
4. **Distinction between *cis* and *trans* isomers** Mass spectrometry is successful in making distinction between *cis* and *trans* isomers. Both *cis* and the *trans* isomers yield similar mass spectra but the molecular ion peak for the *trans* isomer is more intense than the *cis* isomer. Another distinction is that the dehydration fragment for the *cis* isomer is much stronger than the *trans* isomer.
5. **Bonding** Bonding information may be frequently interpreted from fracture patterns. A few of the rules used in this regard are given below:
   i. Branched compounds tend to rupture at the branched carbon atom. The positive charge remains with the branched fragment. The greater the degree of branching, the greater is the probability of rupture.
   ii. Beta position cleavage is the most probable rupture for the double and hetero bonds.
   iii. Ring compounds generally exhibit mass numbers which are characteristic of the ring.

   iv. Side chains are ruptured from saturated ring compounds at the alpha position.
   v. Carbonyl compounds break at the carbonyl bond. The positive charge remains with the carbonyl group.

6. **Reaction kinetics** Mass spectrometry is very frequently used in kinetic and mechanistic reaction studies. As unstable intermediates can be detected in a mass spectrometer, the technique is used in the study of radical reactions.
7. **Impurity detection** Mass spectrometry is one of the best methods to detect impurities. The detection of impurities in parts per million is only possible if their structures differ considerably from those of the major components. If the molecular weights of the impurities are much larger than major components, their detection is more easy because their higher mass peaks are free from contributions by those of the major components. On the other hand if the molecular weights of the impurities are much lower than the major components, their detection is not an easy job because of the formation of common fragmentations.
8. **Characterization of polymers** Mass spectrometry is used for the characterization of polymers. First of all, the polymer is pyrolysed (i.e., decomposed by heating) and then the pyrolized products are fed into the inlet of a mass spectrometer and identified. From the results, one may obtain much information concerning the structure of the polymer. For example, mass spectrometry will be able to distinguish chlorinated polymers to know whether the chlorine occurs randomly or in blocks in the structure of polymers.
9. **Identification of unknown compound** In order to identify the mass spectrum of an unknown substance the following points are to be kept in mind:
   i. A background spectrum is run before a sample is introduced. This is done because weak peaks are often present at m/e 41, 43, 55 and 57 (hydrocarbon background). Then the sample is introduced into the ionization chamber and a sample spectrum is run.

ii. The next job is to check whether the peak at the highest mass is likely to be a molecular ion. Then, one has to check whether the peaks immediately below an assumed molecular ion correspond to the loss of plausible neutral molecules or not. Further, one has to find the nature of bonds in molecular ion from molecular ion abundance. Finally, one has to note whether the molecular weight is odd or even.

iii. From the fragmentation pattern, one may find out very easily a preliminary indication about functional groups and also partial structural information about the compound. One must also assign any metastable peaks.

After molecular diagnosis as described above, the final identification of an unknown compound may be done by comparing its mass spectrum with that of an authentic sample. One important point to remember is that the two mass spectra should be recorded under identical conditions.

The various advantages of identification of an unknown compound by mass spectrometry are as follows:

i. The mass spectrum is always recorded in the gaseous state. It means that complications due to solvents and change of physical state do not affect the nature of mass spectrum.

ii. In mass spectrometry, the quantity of the sample required is much less as compared with other techniques.

One main disadvantage of this method is that even small amounts of impurities make the spectra very complicated. Thus, its interpretation is difficult.

REVIEW YOUR LEARNING

1. Define/explain briefly the following:

    i. Electromagnetic radiation

    ii. Wave theory of electromagnetic radiation

    iii. Quantum theory of electromagnetic radiation

iv. Molar extinction coefficient
v. Specific extinction coefficient
vi. Wavelength
vii. Absorption spectrum
viii. Emission spectrum
ix. Visible spectrum
ix. Optical rotatory dispersion (ORD)
x. Circular dichroism (CD)

2. Write short notes on the following:
   i. Electromagnetic spectrum
   ii. γ-ray spectroscopy
   iii. Mössbauer spectroscopy
   iv. X-ray absorption methods
   v. X-ray fluorescence methods
   vi. X-ray diffraction
   vii. X-ray crystallography
   viii. Nebulizer
   ix. Gas chromatography–mass spectrometry (GC–MS)
   x. Liquid chromatography–mass spectrometry (LC–MS)
   xi. Ion mobility spectrometry–mass spectrometry (IMS–MS)
   xii. Tandem mass spectrometry
   xiii. Isotope ratio mass spectrometry (IR–MS)
   xiv. Fourier transform mass spectrometry (FTMS)
   xv. FT-NMR spectroscopy
   xvi. Atomic fluorescence spectrophotometry
3. Explain Beer–Lambert's law.
4. Explain the principle and applications of the following:
   i. Colorimeter
   ii. Micromotility meter
   iii. Ultraviolet and visible spectroscopy (UV-VIS)

iv. Flame photometry
v. Nephelometry
vi. Turbidimetry
vii. Fluorimeter
viii. Phosphorimeter
ix. Spectrofluorimeter
x. Spectrophosphorimeter
xi. Optical rotary dispersion (ORD) and circular dichroism (CD) spectroscopy
xii. Infrared spectrophotometer
xiii. Raman spectroscopy
xiv. Electron spin resonance spectrometry (ESR)
xv. Nuclear magnetic resonance (NMR)
xvi. Atomic emission spectroscopy
xvii. Atomic absorption spectroscopy
xviii. Mass spectrometry

RADIOISOTOPIC TECHNIQUES

Radioactivity is the spontaneous disintegration of the nuclei of some of the isotopes of certain elements, with the emission of alpha (α) particles, beta (β) particles or gamma (γ) rays. Such nuclei are said to be radioactive. The phenomenon is due to spontaneous nuclear changes and is called as **radioactive decay.**

In 1896, **Henri Becquerel** worked with compounds containing the element uranium. To his surprise, he found that photographic plates that were covered to keep out light became fogged or partially exposed, when these uranium compounds were placed near the plates. This fogging suggested that some kind of ray had passed through the plate coverings. Several materials other than uranium were also found to emit these penetrating rays. Materials that emit this kind of radiation are said to be **radioactive**, and they undergo **radioactive decay**. In 1899, **Ernest Rutherford** discovered that uranium compounds produced three different kinds of radiation. He separated the radiations according to their penetrating abilities and named them alpha (α), beta (β), and gamma (γ) radiation, after the first three letters of the Greek alphabet. The α radiation can be stopped by a sheet of paper. Rutherford later showed that an alpha particle is the nucleus of a helium atom, ^4He. Beta particles were later identified as high-speed electrons. Six millimetres thickness of aluminium are needed to stop most β particles. Several millimetres thickness of lead are needed to stop γ rays, which proved to be high-energy photons. Alpha particles and γ rays are emitted with a specific energy that depends on the radioactive isotope. Beta particles, however, are emitted with a continuous range of energies from zero up to the maximum energy allowed by the particular isotope.

RADIOISOTOPES

An atom consists of an extremely small, positively charged nucleus surrounded by a cloud of negatively charged electrons. Nuclei consist of positively charged protons and electrically neutral neutrons held together by nuclear force (Figure 12.1).

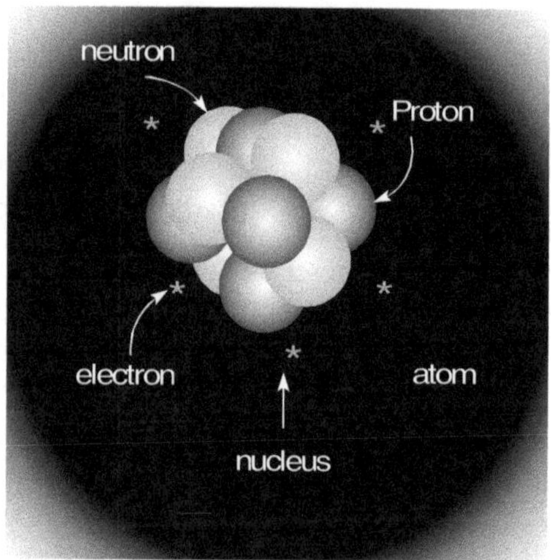

Figure 12.1 Nuclear structure

The number of protons in the nucleus, Z, is called the **atomic number**. The **atomic mass** of the nucleus, A, is equal to the total number of protons (Z) and neutrons (N) in the nucleus (Figure 12.2).

$$A = Z + N$$

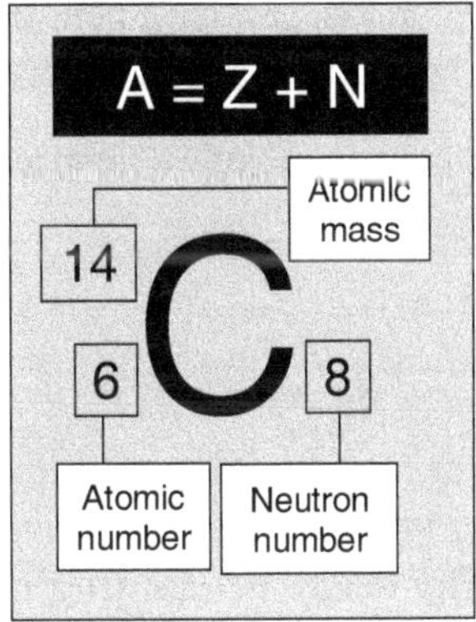

Figure 12.2 Atomic number and atomic mass

Atoms with same atomic number but different mass number are known as **isotopes**. For a given element it may vary from two to as many as 20. For example, hydrogen has three isotopes, $^{1}_{1}H$ (hydrogen), $^{2}_{1}H$ (deuterium) and $^{3}_{1}H$ (tritium). The superscript stands for the mass number. In general, an element is represented $^{Z}_{A}E$, where A is the mass number and Z the atomic number. In an atomic nucleus when the protons and neutrons are equal in number, the isotope is generally a stable one. However, elements with higher atomic numbers frequently have a neutron to proton ratio of more than one. These isotopes are unstable and are known as **radioisotopes**. Radioisotopes are found in nature (e.g. uranium, plutonium, thorium and radium) though they are more often made artificially. Radioactive isotopes emit particles and/or electromagnetic radiation to become a stable isotope. This process is known as **radioactive decay**. A neutron can emit an electron (also known as a negatron to indicate its nuclear origin) to become a proton.

$$\text{Neutron} \rightarrow \text{proton} + \text{negatron } (\beta \text{ –ve})$$

An isotope frequently used in biological work, which decays by negatron emission is $^{14}_{6}C$.

$$^{14}_{6}C \rightarrow {}^{7}_{14}C + \beta \text{ –ve}$$

Similarly, a proton can become a neutron by emitting a positron. A positron has the same mass as an electron but is positively charged.

$$\text{Proton} \rightarrow \text{neutron} + \text{positron}\ (\beta\ +\text{ve})$$

An example of an isotope decaying by positron emission is $^{22}_{11}Na$.

$$^{22}_{11}\text{Na} \rightarrow\ ^{22}_{10}\text{Ne} + \beta\ +\text{ve}$$

Radioactive decay can also result in the emission of α particles and γ-rays. Isotopes of elements with high atomic numbers frequently decay by emitting *alpha* particles (α). An α-particle is a helium nucleus in that it consists of two protons and two neutrons ($^{4}_{2}He^{2+}$). Emission of α-particles results in a considerable lightening of the nucleus, a decrease in atomic number of two and a decrease in the mass number of four. Isotopes which decay by α-emission are not frequently encountered in biological work. $^{226}_{88}$Radium ($^{226}_{88}Ra$) decays by α-emission to $^{222}_{86}$Radon ($^{222}_{86}Rn$) which itself is radioactive. Thus a complex decay series begins which culminates in the formation of $^{206}_{82}Pb$.

$$^{226}_{88}\text{Ra} \rightarrow\ ^{4}_{2}\text{He}^{2+} +\ ^{222}_{86}\text{Rn} \rightarrow \rightarrow \rightarrow\ ^{206}_{82}\text{Pb}$$

In contrast to emission of α- and β-particles, gamma (γ) emission involves electromagnetic radiation similar to, but with a shorter wavelength than, X-rays. These γ-rays result from a transformation in the nucleus of an atom (in contrast to X-rays which are emitted as a consequence of excitation involving the orbital electrons of an atom) and frequently accompany α- and β-particle emission. Emission of γ-radiation in itself leads to no change in atomic number or mass.

BIOCHEMICAL APPLICATIONS OF RADIOISOTOPES

1. Radioisotopes are frequently used for identifying metabolic pathways.
2. Radioisotopes are used in clinical and biochemical fields for quantitative analysis of hormones like thyroxine and steroids.

3. The age of rocks, fossils or sediments can be determined using radioisotopes.
4. Isotopes are widely used in the development of new drugs.
5. Radioactive isotopes are widely used for identifying various diseases. Thyroid function tests using ^{131}I are employed in the diagnosis of hypo- and hyperthyroidism. Xenon-133 (^{133}Xe) is useful in the diagnosis of malfunctioning of lung ventilation. Kidney function tests using ^{131}I-iodohippuric acids are useful in identifying kidney infections, kidney blockage or imbalance of function between two kidneys. Further, half-life period of RBC, blood volume and blood circulation time are determined using radioisotopes.
6. Radioisotopes are widely used in various research laboratories to study the mechanism and rate of absorption, accumulation and transportation of inorganic and organic compounds by both plants and animals.
7. Many radioisotopes are used in recent techniques in molecular biology such as DNA and RNA sequencing, DNA replication, transcription, synthesis of cDNA and recombinant DNA technology.
8. Radioisotopes are used in industrial microbiology.
9. The migratory patterns and behavioural patterns of many animals (e.g. birds and fishes) are monitored using radiotracers.
10. Transport of ions into the tissues of the body is extensively studied with the help of isotopes.

RADIATION DOSIMETRY

Two common units to measure the activity of a substance are the Curie (Ci) and Rutherford (rd). In SI units, the unit of radioactivity is Becquerel (Bq).

Curie is the international unit of radioactivity based on the radioactivity of 1 g of radium. One curie (Ci) equals 3.7×10^{10} nuclear disintegrations per sec (3.7×10^{10} dps).

1 curie (1 Ci) $= 3.7 \times 10^{10}$ dps

1 millicurie (1mCi) $= 3.7 \times 10^{7}$ dps

1 microcurie(1 μ Ci) $= 3.7 \times 10^{4}$ dps

A Becquerel is a more fundamental unit of measure of radioactive decay that is equal to 1 disintegration per second. Currently, the curie is more widely used in the United States, but usage of Becquerel can be expected to broaden as the metric system slowly comes into wider use.

$$1 \text{ curie} = 3.7 \times 10^{10} \text{ becquerels}$$

Roentgen (R) is the amount of γ- or X-rays, which deposits 8.33×10^{-6} J of energy per g of air to produce 1.61×10^{12} ion pairs g^{-1} (or 2.08×10^{9} ion pairs cm^{-3} in dry air at STP).

Roentgen-equivalent physical (Rep) is the amount of γ- or X-rays, which deposits in water or biological tissues. It is identical with R in value.

Radiation absorbed dose (rad) is the quantity of γ- or X-rays that can deposit 100 ergs or 1×10^{-5} J g^{-1} in matter including biological tissues.

Gray (GY) equals 100 rads and is the dose of γ -rays depositing 10^{4} ergs g^{-1} or 1 J kg^{-1}.

Relative biological effectiveness (RBE) is the ratio between the doses (rad) of the radiation under investigation and of γ- rays, producing identical biological effects. RBE is 1 for γ-rays.

Roentgen-equivalent mammals (Rem) measures a radiation in terms of its effects on mammalian tissues. Rem = rad $\times$ RBE.

RADIOISOTOPIC TECHNIQUES

Various techniques have been developed using radioisotopes to quantify the compounds present in living organisms in trace amounts, to determine the age of rock, fossils and sediments, quantitative analysis of drugs, hormones, etc.

Isotope Dilution Technique

Many compounds present in living organisms cannot be accurately measured because they are present in such low amounts. Isotope dilution technique offers a convenient and accurate way to overcome this problem.

In principle, the method consists of adding to the system an accurately weighed amount of radioactive isotope of known specific activity and determination of the size of that system by the dilution of the radioactivity added. It is assumed that there is no loss of isotope from the system. The amount of unlabelled compound originally present in the mixture is calculated using the equation,

$$x = a\left(\frac{S_1}{S_2} - 1\right)$$

where,

x is the amount of unlabelled compound,
a is the amount of labelled compound added,
S_1 is the specific activity of the compound before adding and
S_2 is the specific activity of the isolated material.

Isotope dilution analysis is widely used in the field of biochemistry and medicine especially in the measurement of (a) erythrocyte volume with ^{51}Cr-labelled erythrocytes, (b) plasma volume with ^{131}I-labelled serum albumin, (c) Body water with tritiated water and, (d) extracellular space with ^{82}Br.

Radiodating

The age of rocks, fossils or sediments can be determined using radioisotopes by a method known as **radiodating**. In this technique, it is assumed that the proportion of an element that is naturally radioactive is the same throughout time. The radioactive isotope will start decaying from the time of fossilization or deposition. By determining the amount of radioisotope remaining and from knowledge of the half-life, it is possible to date the sample. For instance, if the radioisotope normally consists of 1% of the element and if it is found that the sample actually contains 0.25 percent, then two half-lives can be assumed to have elapsed since deposition. If the half-life is one million years then the sample can be dated as being two million years old. For long-term dating, isotopes with long-lives such as ^{235}U, ^{238}U and ^{40}K are necessary, whereas for short-term dating ^{14}C is widely used. It is not a very accurate method since it gives only the approximate age of the samples.

Radioactive Tracer Technique

In the tracer technique, a radioactive element is always used in trace amounts. The chemical properties of radioactive elements are similar to corresponding non-radioactive elements. Therefore, the emitted radiation alone acts as a means of identifying or tracing the radiating atom. This technique is extensively used in chemical research and analytical chemistry. Radioactive compounds used in tracer techniques are called radioactive tracers.

In this technique, a particular compound is tagged with trace amounts of a radioactive isotope of an element such as $^{3}H_{1}$ or $^{14}C_{6}$. The tagged compound is introduced into a biological system and the experiment is carried out as usual. The resulting products can be analysed radiochemically.

Radioimmunoassay (RIA)

Radioimmunoassay is one of the most important techniques used in clinical and biochemical fields for the quantitative analysis of hormones, steroids and drugs. It combines the specificity of the immunoreaction with the sensitivity of the radioisotope techniques. This technique was introduced in 1960 by Berson and Yalow as an assay for the concentration of insulin in plasma. Radioimmunoassay is based on the competition between unlabelled antigen (Ag) and a finite amount of the corresponding radiolabelled antigen (Ag^*) for a limited number of antibody-binding sites in a fixed amount of antiserum. In this technique, a mixture of radioactive antigen and antibodies against that antigen is prepared. Because of the ease with which iodine atoms can be introduced into tyrosine residues in a protein, the radioactive isotopes ^{125}I or ^{131}I are often used. A known amount of unlabelled ("cold") antigen is added to the samples of the mixture. These compete for the binding sites of the antibodies (Figure 12.3).

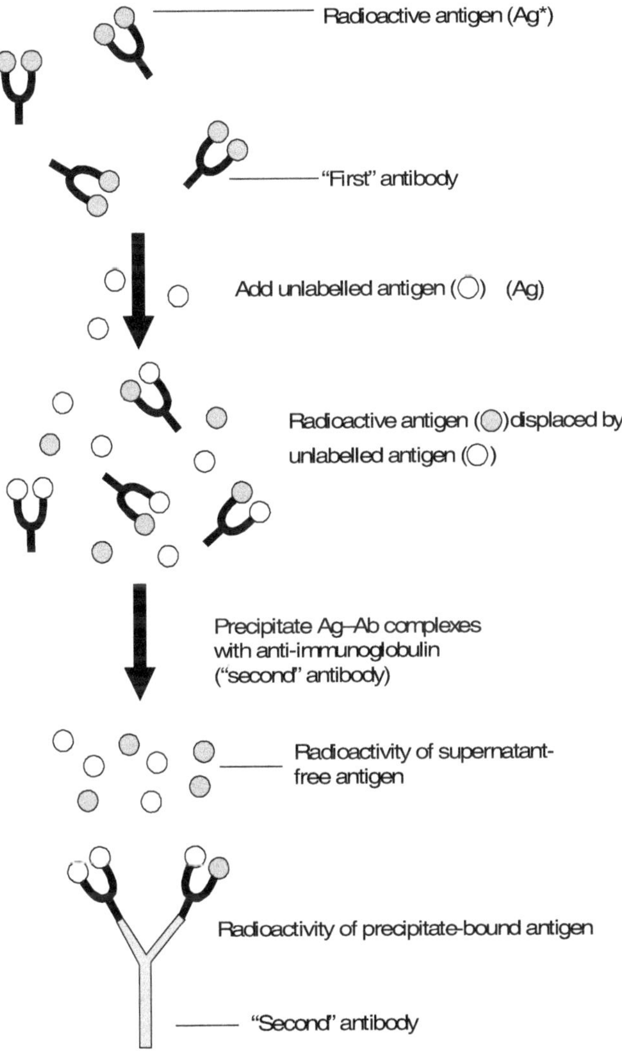

Figure 12.3 Radioimmunoassay

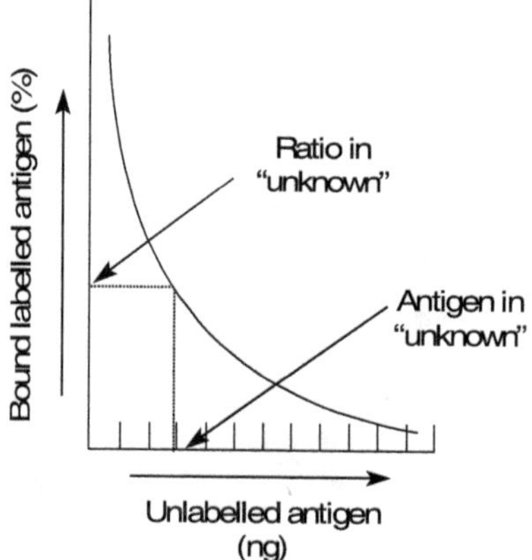

Figure 12.4 Radioimmunoassay calibration curve

At equilibrium, there will be free unlabelled antigen, labelled antigen, antibody bound to unlabelled antigen, and antibody bound to labelled antigen (Ag, Ag*, AgAb and Ag*Ab). The bound and free forms are separated by protein precipitating agents such as polyethylene glycol. Under standard conditions, the amount of labelled antigens bound to antibodies will decrease as the amount of unlabelled antigen in the sample increases. A calibration curve is prepared by taking the concentration of unlabelled antigens antigen in the X-axis and % of bound labelled antigen in the Y-axis (Figure 12.4). From this graph, an unknown amount of antigen can be determined. RIA is used to assay any compound that is immunogenic and it has very high sensitivity.

DETECTION AND MEASUREMENT OF RADIOACTIVITY

The detection and measurement of radioactivity depend upon the ionizing effects of radioactivity.

Fountain-pen Dosimeter

This instrument, a modification of pocket electroscope, consists of an earthed and gas-filled tube closed at one end by an eyepiece lens, and at the other end by an insulated plug.

A metal rod passes into the tube through the insulated plug, carries two thin movable gold foils at its farthest end while its other end terminates in a charging terminal outside the tube. On charging the metal rod, the gold leaves carry like charges and divert at an angle proportional to the potential, the divergence being seen through the lens. In the presence of radioactivity, γ rays and β particles (but not α -particles) penetrate through the tube wall and ionize the gas molecules inside. These ions discharge the gold leaves to reduce the angle of their divergence. The change in divergence, read from an incorporated scale, gives an estimate of ionizing radiation. Though sensitive to γ-rays and β-particles, the instrument cannot detect neutrons or α -particles and cannot enumerate individual emitted particles, and can be used only for short periods after which it needs recharging.

Geiger–Müller Counter

This instrument is sensitive to β-particles, γ-rays and also α -particles. It can also detect particles with weak ionizing effects, can count individual particles at a very high rate and can thus precisely enumerate particles at high radiation levels. It can be designed to curtail the low-ionizing γ-rays. But though very efficient in counting high-energy β-particles, its efficiency is poor for low-energy β-particles. It is less efficient than scintillation counters in detecting γ-rays, nor does its output indicate the type of emission.

A typical Geiger–Müller (GM) counter consists of a GM tube, a high voltage supply for the tube, a scaler to record the number of particles detected by the tube, and a timer which will stop the action of the scaler at the end of a preset interval (Figure 12.5).

The GM tube is a cylindrical metal tube, one end of which is closed by a thin mica membrane through which α-particles can enter; the other end is closed by an insulated plug through which a thin metal wire runs inside the tube along its axis. The tube is filled with an inert ionizable gas (He, Ar or Xe) at a low pressure, along with a small volume of an easily dissociable "quenching" substance like ethanol or a polyatomic organic gas. The axial wire serves as the anode and the metal wall of the tube acts as the cathode; they are connected to a source of high potential like 1500 V, and a sensitive electronic indicator is also included in the circuit for recording any current pulse generated in the circuit. The axial wire is maintained highly charged as the anode so that the surrounding electric field stresses the gas inside the tube almost to the breaking point of a spark.

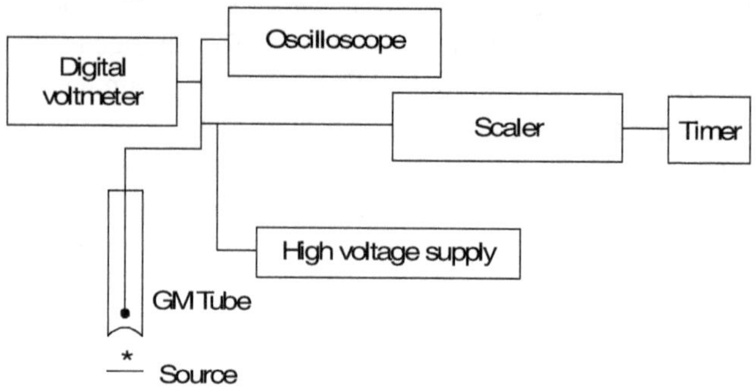

Figure 12.5 Schematic diagram of Geiger–Müller counter

To detect or estimate radioactivity in a tissue or organ *in situ* (e.g. in radiography), the counter may be placed with its mica window on the skin surface overlying the tissue or organ. To detect or estimate radioactive emissions from a biological sample, the latter is placed on a sample pan held by a holder just below the mica window of a Geiger–Müller counter inside a chamber with a lead shield on its wall for protecting the investigator from exposure. In both methods, radioactive emissions entering the tube through the thin mica window of the counter ionize the gas atoms by knocking off electrons, which are then greatly accelerated by the anodal field to cause further ionizations. This soon causes an avalanche of the electrons to reach the anode, generating a large current pulse between the anode and the cathode to produce audible sounds or visible indications, which are recorded electronically. The quenching agent terminates the current pulse by absorbing the energy of the electronic avalanche.

The Geiger–Müller counter may also be used to estimate or detect a radioisotope in metabolites either in the elute emerging from a chromatographic column or in the electrophoretic bands on a paper strip or a gel block.

Scintillation Counter

Fluorescence due to excitation by radioactivity is known as scintillation. The instrument which is used to measure scintillation is called scintillation counter. This counter carries a solid ionizable material (scintillation or phosphor) instead of gas. Even very high-energy γ-rays are absorbed by the scintillation counter to produce detectable ionizations, making this instrument more sensitive to γ-rays than the Geiger counter. It can also count α- and β-particles at a faster and at higher radiation levels than the Geiger counter. Unlike the latter, it can distinguish between different types of emissions because the height of its output current pulse is proportional to the energy contents of emitted particles or radiations. It is therefore more suitable and sensitive for estimating radioactive material present in a sample.

Scintillation counters are of two types namely solid or external scintillation counters and liquid or internal scintillation counters.

Solid or external scintillation counters Scintillation counter consists of a scintillation chamber connected by a light tube to a photomultiplier tube (Figure 12.6).

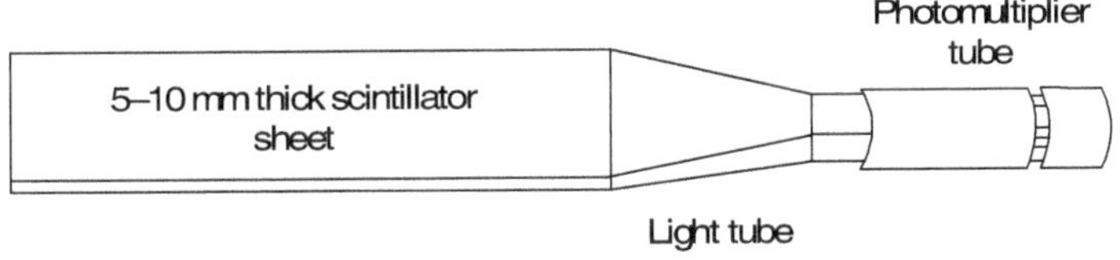

Figure 12.6 Scintillation counter

In solid scintillation counter (Figure 12.7), the sample is placed close to the fluor crystal which in turn is placed adjacent to a photomultiplier. This photomultiplier is connected to a high voltage supply and a scaler. Radiations entering the chamber activate the scintillator to emit visible light photons, which pass through the light tube to fall on the photoelectric cathode surface of the photomultiplier. The photomultiplier converts the optical signal to an electrical one, and provides a large degree of amplification. This consists of an evacuated glass envelope coated at one end with a

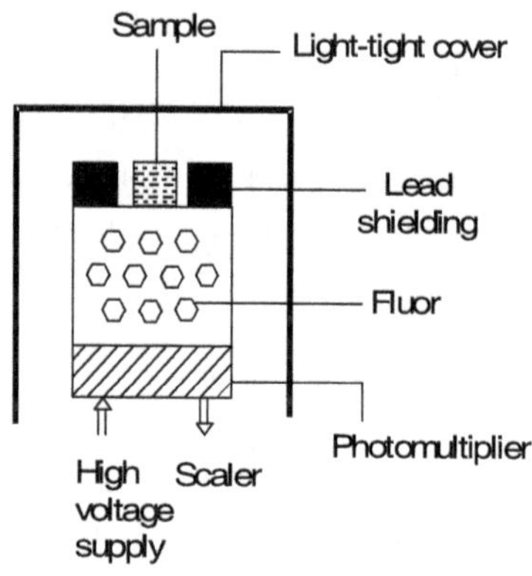

Figure 12.7 Solid scintillation counter

photocathode made of alkali metals, which is maintained at a large negative potential. Photons liberate electrons from the cathode, which are accelerated towards a (less negative) dynode, and here knock out a number of secondary electrons. This process continues down the dynode chain, until eventually a large signal is collected at the anode (Figure 12.8).

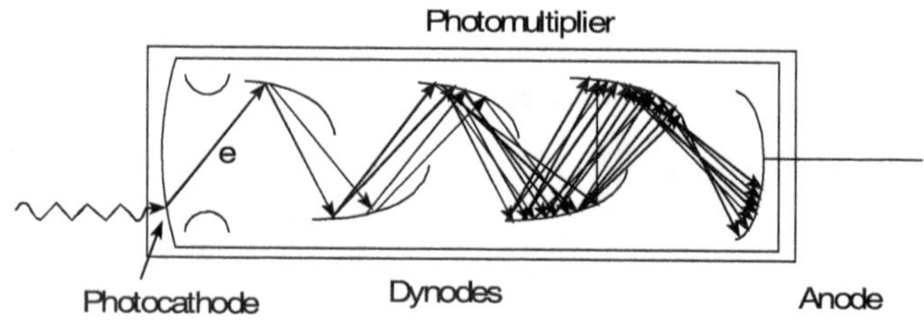

Figure 12.8 Photomultiplier

The vast number of electrons finally impinges on a collector plate to generate a current pulse, which can be visually observed and electronically recorded. Solid scintillation counting is useful for measurement of γ-emitting isotopes.

Liquid or internal scintillation counters Liquid scintillation counter is used to estimate radioactive metabolites separated from tissue samples either in the elute emerging from a chromatographic column or in the fractions obtained by differential or density gradient ultracentrifugation. The fluid containing the radioactive material is thoroughly mixed with a solution of the scintillator for very close contact between the molecules of the two. The mixed liquid is taken in a plastic vial, which is then inserted in a well of the scintillation counter (Figure 12.9). The β-particles and α -rays from radioactive molecules activate the scintillator molecules to emit light, which falls on the photoelectric cathode of the multiplier tube, causing an emission of electrons from the cathode. The emitted electrons, progressively accelerated and multiplied in number in the photomultiplier tube, finally impinge on the plate of the counter to generate current pulses which are recorded.

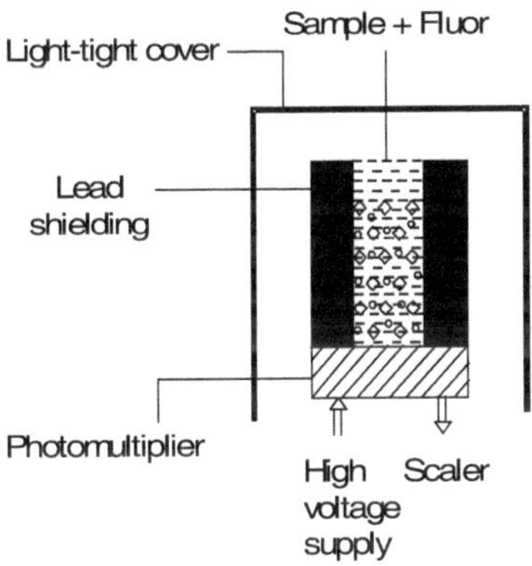

Figure 12.9 Liquid scintillation counter

The greatest disadvantage of scintillation counting is quenching. Reduction in the efficiency of transferring energy from the β-particles to the photomultiplier is called **quenching**. It results in a decreased number of photons per β-particle and the production of pulse of reduced voltage. Quenching may be due to the usage of inappropriate scintillation vials which absorbs some of the light emitted before it reaches photomultiplier. This type of quenching is called **optical quenching**. Coloured samples may cause colour quenching in which the emitted light is absorbed within the scintillation cocktail (the solution of primary and secondary fluor in an organic solvent is called scintillation cocktail). This can be avoided by bleaching the sample before counting. **Chemical quenching** may also occur when the substance in the sample interferes with the transfer of energy from the solvent to the primary fluor or from the primary fluor to the secondary fluor. This is overcome by determining the counting efficiency of each sample by converting the counts per minute to absolute counts (i.e., disintegration per minute).

Scintillation counters are now automated. Hundreds of samples can be counted in a short time and built-in computer facilities carry out data analysis. In liquid scintillation counting, any type of sample (liquids, solids, suspensions and gels) can be counted. The sample preparation is easy. Dual labelling experiments can be carried out. The radioactivity of two different isotopes can be counted in the same sample.

Cerenkov Counting

The Cerenkov effect occurs when a particle passes through a substance with a speed higher than that of light passing through the same substance. If a β-emitter has decay energy in excess of 0.5 MeV, then this causes water to emit a bluish white light usually referred to as Cerenkov light. It is possible to detect this light using a typical liquid scintillation counter. Since there is no requirement for organic solvents and flours, this technique is relatively cheap, sample preparation is very easy, and there is no problem of chemical quenching. Table 12.1 shows some isotopes which are suitable for this detection method. Most work has been done on ^{32}P which has 80% of its energy spectrum above the Cerenkov threshold and which can be detected at around 40% efficiency. It will be noted from Table 12.1 that, as the proportion of the energy spectrum above 0.5 MeV increases, the detection efficiency also increases. So far, few workers have exploited the possibility of combining Cerenkov counting and scintillation counting for pairs of isotopes in a single sample, but this is obviously a possibility. It would simply be necessary to choose two isotopes, one of which has an E_{max} below 0.5 MeV, while the other has much of its energy above this threshold. In particular, it would seem useful to combine a soft β-emitter with a higher energy γ-emitter, since γ-emitters are readily detected by this method, whereas they are not usually efficiently counted by liquid scintillation counting.

Table 12.1 Isotopes suitable for Cerenkov counting

Radioisotope	E_{max} (MeV)	% of spectrum above 0.5 MeV	Counting efficiency
^{22}Na	1.36	60	30
^{32}P	1.71	80	40
^{36}Cl	0.71	30	10
^{42}K	3.50	90	80

Autoradiography

This method can locate radioisotopically labelled molecules in the sections of cells or tissues.

After incubation with the specific radioisotopic compound (e.g. $_1H^3$-labelled thymidine), cells are fixed in a suitable fixative, dehydrated, and embedded in either paraffin wax or plastic. Sections are put in contact with the photographic emulsion for a certain period. During this period, β-particles from the radioisotopes, located at specific sites of the cell, strike the AgBr crystals of the overlying photographic emulsion to change them to silver. During subsequent developing of the photographic emulsion, all unexposed AgBr crystals are dissolved out leaving the dark silver grains in the emulsion layer. Microscopic re-examination of the processed grid or slide reveals the cellular locations of the radioactive compound.

Autoradiography is also used to locate radiolabelled molecules in the chromatographic or electrophoretic bands on polyacrylamide gel blocks.

SAFETY ASPECTS

All ionizing radiations can have an effect on the human body. When radiation damages living cells, it can destroy or mutate the cells, possibly causing a cancerous growth. There is a great difference in the penetrating powers of alpha (α) particles, beta (β) particles and gamma (γ) rays. Of the three types of radiation, alpha (α) particles are the easiest to stop. A sheet of paper is all that is needed for their absorption. However, it may require a material with a greater thickness and density to stop beta (β) particles. Gamma (γ) rays have the most penetrating power of the three radiation types. Hence the most important safety aspect when dealing with radioactive sources is that of shielding—often simply keeping a reasonable distance from the source will be sufficient as the air acts as a shield. When working with more intense sources, some form of shielding may be required. Sealed sources should be handled with tongs or a special source holder—never with the fingers. One should not probe inside sealed sources or allow them to come into contact with any substance, which might attack or dissolve the source or its container. When not in use, sealed sources should always be returned to their lead-lined storage boxes. Radioactive substances should be handled with the same care and respect as concentrated acids. Washing of hands thoroughly after using radioactive sources is very essential.

REVIEW YOUR LEARNING

1. Define the following:
    i. Radioactivity
    ii. Radioactive decay
    iii. Atomic number
    iv. Atomic mass
    v. Radioisotopes
    vi. Curie
    vii. Becquerel

viii. Roentgen
ix. Radiodating
x. Cerenkov effect

2. Explain the following:
   i. Radioisotopes
   ii. Biochemical applications of radioisotopes
   iii. Radiation dosimetry
   iv. Isotope dilution technique
   v. Radiodating
   vi. Radioactive tracer technique
   vii. Radioimmunoassay (RIA)
   viii. Fountain-pen dosimeter
   ix. Geiger–Müller counter
   x. Scintillation counter
   xi. Cerenkov counting
   xii. Autoradiography

BIOSENSORS

A biosensor is an analytical device consisting of a biocatalyst (enzyme, cell or tissue) and a transducer, which can convert a biological or biochemical signal or response into a quantifiable electrical signal. The term 'biosensor' is often used to cover sensor devices used in order to determine the concentration of substances and other parameters of biological interest. Number of biological systems is utilized by biosensors; tissues, bacteria, yeast, enzymes, algae, whole cell metabolism, ligand binding and the antibody–antigen reaction can all be used as the basis of biosensors. Research and development in this field is wide and multidisciplinary, which encompass biochemistry, bioreactor science, physical chemistry, electrochemistry, electronics and software engineering. Most of this current endeavour concerns potentiometric and amperometric biosensors and colorimetric paper enzyme strips.

DEFINITION

A biosensor is defined as a compact analytical device incorporating a biological or biologically-derived sensing element either integrated within or intimately associated with a physico-chemical transducer. The aim of a biosensor is to produce either discrete or continuous digital electronic signals, which are proportional to a single analyte or a related group of analytes.

HISTORY

Professor **Leland Clark** is the father of the biosensor concept. In 1956, **Clark** published his paper on the oxygen electrode. Based on this experience and addressing his desire to expand the range of analytes that could be measured in the body, he made a landmark address in 1962 at a New York Academy of Sciences Symposium in which he described how 'to make electrochemical sensors (pH, polarographic, potentiometric or conductometric) more intelligent' by adding 'enzyme transducers as membrane enclosed sandwiches.' The concept was illustrated by an experiment in which glucose oxidase was entrapped at a Clark oxygen electrode using dialysis membrane. The decrease in measured oxygen concentration was proportional to glucose concentration. In the published paper, **Clark** and **Lyons** coined the term *enzyme electrode*. **Guilbault** and **Montalvo** were the first to detail a potentiometric enzyme electrode. They described a urea sensor based on urease, immobilized at an ammonium-selective liquid membrane electrode. Clark's ideas became commercial reality in 1975 with the glucose analyser based on the amperometric detection of hydrogen peroxide. The use of thermal transducers for biosensors was proposed in 1974 and the new devices were named as *thermal enzyme probes* and *enzyme thermistors*, respectively.

The biosensor took a further fresh evolutionary route in 1975, when **Divis** suggested that bacteria could be harnessed as the biological element in *microbial electrodes* for the measurement of alcohol. This paper marked the beginning of a major research effort in Japan and elsewhere into biotechnological and environmental applications of biosensors. **Lubbers** and **Opitz** coined the term *optode* in 1975 to describe a fibre-optic sensor with immobilised indicator to measure carbon dioxide or oxygen. They extended the concept to make an optical biosensor for alcohol by immobilizing alcohol oxidase on the end of a fibre-optic oxygen sensor. Commercial optodes are now showing excellent performance for *in vivo* measurement of pH, pCO_2 and pO_2, but enzyme optodes are not yet widely available. In 1976, **Clemens** *et al.* incorporated an electrochemical glucose biosensor in an artificial pancreas and this was later marketed by Miles as the Biostator. Although the Biostator is no longer commercially available, based on this a new semi-continuous catheter-based blood glucose analyser has recently been introduced. In the same year, La Roche (Switzerland) introduced the Lactate Analyser, LA 640, in which the soluble mediator, hexacyanoferrate, was used to shuttle electrons from lactate dehydrogenase to an electrode. Although this was not a commercial success at the time, it turned out in an important forerunner of a new generation of mediated-biosensors and of lactate analysers for sports and clinical applications. A major advance in the *in vivo* application of glucose biosensors was reported by Shichiri *et al.*, who described the first needle-type enzyme electrode for subcutaneous implantation in 1982. In 1984, Cass *et al.* published a paper on the use of ferrocene and its derivatives as an immobilised mediator for use with oxidoreductases in the construction of inexpensive enzyme electrodes. This formed the basis for the screen-printed enzyme electrodes launched by MediSense (Cambridge, USA) in 1987 with a pen-sized meter (blood glucose meter) for home blood-glucose monitoring (Figure 13.1).

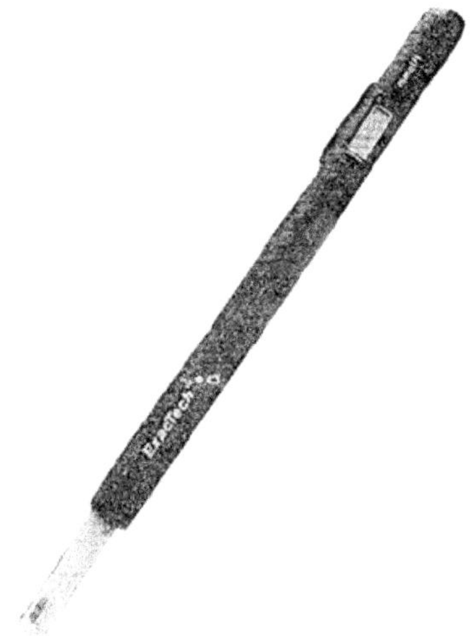

Figure 13.1 Blood glucose meter

The significance of biosensors is likely to increase as the technology develops. This is because they can be made to respond specifically and at high sensitivity to a wide range of molecules, including those of industrial and clinical significance. Pollution by pesticides can be monitored using biosensors. They can also be used for industrial quality control and there are agricultural applications. In the clinical area, one of the most important research areas is the assay of glucose for the management of diabetes. Continued development of this kind of biosensor led to the commercialization of various devices for such applications as the measurement of glucose in blood and the detection of glutamate, aspartate, sulphite, lactose, and ethanol in food products.

COMPONENTS OF BIOSENSORS

The main components of biosensors are the biocatalyst, transducer, amplifier, processor and output (Figure 13.2). The biocatalyst converts the substrate to product. This reaction is determined by the transducer, which converts it to an electrical signal. The output from the transducer is amplified by the amplifier and processed by the processor and displayed.

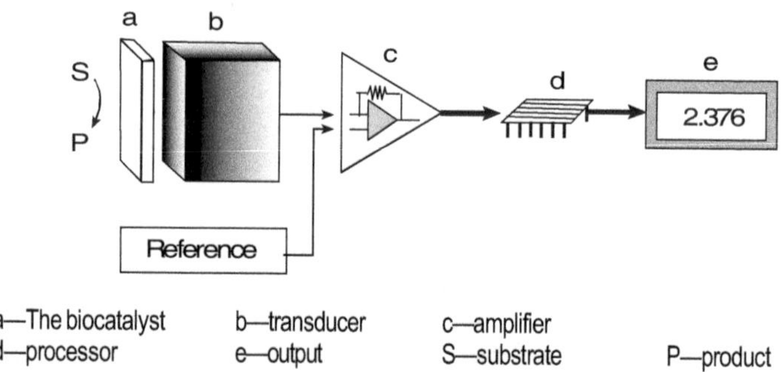

a—The biocatalyst b—transducer c—amplifier
d—processor e—output S—substrate P—product

Figure 13.2 Schematic diagram showing the main components of a biosensor

The key part of a biosensor is the transducer, which makes use of a physical change accompanying the reaction. This may be

1. The heat output (or absorbed) by the reaction (calorimetric biosensors)

2. Changes in the distribution of charges causing an electrical potential to be produced (potentiometric biosensors)
3. Movement of electrons produced in a redox reaction (amperometric biosensors)
4. Light output during the reaction or a light absorbance difference between the reactants and products (optical biosensors)
5. Effects due to the mass of the reactants or products (piezoelectric biosensors).

Types of transducer used in biosensors are given in the Table 13.1.

Table 13.1 Types of transducers used in biosensors

Type	Operating principles	Reactions or molecules detected
Amperometric	A voltage is set and if the molecule is present, a current flows	O_2 (using Pt at –0.8 V) H_2O_2 (using Pt at +0.8 V) I_2, NADH
Ion-selective electrode/pH electrode	Potential depends on ion concentration	H^+; Na^+; Cl^-
Gas-sensing electrode	Potential depends on gas concentration	CO_2; NH_3
Photomultipliers with fibre optics for bioluminescence	Light emission detected	If luciferase present, ATP detected
Photomultipliers with photodiode	Light absorption detected, e.g. pH changes cause a change in colour of a dye	Penicillin; urea

(Contd.)

Table 13.1 (Continued)

Type	Operating principles	Reactions or molecules detected
Thermistor	Heat of reaction detected	Almost universal
Piezoelectric crystal	Mass absorption detected	Reactions involving volatile gases and vapours
Miniature field effect transistor (FET)	Microelectronic device	Can be made into an ion-selective field effect transistor (ISFET) to detect ions and a chemically sensitive field effect transistor (CHEMFET) to detect molecules
Non-specific ionic conductance	Measures increase in total number of ions	Many reactions, e.g. urease

A successful biosensor must possess at least some of the following beneficial features:

1. The biocatalyst must be highly specific for the purpose of the analyses and stable under normal storage conditions.
2. The reaction should be independent of physical parameters such as stirring, pH and temperature. This would allow the analysis of samples with minimal pretreatment. If the reaction involves cofactors or coenzymes these should, preferably, also be co-immobilized with the enzyme.
3. The response should be accurate, precise, reproducible and linear over the useful analytical range, without dilution or concentration. It should also be free from electrical noise.
4. If the biosensor is to be used for invasive monitoring in clinical situations, the probe must be tiny and

biocompatible, having no toxic or antigenic effects. If it is to be used in fermenters it should be sterilizable. This is preferably performed by autoclaving but no biosensor enzymes can presently withstand such drastic wet-heat treatment. In either case, the biosensor should not be prone to fouling or proteolysis.

5. The complete biosensor should be cheap, small, portable and capable of being used by semi-skilled operators.

Biosensors may be categorized as first-, second- or third-generation instruments according to the degree of intimacy between the biocatalyst and transducer. In first-generation instruments the two components, biocatalyst and transducer, may be easily separated and both may remain functional in the absence of the other. In second-generation instruments the two components interact in a more intimate fashion and removal of one of the two components affects the usual functioning of the other. In third-generation instruments the biochemistry and electrochemistry are even more closely linked and where the electrochemistry occurs at a semiconductor the term biochip may be applied to describe such instruments.

TYPES OF BIOSENSORS

A wide variety of devices have been developed exploiting enzymes, nucleic acids, cell receptors, antibodies and intact cells, in combination with electrochemical, optical, piezoelectric and thermometric transducers. Within each permutation lies a myriad of alternative transduction strategies and each approach can be applied to numerous analytical problems in health care, food and drink, the process industries, environmental monitoring, defense and security.

Enzyme Electrodes

It is possible to exploit the specificity and diversity of enzymes to estimate the concentration of many different substrates. Some examples are given in the Table 13.2. Enzyme electrodes can either be made to dip into a solution (this type is called a bioprobe) or can be made into a column (or sometimes a porous pad) and the solution is passed through it (forming a bioreactor or a flow system).

In bioprobes, the enzyme is immobilized by entrapment in a gel (e.g. polyacrylamide), by encapsulation, by ionic, adsorptive or covalent binding to carriers or by cross-linking either to another molecule of the same enzyme or to another protein, e.g. serum albumin. Glutaraldehyde, a bifunctional reagent, can be used to form cross-links or to give covalent binding, e.g. to a teflon or nylon membrane. The substance to be measured, which is usually the substrate for the enzyme, diffuses to the enzyme where a product is formed which then affects the electrode. For example, a penicillin electrode uses penicillinase, which degrades penicillin and causes a pH change that is recorded using a pH electrode.

Enzyme electrodes normally have to be kept refrigerated to keep the lifetime as long as possible; even so, few have a lifetime of over a month. With some electrodes there is a problem of choosing the operating pH because the optimum pH for the enzyme reaction may be different from the optimum pH for detecting the product. A compromise pH may have to be selected.

Enzyme electrodes do not give an instant response. This is because the substrate has to diffuse through the membrane, and the product then has to diffuse to the transducer (sensing device). However, the response using immobilized enzymes is usually faster than that using immobilized cells where the cell membrane forms an additional barrier.

Bacterial Electrodes (Cell-based Biosensors)

Some of the disadvantages of enzyme electrodes may be overcome by using bacterial electrodes (cell-based biosensors). For example, bacterial electrodes are less sensitive to inhibition by solutes and are more tolerant of sub optimal pH and temperature values than the enzyme electrodes. Bacterial electrodes also tend to have a longer lifetime than enzyme electrodes (20 days or more compared with an average of 14 days). They are cheaper because an active enzyme

Table 13.2 Biosensors

Compound detected	Biological material used	Sensor	Immobilization	Stability	Response time
Alcohol	Alcohol oxidase	O_2	Glutaraldehyde	2 weeks	1–2 min.
β-glucose	β-Glucose oxidase	O_2	Chemical	3 weeks	1 min.
Glutamate	Glutamate decarboxylase	CO_2	Glutaraldehyde	1 week	10 min.
Penicillin	Penicillinase	pH	Polyacrylamide	2 weeks	15–30 sec.
Urea	Urease	NH_4	Polyacrylamide	19 days	20–40 sec.
Arginine	*Streptococcus faecalis*	NH_3	Physically entrapped	20 days	20 min.
Cholestrol	*Nocardia erythropolis*	O_2	Physically entrapped	4 weeks	35–70 sec.
Nitrate	*Azotobacter vinelandii*	NH_3	Physically entrapped	2 weeks	7–8 min.
NAD^+	NADase + *Escherichia coli*	NH_3	Dialysis membrane	1 week	5–10 min.

does not have to be isolated. However, cells contain many enzymes and care has to be taken to ensure selectivity, e.g. by optimizing storage conditions or adding specific enzyme or transport inhibitors to stop undesirable enzyme reactions. Mutant bacteria lacking enzymes can be used. Bacterial electrodes do have some disadvantages. For example, some have a longer response time than enzyme electrodes, but a more serious problem is the time taken for cell-based sensors to return to a baseline potential after use.

When bacterial electrodes were first developed, they involved the use of rather harsh techniques for the immobilization such as polyacrylamide gel, or used cells whose permeability had been increased. Thus the cells were generally not viable. However, the enzymes within them were still active. More recent immobilization techniques have tended to use more gentle physical methods so that viability is retained. The advantage of this is that such cells may be involved in converting substrate into a product via a multi enzymes pathway, without having to immobilize each of the enzymes and then provide them with expensive coenzymes.

More than one kind of bacterium can be incorporated into one electrode, which increases the number of potential applications. Thus biochemical oxygen demand (BOD) due to organic matter in wastewater is detected by a mixed culture of bacteria obtained from soil since a single microorganism species would be unable to use all the organic compounds likely to be found in the sample.

It is also possible to combine an enzyme preparation with a microorganism. It is a detector for NAD^+. *Escherchia coli* cells provide the enzyme nicotinamide deaminase and it is combined with NADase from *Neurospora crassa*. The following reaction occurs.

$$\text{NAD} + \text{H}_2\text{O} \xrightarrow{\text{NADase}} \text{nicotinamide} + \text{ADP–ribose}$$

$$\text{Nicotinamide} + \text{H}_2\text{O} \xrightarrow{\text{nicotinamide deaminase}} \text{nicotinic acid} + \text{NH}_3$$

A gas-sensing electrode detects ammonia released.

The detection of cholesterol is of great clinical importance, and can be carried out by using *Nocardia erythropolis* immobilized in polyacrylamide on agar on an oxygen electrode. The reaction is carried out as follows:

$$\text{Cholesterol} + O_2 \xrightarrow{\text{Cholesterol oxidase}} \text{Cholest - 4 - en - 3 - one} + H_2O_2$$

The oxygen electrode measures the rate of oxygen uptake and this can be related to the cholesterol content of a biological sample (plasma).

Like enzyme electrodes, bacterial electrodes can be made into bioprobes or into bioreactors. Bioreactors are very suitable for the commercial production of metabolites, but can also be used as biosensors. In future, it may be possible to increase the life span of bacterial electrodes by stimulating growth of fresh cells on the membrane surface.

Enzyme Immunosensors

Enzyme-immunosensors that combine the molecular recognition properties of antibodies with high sensitivity of enzyme based analytical methods of several kinds have been developed. The enzyme is used as a marker as it reacts with its substrate, giving a change, which can be detected by a transducer of some kind. There is a similarity between this method and ELISA.

The enzyme-immunosensor for IgG consists of anti-IgG antibody bound to a membrane and linked to an oxygen electrode. Catalase is bound to the IgG and so forms a label for it. This complex is then mixed with the sample that contains an unknown amount of unlabelled IgG. The labelled and unlabelled IgG compete for the antibody on the membrane. After exposure to test an IgG solution, the sensor is rinsed to remove any non-specifically associated IgG and is then immersed in H_2O_2 solution as a substrate for the catalase. The more unlabelled IgG present, the lower the amount of catalase present and therefore the lower the rate of oxygen evolution. A similar assay can be carried out for Human Chorionic Gonadotropin (HCG), a diagnostic hormone for pregnancy. Catalase is again used to label the HCG, and oxygen evolution is followed as before.

In attempts to construct enzyme immunosensors, bioluminescence, chemiluminescence and fluorescence principles are being explored, because of their great sensitivity. A luminescent immunoassay with catalase has been used to detect human serum albumin at 1 ng cm^{-3}.

Biosensors to Detect Cancer and Health Abnormalities

In the past, determining accurately whether a patient has cancer of the oesophagus has required surgical biopsy. The laser-based fluorescence method developed by **Tuan Vo-Dinh** *et al.* has eliminated the need for biopsy, reducing pain and recovery time for patients. In laser-based fluorescence method, laser light of the appropriate wavelength is directed to the inner surface of the oesophagus by means of a fiber-optic device that is swallowed by the patient. The epithelial cells and tissue inside the oesophagus fluoresce when excited by the laser light. When the oesophagus interior is illuminated with blue light (410 nm), the normal tissue emits light at wavelengths different from those emitted by the cancer cells. Emissions from normal cells and cancer cells can be distinguished quite accurately.

Biosensors to Monitor the Status of Diabetes without using Blood Samples

Another biosensor developed by **Vo-Dinh**, provides a way to monitor the status of diabetes without using blood samples. In this case, light is used to illuminate the eyeball and stimulate certain substances, including proteins, to emit fluorescent light. This fluorescence changes in intensity and wavelength when the distribution and status of proteins in the eye change. This truly non-invasive method depends on a relatively new development for selecting the wavelength of light for illumination. Instead of prisms or gratings to refract the light into different wavelengths, a device called the acousto-optic tunable filter (AOTF) is used. One AOTF selects the wavelength of light to shine on the eyeball and another selects the wavelength of fluorescent light emitted from the eyeball. Both AOTF wavelengths are scanned simultaneously using the synchronous luminescence technique developed previously for environmental screening. The AOTFs, which are manipulated with a radio-frequency signal, can scan the entire visible spectrum and portions of the ultraviolet and infrared spectra in milliseconds to select the appropriate wavelengths to use to illuminate the eyeball. They can also select the correct wavelength to use to read the fluorescence signal from the patient's eye instantaneously. In this way many spectral scans can be taken and averaged in a computer to obtain the accuracy required to measure the status and changes in the eye proteins of diabetics.

Environmental Biosensors

Recently, researchers have investigated biosensors and ancillary technologies for environmental applications, especially those that require continuous monitoring. The potential for environmental applications lies in the ability of biosensors to measure the interaction of pollutants with biological systems through a biomolecular recognition capability. Biosensors are currently available for monitoring biochemical oxygen demand (BOD) and

Figure 13.3 A new biosensor to monitor explosives such as TNT and RDX has been developed by the U.S. Naval Research Laboratory. The continuous-flow immunosensor uses a clear plastic disposable immunoassay cartridge that is inserted into an analytical system (black box); measurements are converted into database format with associated signal acquisition and analysis software.

are in use at water treatment facilities in Europe and Japan. Recently, biosensors for 2, 4, 6-trinitrotoluene (TNT) and RDX (hexahydro-1, 3, 5-trinitro-1, 3, 5-triazine) have been used at the U.S. Naval Research Laboratory (Figure 13.3). Other promising applications for environmental biosensors include groundwater monitoring, drinking-water analysis, and the rapid analysis of extracts of soils and sediments at hazardous waste sites.

Bioreporters

Yet another example of a biosensor is based on detection of light emitted by a specially engineered microorganism that is involved in bioremediation. This type of biosensor is called bioreporters (Figure 13.4). In this case the light originates from a particular protein that has been installed in certain bacteria by modern molecular genetic methods. In one case, the gene for luciferase is placed in the operon (a sequence of genes that specify enzymes that carry out a related series of metabolic steps) that is responsible for

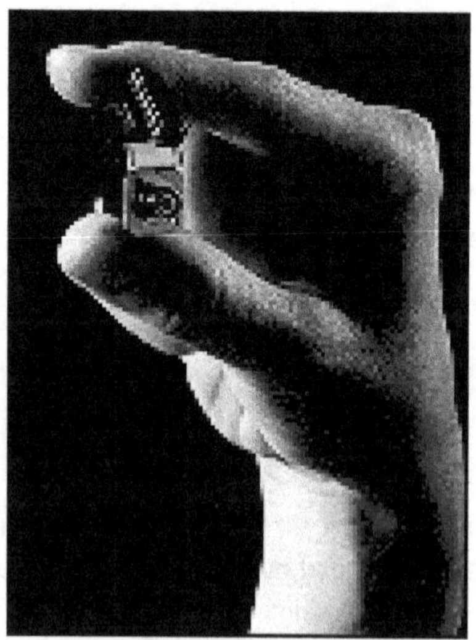

Figure 13.4 Bioreporter: "critters on a chip" in which bioluminescent bacteria signal the presence of pollutants

degrading unwanted chemicals such as toluene, an organic solvent. When the bacteria are metabolizing the toluene, the genetic control mechanism also turns on the synthesis of the enzyme luciferase, which produces light in the presence of oxygen. To deal with situations in which bacteria degrade an organic solvent under conditions of very limited oxygen a "green fluorescent protein," is used. In this case, a "green fluorescent protein," which emits green light (with a wavelength of 509 nm) when excited by blue light (395 nm), is installed in the operon. No oxygen is required for the light emission. Again, as the toluene is metabolized by the enzymes synthesized for that activity, the green fluorescent protein is also produced. Because it is active, it can be monitored by remote light activation and spectral emission analysis.

Miniaturized Devices

Another class of biosensors uses various techniques to turn a biological system into a tiny electronic device, to analyse biological or physiological processes, or to detect and identify bacteria. Some of these techniques produce or are carried out in miniaturized devices.

Leaf disc electrode The site for photosynthesis in a green leaf contains a complex set of enzymes and proteins that capture light energy and convert carbon dioxide into compounds that help the plant to grow. If a platinum salt in a certain oxidation state is supplied to one of two photosynthetic systems in plant chloroplasts, one photosynthetic reaction system will use light energy to provide electrons that will reduce platinum to the metal form. The metal is deposited on the photosystem complex to form a tiny platinum centre that can be employed in sophisticated diode-based microelectronics for measurements of light energy at extremely high sensitivity, resolution, and speed. Such a biomolecular optoelectronic sensor has been demonstrated by Eli Greenbaum, James Lee, Ida Lee, and Steve Blankinship.

Medical telesensors Blood pressure and pulse rate may be measured by chips designed to detect pressure changes. If a silicone fibre on a chip can sense pressure at various positions in the body, it may be used for monitoring blood pressure, pulse rate, breathing (chest expansion), knee bending during physical rehabilitation, and foot pressure distribution.

A chip (medical telesensor) on the fingertip measures and transmits data on the body temperature (Figure 13.5). An array of chips attached to your body may provide additional information on blood pressure, oxygen level, and pulse rate. This type of medical telesensor, which is being developed for military troops in combat zones, will report measurements of vital functions to remote recorders. The goal is to develop an array of chips to collectively monitor bodily functions. These chips may be attached at various points on a soldier using a non-irritating adhesive like that used in waterproof band-aids. These medical telesensors would send physiological data by wireless transmission to an intelligent monitor on another soldier's helmet. The monitor could alert medics if the data showed that the soldier's condition fit one of five levels of trauma. The monitor also would receive and transmit global satellite positioning data to help medics locate the wounded soldier.

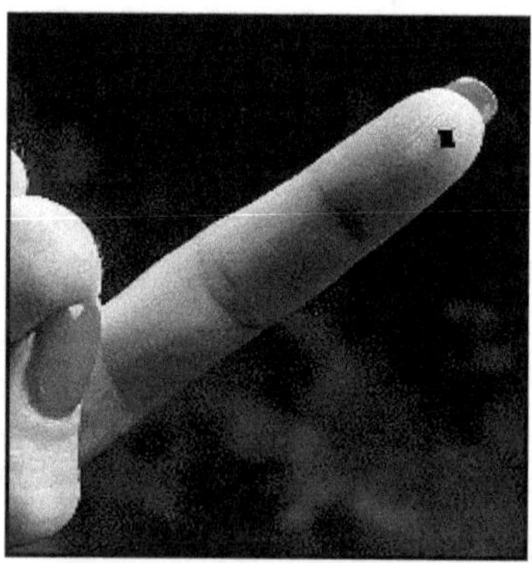

Figure 13.5 Medical telesensor

DNA can be used to identify organisms ranging from humans to bacteria and viruses. DNA sequence information would be used to determine whether a person is a carrier of a disease gene known to exist in their family, whether blood at a crime scene was that of the arrested suspect, or whether a biowarfare bacterium was descending on a battlefield.

REVIEW YOUR LEARNING

1. Explain the principle and application of the following:
    i. Biosensors
    ii. Enzyme electrodes
    iii. Cell-based biosensors
    iv. Environmental biosensors
    v. Bioreporters
    vi. Medical telesensors
2. What are various types of transducers used in biosensors?
3. Explain the use of biosensors in the field of medicine.

DNA SEQUENCING

DNA sequencing, first devised in 1975, has become a powerful technique in molecular biology, allowing analysis of genes at the nucleotide level. DNA sequencing is the determination of all or part of the nucleotide sequence of a specific deoxyribonucleic acid (DNA) molecule. For this reason, this tool has been applied to many areas of research. For example, the polymerase chain reaction (PCR), a method which rapidly produces numerous copies of a desired piece of DNA, requires first knowing the flanking sequences of this piece. Another important use of DNA sequencing is identifying restriction sites in plasmids. Knowing these restriction sites is useful in cloning a foreign gene into the plasmid. Before the advent of DNA sequencing, molecular biologists had to sequence proteins directly; now amino acid sequences can be determined more easily by sequencing a piece of cDNA and finding an open reading frame. In eukaryotic gene expression, sequencing has allowed researchers to identify conserved sequence motifs and determine their importance in the promoter region. Furthermore, a molecular biologist can utilize sequencing to identify the site of a point mutation. These are only a few examples illustrating the way in which DNA sequencing has revolutionized molecular biology.

The fundamental reasons for which the DNA molecule is sequenced are:

i. To make predictions about its function
ii. To facilitate manipulation of the molecule

Several methods are available now for sequencing nucleic acids. All these methods involve the following basic aspects.

POLYACRYLAMIDE GEL ELECTROPHORESIS (PAGE)

Sequencing is carried out in PAGE. Generally 5–8% polyacrylamide gel capable of resolving fragments which vary in length by only one nucleotide, is used. Depending on the requirement of experiment, as many as 1000 nucleotide fragments are analysed in a single gel.

Unlike proteins, DNA molecules have lesser charge density (net charge per unit length). Secondly, as they are subjected to a long separation distance, higher electrical field strength is required to run desirable time. So that, power supply capable of delivering up to 5000 volts with extreme safety devices is required.

Passing electricity at high volts in sequencing gel generates enormous heat. Excess heat affects not only the gel but also the separation pattern. So the sequencing apparatus must have an efficient cooling system and/or uniform heat-dissipating devices.

Sequencing reactions are often carried out with radiolabelled (^{32}P, ^{35}S) nucleotides. When sample is in contact with the buffer, it is likely to be contaminated with radioactive materials. So in any circumstances there should not be any leakage of buffer from the top chamber.

METHODS OF GENE SEQUENCING

Several methods are adopted for gene sequencing.

Manual Methods of DNA Sequencing

There are two methods in use, the deoxy or chain termination method of F. Sanger and the chemical cleavage method of A. Maxam and W. Gilbert. Both methods are based on high-resolution electrophoresis of four sets of radioactive oligonucleotides produced from the DNA to be sequenced, but they differ in the procedures used to generate the oligonucleotides.

Sanger's Method for DNA Sequencing Dideoxynucleotide sequencing is the method of sequencing of DNA devised by Sanger. This technique utilizes 2´,3´-dideoxynucleotide triphosphates (ddNTPs), molecules that differ from deoxynucleotides by having a hydrogen atom attached to the 3´ carbon rather than an OH group (Figure 14.1). These molecules terminate DNA chain elongation because they cannot form a phosphodiester bond with the next deoxynucleotide.

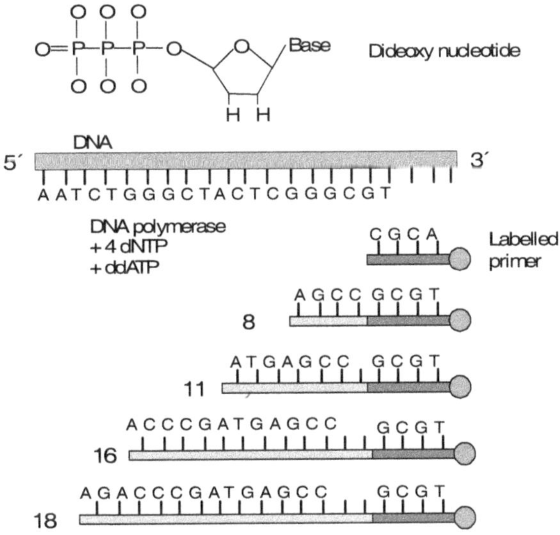

Figure 14.1 Structure of a dideoxynucleotide (notice the H atom attached to the 3´ carbon), the ingredients for a Sanger reaction. Notice the different lengths of labelled strands produced in this reaction.

In order to perform the sequencing, the double-stranded DNA should be converted into single-stranded DNA. This can be done by denaturing the double-stranded DNA with NaOH. A Sanger reaction mixture consists of the following: a strand to be sequenced (one of the single strands which was denatured using NaOH), DNA primers (short pieces of DNA that are both complementary to the strand which is to be sequenced and radioactively labelled at the 5´ end), a mixture of a particular ddNTP (such as ddATP) with its normal dNTP (dATP in this case), and the other three dNTPs (dCTP, dGTP, and dTTP) (Figure14.1). The concentration of ddATP should be 1% of the concentration of dATP. The logic behind this ratio is that after DNA polymerase is added, the polymerization will take place and will terminate whenever a ddATP is incorporated into the growing strand. If the ddATP is only 1% of the total concentration of dATP, a whole series of labelled strands will result.

This reaction is performed four times using a different ddNTP for each reaction. When these reactions are completed, a polyacrylamide gel electrophoresis (PAGE) is performed (Figure 14.2). The gel is transferred to a nitrocellulose filter and autoradiography is performed so that only the bands with the radioactive label on the 5´ end will appear. In PAGE, the shortest

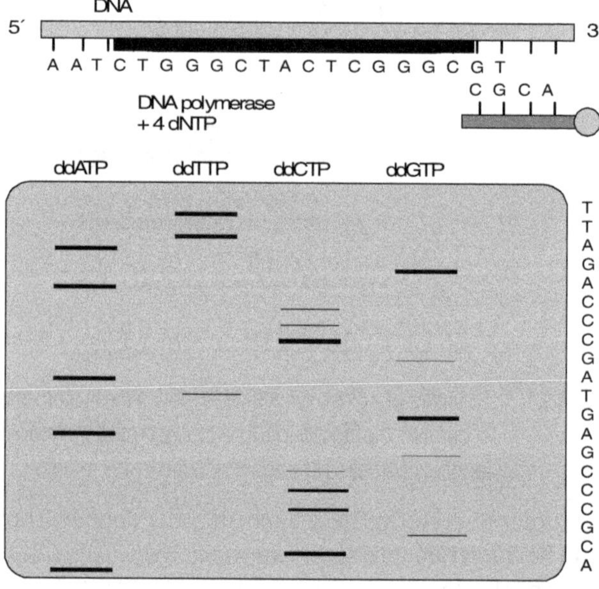

Figure 14.2 Gene sequencing in polyacrylamide gel

fragments will migrate the farthest. Therefore, the bottom-most band indicates that its particular dideoxynucleotide was added first to the labelled primer. In Figure 14.2, for example, the band that migrated the farthest was in the ddATP reaction mixture. Therefore, ddATP must have been added first to the primer, and its complementary base, thymine, must have been the base present on the 3′end of the sequenced strand. One can continue reading in this fashion.

The sequence of the strand of DNA complementary to the sequenced strand is 5′ to 3′ ACGCCCGAGTAGCCCAGATT while the sequence of the sequenced strand, 5′ to 3′, is AATCTGGGCTACTCGGGCGT.

Maxam and Gilbert sequencing The chemical cleavage method of Maxam and Gilbert usually starts with the enzymic addition of a radioactive label to either the 3′ or the 5′ ends of a double-stranded DNA preparation. The strands are then separated by electrophoresis under denaturing conditions, and analysed separately. DNA labelled at one end is split into four portions and each is treated with chemicals that will act on a specific base (or, in some cases, either of two bases) by methylation or removal of the base. Conditions are chosen so that, on average, each molecule is modified at only one position along its length; every base in the DNA strand has an equal chance of being modified. After the modification reactions, the separate samples are cleaved by piperidine, which breaks phosphodiester bonds exclusively at the 5′ side of nucleotides whose base has been modified. The result is similar to that produced by Sanger method, since each sample now contains radioactive molecules of various lengths, all with one end in common (the labelled end), and with the other end cut at the same type of base. Analysis of the reaction products by electrophoresis is as described for the Sanger method.

Dye Terminator Sequencing

An alternative to labelling the primer is to label the terminators instead, commonly called **dye terminator sequencing**. The major advantage of this approach is that the complete sequencing set can be performed in a single reaction, rather than the four needed with the labelled-primer approach. This is accomplished by labelling each of the dideoxynucleotide chain terminators with a separate fluorescent dye, which fluoresces at a different wavelength. This method is easier and quicker than the dye primer approach, but may produce more uneven data peaks (different heights), due to a template-dependent difference in the incorporation of the large dye chain-terminators. This problem has been significantly reduced with the introduction of new enzymes and dyes that minimize incorporation variability. This method is now used for the vast majority of sequencing reactions, as it is both simpler and cheaper.

To sequence DNA by this method, a primer is annealed to the single strand template near the 3´ end of the inserted foreign DNA. The DNA to be sequenced is prepared as a single strand.

This template DNA is supplied with

- a mixture of all four normal (deoxy) nucleotides in ample quantities—dATP, dGTP, dCTP, dTTP.
- a mixture of all four dideoxynucleotides, each present in limited quantities and each labelled with a "tag" that fluoresces a different colour—ddATP, ddGTP, ddCTP, ddTTP (chemical structures of dTTP and ddTTP are shown in the Figure 14.3).
- DNA polymerase I

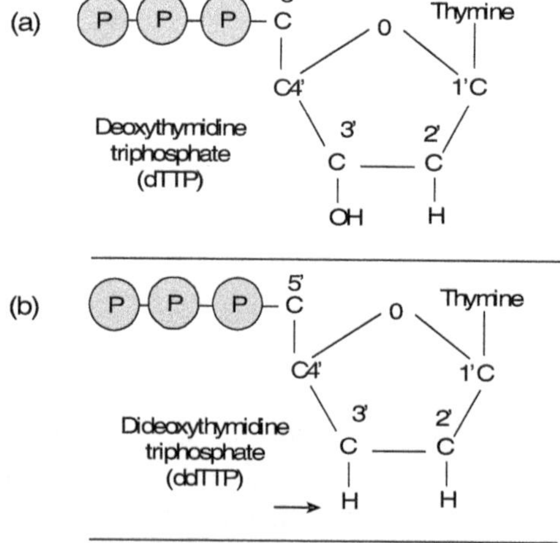

Figure 14.3 (a) Deoxythymidine triphosphate (dTTP), (b) Dideoxythymidine triphosphate (ddTTP)

Since all four normal nucleotides are present, chain elongation proceeds normally until, by chance, DNA polymerase inserts a dideoxynucleotide, instead of the normal deoxynucleotide. If the ratio of normal nucleotide to the dideoxy versions is high enough, some DNA strands will succeed in adding several hundred nucleotides before insertion of the dideoxy version halts the process.

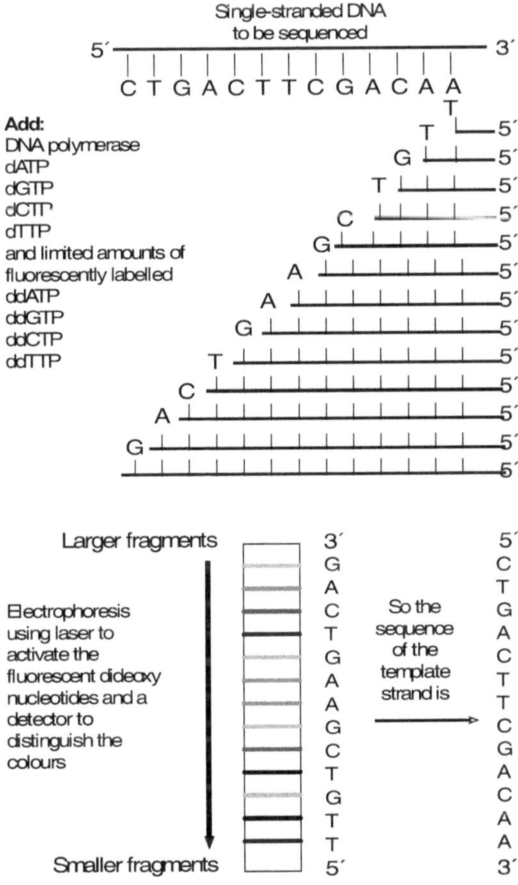

Figure 14.4 Dye terminator sequencing

At the end of the incubation period, the fragments are separated by length from longest to shortest. The resolution is so good that a difference of one nucleotide is enough to separate that strand from the next shorter and next longer strand. Each of the four dideoxynucleotides fluoresces a different colour when illuminated by a laser beam and an automatic scanner provides a printout of the sequence (Figure 14.4).

Sequencing by Direct Blotting Electrophoresis (Semi-automatic)

Beck (1986) described a method called 'Direct blotting electrophoresis'. This is a simultaneous method for both electrophoretic separation and blotting of nucleotides. As the nylon membrane moves in contact with the bottom of the gel, nucleotides are automatically blotted on it. The membrane is then accessible for different detection methods like colorimetric or chemiluminescent methods or autoradiography. Autoradiogram results obtained with blotted membranes are sharper than those made with a dried or wet gel because fragments blotted on the membrane have a direct contact with the film. The exposure time is also less. For example, in a six-hour time, as many as 400 bases can be read with 95% accuracy.

Automated DNA Sequencer

Gene and genome sequencing involve a variety of computers, software programs, automated sequencing machines, fluorescent dyes, lasers, and other tools. The development of machines that can quickly chop up, separate, realign, and read bits of DNA have greatly speeded up the sequencing process. Using automation and high technology, gene-sequencing machines have compressed the time to sequence genes into hours, which was once a tedious, painstaking process, and took even a year to sequence the gene.

Automated DNA sequencer was employed by Middendorf *et al.* (1992) for a highly accurate DNA sequencing and fragment analysis. In this method DNA fragments are labelled with a carbocyanine fluorescent dye, Cy5. While they are separating in the gel, a laser beam is passed on to induce the fluorescence. Detection of fluorescence by photodiodes in each lane indicates the presence of labelled fragment. The data is then processed by a computer to get maximum sequence information. The automated system has the following essential components:

1. Sequencing apparatus for running PAGE.
2. Detector, long-life red helium, neon laser (633 nm) which penetrates the entire width of the gel.
3. Photodiodes, fixed one for each lane located at right angles to the laser beam measures the fluorescence emitted by the Cy5-labelled DNA fragments as they pass the detector.
4. A computer and suitable software to analyse the data obtained from the detector.

Recently, by using automated infrared DNA sequencer, analysis of DNA fragments up to 100 bp is possible. Applications of automated sequencer include, fragment analysis for genetic and mutational analyses such as detection of restriction fragment length polymorphisms (RFLP), heterozygosity and various other sequence polymorphisms.

Sequencing on an automated sequencer The purpose of sequencing is to determine the order of the nucleotides of a gene. For sequencing, PCR fragments or cloned genes are used (Figure14.5).

1. *The sequencing reaction* There are three major steps in a sequencing reaction (like in PCR), which are repeated for 30 or 40 cycles.

   i. *Denaturation at 94°C* During denaturation, the double-stranded DNA melts open to single-stranded DNA.

   ii. *Annealing at 50°C* In sequencing reactions, only one primer is used, so there is only one strand copied (in PCR : two primers are used, so two strands are copied). The primer is jiggling around, caused by the Brownian motion. Ionic bonds are constantly formed and broken between the single-stranded primer and the single-stranded template. The more stable bonds last a little bit longer (primers that fit exactly) and on that little piece of double-stranded DNA (template and primer), the polymerase can attach and starts copying the template. Once there are a few bases built in, the ionic bond is so strong between the template and the primer that it does not break anymore.

iii. *Extension at 60°C* This is the ideal working temperature for the polymerase. Since it has to incorporate ddNTP which are chemically modified with a fluorescent label, the temperature is lowered so that it has time to incorporate the 'strange' molecules. The primers, where there are a few bases built in, already have a stronger ionic attraction to the template than the forces breaking these attractions. Primers that are on positions with no exact match, come loose again and do not give an extension of the fragment. The bases (complementary to the template) are coupled to the primer on the 3´ side (adding dNTPs or ddNTPs from 5´ to 3´, reading from the template from 3´ to 5´ side, bases are

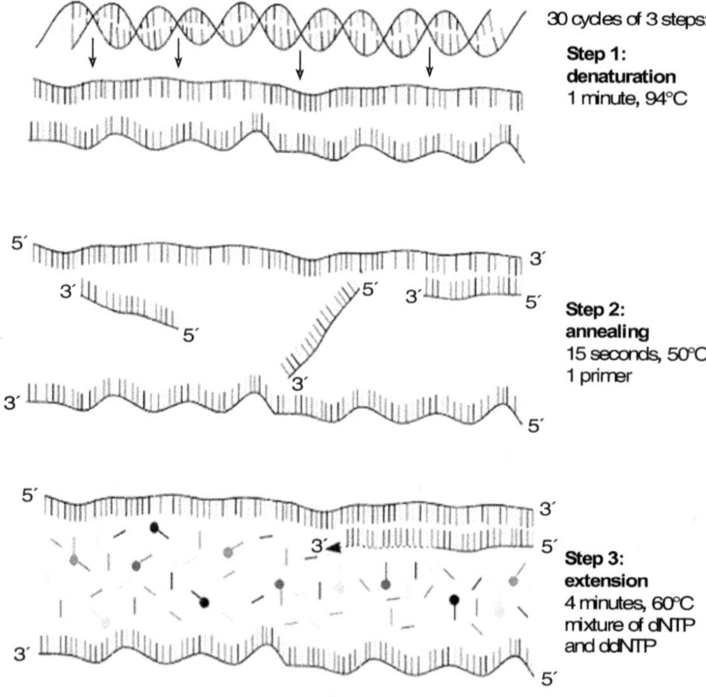

Figure 14.5 The different steps in sequencing

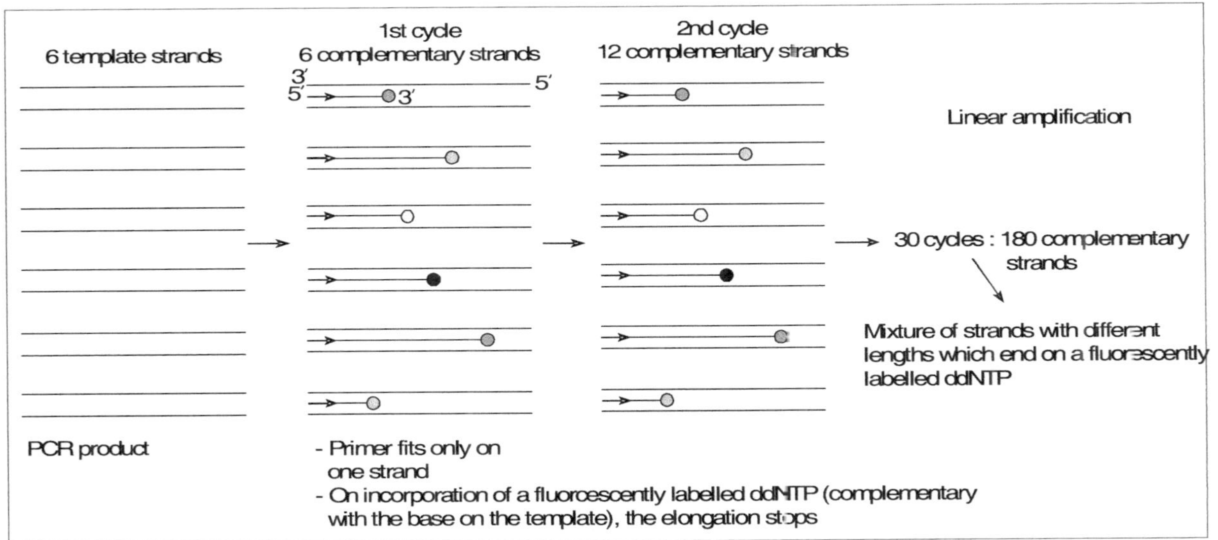

Figure 14.6 The linear amplification of the gene in sequencing

added complementary to the template). When a ddNTP is incorporated, the extension reaction stops because a ddNTP contains a H-atom on the 3rd carbon atom (dNTPs contain an OH atom on that position). Since the ddNTPs are fluorescently labelled, it is possible to detect the colour of the last base of this fragment on an automated sequencer.

Because only one primer is used and only one strand is copied during sequencing, there is a linear increase of the number of copies of one strand of the gene (Figure 14.6). Therefore, there has to be a large amount of copies of the gene in the starting mixture for sequencing. Suppose there are 1000 copies of the wanted gene before the cycling starts, after one cycle, there will be 2000 copies: 1000 original templates and 1000 complementary strands with each one fluorescent label on the last base, after two cycles, there will be 2000 complementary strands, three cycles will result in 3000 complementary strands and so on.

2. *Separation of the molecules* After the sequencing reactions, the mixture of strands, all of different lengths and all ending on a fluorescently labelled ddNTP have to be separated. This is done onan acrylamide gel which is capable of separating a molecule of30 bases from one of 31 bases, and also a molecule of 750 bases

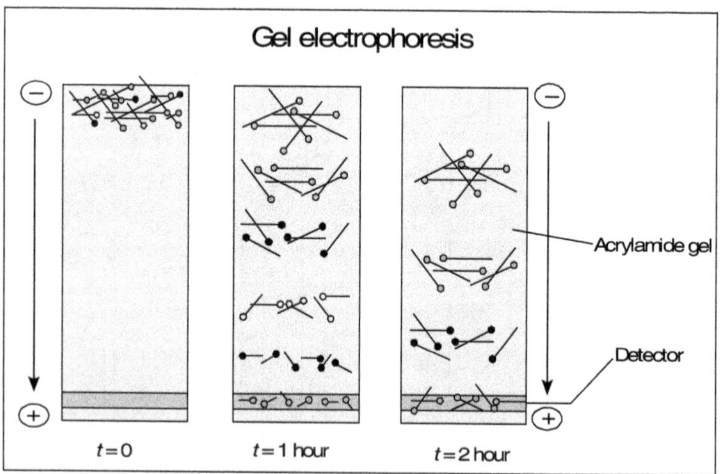

Figure 14.7 The separation of the molecules with electrophoresis

from one of 751 bases. All this is done with gel electrophoresis (Figure 14.7). DNA has a negative charge and migrates to the positive side. Smaller fragments migrate faster, so the DNA molecules are separated on their size.

3. *Detection on an automated sequencer* The fluorescently labelled fragments that migrate through the gel pass through a laser beam at the bottom of the gel. The laser excites the fluorescent molecule which sends out light of a distinct colour. That light is collected and focused by lenses into a spectrograph. Based on the wavelength, the spectrograph separates the light across a CCD camera (charge-coupled device) (Figure 14.8). Each base has its own colour, so the sequencer can detect the order of the bases in the sequenced gene (Figure 14.9).

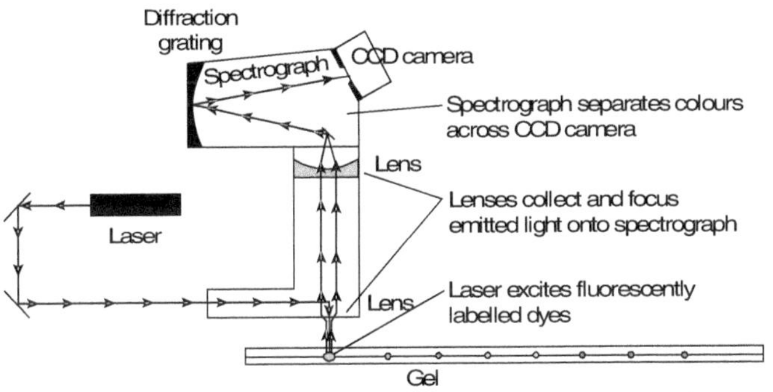

Figure 14.8 The scanning and detection system on the automated sequencer

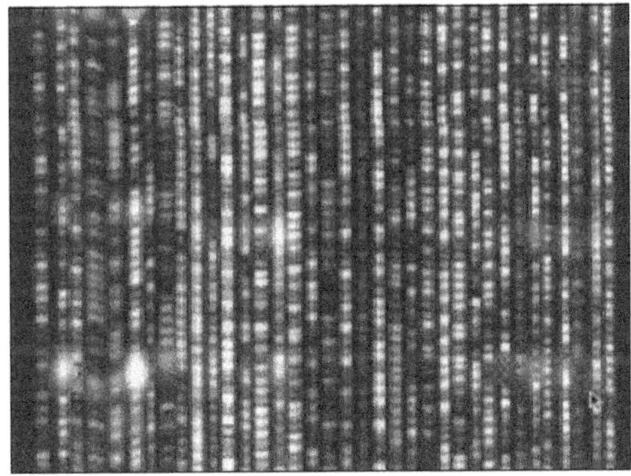

Figure 14.9 A snapshot of the detection of the molecules on the sequencer

4. ***Assembling of the sequenced parts of a gene*** For publication purposes, each sequence of a gene has to be confirmed in both directions. To accomplish this, the gene has to be sequenced with forward and reverse primers. Since it is only possible to sequence a part of 750 till 800 bases in one run, a gene of, for example, 1800 bases has to be sequenced with internal primers. When all these fragments are sequenced, a computer program tries to fit the different parts together and assembles the total gene sequence (Figure 14.10).

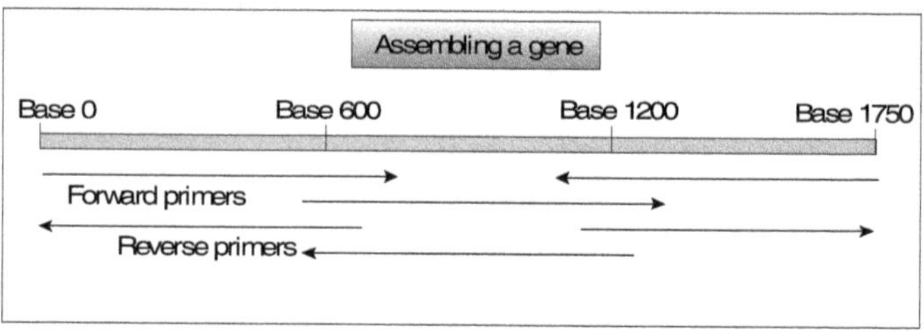

Figure 14.10 Assembling of the sequenced parts of a gene

RNA Sequencing

RNA is less stable in the cell, and also more prone to nuclease attack experimentally. As RNA is generated by transcription from DNA, the information is already present in the cell's DNA. However, there are many occasions when it may be necessary to sequence RNA directly, especially to determine the position of modified nucleotides present in it. This is normally done for tRNA and rRNA. This is achieved by base-specific cleavage of 5′ end labelled RNA using RNases. (RNase T_1 cleaves after G, RNase U_2 cleaves after A, RNase Phy M cleaves after A and U and *Bacillus cereus* RNase cleaves after U and C). Limited amount of enzymes and time of digestion are employed to generate a ladder of cleaved products which are analysed by PAGE.

In eukaryotes, RNA molecules are not co-linear with their DNA template, as introns are excised. To sequence these RNA, the usual method is to first reverse transcribe the sample to generate DNA fragments using reverse trancriptase. This DNA fragment can then be sequenced as described for DNA.Thus RNA viruses such as polio, influenza and AIDS virus HIV can be sequenced.

Genetic Fingerprinting

Genetic fingerprinting is a technique to distinguish between individuals of the same species using only samples of their DNA. It was invented by **Sir Alec Jeffreys** in 1985.

Two humans will have the vast majority of their DNA sequence in common. Genetic fingerprinting exploits highly variable repeating sequences called microsatellites. Two unrelated humans are likely to have different numbers of microsatellites at a given locus. By using PCR to detect the number of repeats at several loci, it is possible to establish a match that is extremely unlikely to have arisen by coincidence. Genetic fingerprinting is used in forensic science, to match samples of blood, hair, saliva or semen of suspects. It is also used in such applications as studying populations of wild animals, paternity testing, identifying dead bodies, etc.

Sometimes, a point mutation (deletion or insertion) occurs within a DNA sequence resulting in DNA polymorphism. This minor difference in sequence can be detected by using restriction enzymes and gel electrophoresis. For example, if the DNA sequence on a pair of homologous chromosomes is slightly different when a restriction enzyme is used to cut the DNA, the fragments will be of different lengths because restriction enzymes were used to cut these polymorphs and the fragment length varies. The technique is called **restriction fragment length polymorphism** or **RFLP**.

Recently, an additional technique for genetic fingerprinting has been introduced—**amplified fragment length polymorphism** or **AFLP**. This new technique is similar to RFLP analysis, but introduces a few other features, like two rounds of amplification and specially made primers. AFLP analysis is now highly automated, and allows for easy creation of phylogenetic trees based on comparing individual samples of DNA.

DNA Footprinting

DNA footprinting is a technique for identifying exactly where a protein binds to DNA. Knowing where a protein binds to DNA often aids in understanding how gene expression is regulated. Consequently, DNA footprinting is often part of a larger study to determine how a particular gene is controlled.

DNA footprinting is based on the observation that when a protein binds to DNA, the DNA is protected from chemicals that would otherwise cleave it (Figure 14.11). In a typical DNA footprinting experiment, a DNA fragment with a suspected protein-binding site is first isolated, and then labelled with a radioactive nucleotide or other chemical that will allow it to be detected later on. Once labelled, the DNA is mixed in a test tube with a DNA-binding protein and a chemical that cleaves DNA, such as the enzyme DNase I. In a separate test tube, more of the same labelled DNA is mixed with the same cleaving chemical, but without the binding protein. The DNA fragments in each tube are allowed to incubate long enough for the molecule to cleave once, and then are separated by size (fractionated) in a DNA sequencing gel. The reactions in the two test tubes (one with the binding protein and one without) are then compared. If the DNA actually contains protein-binding sites, these will have been protected from cleaving in the test tube that contains DNA-binding protein, and a "footprint" of those sites where no DNA

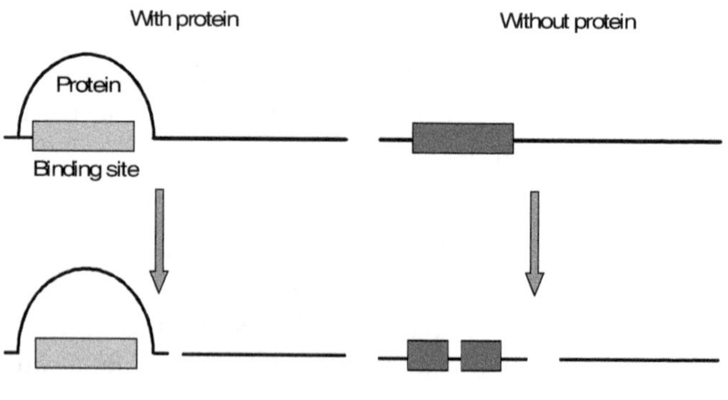

Figure 14.11 DNA footprinting

cleavage occurred will be observed. By comparison with a sequencing reaction run on the same gel, one can determine the exact location where a protein has been bound to the DNA. A related technique, called gel retardation, can also be used to test for protein binding to DNA, but this method is less precise than DNA footprinting.

Uses in research DNA footprinting is often used to locate the binding site for proteins that regulate transcription. For example, a researcher may suspect that a particular protein binds to a particular DNA fragment and inhibits transcription. After conducting a DNA footprinting experiment, the researcher will know the location of the exact sequence of DNA bound by that protein. If that sequence matches the sequence of a promoter, the DNA footprinting experiment can help explain how that DNA-binding protein carries out its function.

Modified DNA footprinting experiments can also be performed to detect where proteins bind to DNA in a living cell. In these experiments, cells are grown under conditions where the protein of interest would be expected to bind to DNA. The cells are then treated with a chemical that causes proteins bound to DNA to become permanently attached to the DNA. The resulting DNA–protein complexes are then purified from the cell, and the DNA sequences are identified.

Since DNA footprinting is used to identify specific sequences in DNA where a protein binds, the technique is useful in genetic research. For example, DNA footprinting is used in characterizing the function of proteins identified in the Human Genome Project and other genome projects, making it an important component of a molecular geneticist's toolbox.

REVIEW YOUR LEARNING

1. Define or explain the following:
   i. DNA sequencing
   ii. RNA sequencing

    iii. Genetic fingerprinting
    iv. Restriction Fragment Length Polymorphism (RFLP)
    v. Amplified Fragment Length Polymorphism (AFLP)
    vi. DNA footprinting
2. Explain Sanger's method for DNA sequencing.
3. Explain Maxam and Gilbert's method for DNA sequencing.
4. Explain the method used for sequencing on an automated sequencer.
5. Explain DNA footprinting and its uses.

POLYMERASE CHAIN REACTION (PCR)

There have been a number of key developments in molecular biological techniques; the one that has had the most impact in recent years is the polymerase chain reaction (PCR). One of the reasons for the adoption of PCR is the elegant simplicity of the reaction and relative ease of the practical manipulation steps. The polymerase chain reaction is a technique for quickly "cloning" a particular piece of DNA in the test tube (rather than in living cells like *E. coli*). Unlimited copies of a single DNA molecule can be synthesized from a mixture containing many different DNA molecules.

Kary Mullis invented PCR in 1983. He was awarded the Nobel Prize in Chemistry in 1993 for this achievement. Mullis's aim was to develop a process by which DNA could be artificially multiplied through repeated cycles of duplication driven by an enzyme called DNA polymerase. DNA polymerase occurs naturally in living organisms, where it functions to duplicate DNA when cells divide in mitosis and meiosis. Polymerase works by binding to a single DNA strand and creating the complementary strand. In Mullis's original process, the enzyme was used *in vitro*. The double-stranded DNA was separated into two single strands by heating it to 94°C. At this temperature, however, the DNA polymerase used at the time was destroyed, so the enzyme had to be replenished after the heating stage of each cycle. Mullis's original procedure required a great deal of time, large amounts of DNA polymerase, and continual attention throughout the process.

Later, this original PCR process was greatly improved by the use of DNA polymerase taken from thermophilic bacteria grown in geysers at a temperature of over 110°C. The DNA polymerase taken from these organisms is stable at high temperatures and, when used in PCR, does not break down when the mixture was heated to separate the DNA strands. Since there was no longer a need to add new DNA polymerase for each cycle, the process of copying a given DNA strand could be simplified and automated.

One of the first thermostable DNA polymerases was obtained from *Thermus aquaticus* and was called *Taq*. *Taq* polymerase is widely used in current PCR practice. A disadvantage of *Taq* is that it sometimes makes mistakes when copying DNA, leading to mutations in the DNA sequence, since it lacks 3´ → 5´ proofreading exonuclease activity. Polymerases such as *Pwo* or *Pfu*, obtained from *Archaea,* have "proofreading mechanisms" (mechanisms that check for errors) and can significantly reduce the number of mutations that occur in the copied DNA sequence. However, these enzymes polymerize DNA at a much slower rate than *Taq*. Combinations of both *Taq* and *Pfu* are available nowadays, and

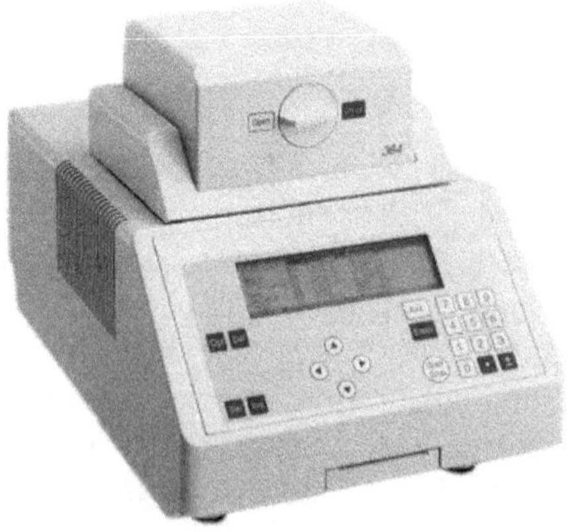

Figure 15.1 Thermal cycler

provide both high processivity (fast polymerization) and high fidelity (accurate duplication of DNA).

The PCR reaction is carried out in a thermal cycler (Figure 15.1). This is a machine that heats and cools the reaction tubes within it to the precise temperature required for each step of the reaction. To prevent evaporation of the reaction mixture, a heated lid is placed on top of the reaction tubes or a layer of oil is put on the surface of the reaction mixture.

A block diagram of the various parts of the thermal cycler is shown in Figure 15.2. It consists of the induction heater and its river circuit, a fan for cooling the reaction tubes, a temperature probe and control circuit, and a micro-controller with a keypad and an LCD display for programming the cycling paramcters.

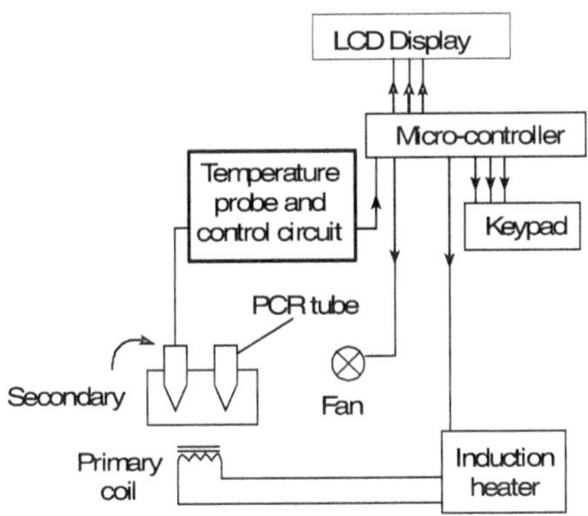

Figure 15.2 Block diagram of the various parts of the thermal cycler

PRINCIPLE AND WORKING MECHANISM

The purpose of a PCR (Polymerase Chain Reaction) is to make a huge number of copies of a gene. This is necessary to have enough starting template for sequencing.

The DNA sample is heated to separate its strands and mixed with the primers. If the primers find their complementary sequences in the DNA, they bind to them. Synthesis begins (5´ – 3´, as always) using the original strand as the template. The reaction mixture must contain all four deoxyribonucleotide triphosphates (dATP, dCTP, dGTP, dTTP) and a DNA polymerase.

DNA Replication (Synthesis)

- A portion of the double helix is unwound by a **helicase**.
- A molecule of a **DNA polymerase** binds to one strand of the DNA and begins moving along it in the 3´ to 5´ direction, using it as a template for assembling a **leading strand** of nucleotides and reforming a double helix. In eukaryotes, this molecule is called DNA polymerase delta (δ).
- Because DNA synthesis can only occur 5´ to 3´, a molecule of a second type of DNA polymerase (epsilon, ε, in eukaryotes) binds to the other template strand as the double helix opens. This molecule must synthesize discontinuous segments of polynucleotides (called Okazaki fragments). Another enzyme, **DNA ligase** I then stitches these together into the **lagging strand**.

Polymerization continues until each newly synthesized strand has proceeded far enough to contain the site recognized by the **other primer**. Two DNA molecules identical to the original molecule are thus synthesized. Each cycle doubles the number of DNA molecules.

The Cycling Reactions

There are three major steps in a PCR, which are repeated for 30 or 40 cycles. This is done on an automated cycler, which can heat and cool the tubes with the reaction mixture in a very short time.

1. **Denaturation at 94°C** During the denaturation, the double strand melts open to single-stranded DNA, all enzymatic reactions stop (for example: the extension from a previous cycle).
2. **Annealing at 54°C** The primers are jiggling around, caused by the Brownian motion. Ionic bonds are constantly formed and broken between the single-stranded primer and the single-stranded template. The more stable bonds last a

little bit longer (primers that fit exactly) and on that little piece of double-stranded DNA (template and primer), the polymerase can attach and start copying the template. Once there are a few bases built in, the ionic bond is so strong between the template and the primer that it does not break anymore.

3. **Extension at 72°C** This is the ideal working temperature for the polymerase. The primers, where there are a few bases built in, already have a stronger ionic attraction to the template than the forces breaking these attractions. Primers that are on positions with no exact match get loose again (because of the higher temperature) and do not give an extension of the fragment. The bases (complementary to the template) are coupled to the primer on the 3´ side (the polymerase adds dNTPs from 5´ to 3´, reading the template from 3´ to 5´ side, bases are added complementary to the template).

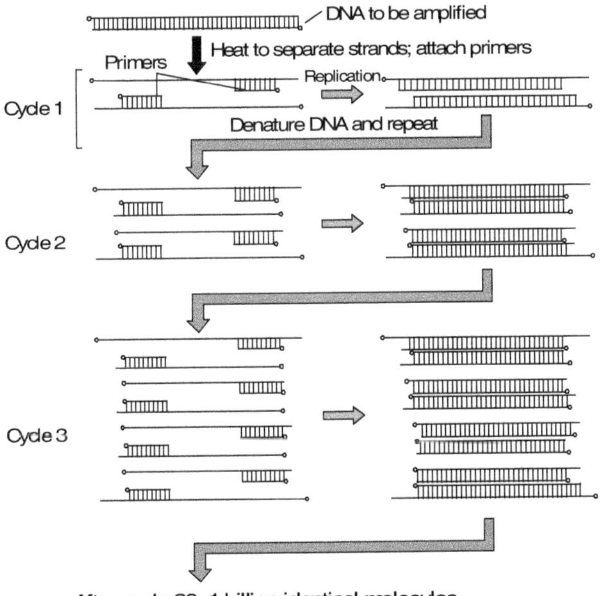

Figure 15.3 The cycling reactions in PCR

After 30 cycles, what began as a single molecule of DNA has been amplified into more than a billion copies (Figure 15.3).

CONSTRAINTS IN PCR

Polymerase Errors

Taq polymerase lacks a 3´ to 5´ exonuclease activity. This makes it impossible for it to check the base it has inserted and remove it if it is incorrect, a process common in higher organisms. This in turn results in a high error rate (approximately 1 in 10000 bases). If an error occurs early, it can alter large proportions of the final product. As a result other polymerases are available for accuracy in vital uses such as amplification for sequencing. Some examples of polymerases with 3´ to 5´ exonuclease activity include Vent® (*Thermococcus litoralis*), Pfu (*Pyrococcus furiosus*) and *Pwo* (*Pyrococcus woesii*).

Size Limitations

PCR works readily with DNA of lengths 2–3 kb, but above this length the polymerase tends to fall off and the typical heating cycle does not leave enough time for polymerization to complete. It is possible to amplify larger pieces (up to 20 kb), with a slower heating cycle and special polymerases.

Non-specific Priming

The non-specific binding of primers is always a possibility due to sequence duplications, non-specific binding and partial primer binding (leaving the 5´ end unattached). This is increased by the use of degenerate sequences or bases in the primer. Manipulation of annealing temperature and magnesium ion (which stabilize DNA and RNA interactions) concentrations can increase specificity.

PRACTICAL MODIFICATIONS OF THE PCR TECHNIQUE

Nested PCR

Nested PCR is intended to reduce the contamination in products due to the amplification of unexpected primer binding sites. Two sets of primers are used in two successive PCR runs, the second set intended to amplify a secondary target within the first run product. This is very successful, but requires more detailed knowledge of the sequences involved.

Inverse PCR

Inverse PCR is a method used to allow PCR when only one internal sequence is known. This is especially useful in identifying flanking sequences to various of genomic inserts. This involves a series of digestions and self ligations before being cut by an endonuclease, resulting in known sequences at either end of the unknown sequence.

RT-PCR

Reverse Transcription PCR is the method used to amplify, isolate or identify a known sequence from the RNA library of a cell or tissues. This is widely used in expression mapping, determining when and where certain genes are expressed.

Asymmetric PCR

Asymmetric PCR is used to preferentially amplify one strand of the original DNA more than the other. It finds use in sequencing where only one of the complementary strands is required for sequencing. PCR is carried out as usual, but with a great excess of the primers for the chosen strand. Due to the slow amplification, many cycles of PCR are required.

Quantitative PCR and Real Time PCR

Quantitative PCR (Q-PCR) is used to rapidly measure the quantity of PCR product (preferably real-time), and thus is an indirect method for quantitatively measuring starting amounts of DNA, cDNA or RNA. This is commonly used for the purpose of determining whether a sequence is present or not, and if it is present, the number of copies in the sample. Quantitative real-time PCR is generally known as QRT-PCR or RTQ-PCR.

Touchdown PCR

Touchdown PCR is a variant of PCR that reduces non-specific primer annealing by variation of primer annealing temperature between cycles.

RECENT DEVELOPMENTS IN PCR TECHNIQUES

- A more recent method which excludes a temperature cycle, but uses enzymes, is helicase-dependent amplification.
- TAIL-PCR, developed by Liu *et al.* in 1995, is the thermal asymmetric interlaced PCR.
- Meta-PCR, developed by Andrew Wallace, allows to optimize amplification and direct sequence analysis of complex genes.

APPLICATIONS

Human Health and the Human Genome Project

PCR has very quickly become an essential tool for improving human health and human life. Medical research and clinical medicine are profiting from PCR mainly in two areas: detection of infectious disease organisms, and detection of variations and mutations in genes, especially human genes. Because PCR can amplify unimaginably tiny amounts of DNA, even that from just one cell, physicians and researchers can examine a single sperm, and track down the elusive source of a puzzling infection.

The method is especially useful to detect the AIDS virus sooner during the first few weeks after infection than the standard ELISA test. PCR looks directly for the virus's unique DNA, instead of the method employed by the standard test, which looks for indirect evidence that the virus is present by searching for antibodies the body has made against it.

PCR can also be more accurate than standard tests. It is making a difference, for example, in a painful, serious, and often stubborn misfortune of childhood, the middle ear infection known as otitis media. The technique has detected bacterial DNA in children's middle ear fluid, signalling an active infection even when culture methods failed to detect it.

Lyme disease, the painful joint inflammation caused by bacteria transmitted through tick bites, is usually diagnosed on the basis of symptom patterns. But PCR can identify the disease organism's DNA contained in joint fluid, permitting speedy treatment that can prevent serious complications.

PCR is the most sensitive and specific test for *Helicobacter pylori*, the disease organism now known to cause almost all stomach ulcers. Unlike other tests, PCR can detect three different sexually transmitted disease organisms on a single swab (herpes, papillomaviruses, and chlamydia) and can even distinguish the particular strain of papillomavirus that predisposes to cancer, which other tests cannot do.

With PCR analysis of cells shed into faeces, premalignant changes in the gastrointestinal tract such as mutations in genes that protect against tumours can be demonstrated. Researchers have also detected potentially metastatic cells in the circulation of patients with newly diagnosed tumours.

PCR can also be used to differentiate mutations in a single gene, each of which can lead to a disorder such as Duchenne muscular dystrophy. It helps doctors track the presence or absence of DNA abnormalities characteristic of particular cancers, so that they can start and stop drug treatments and radiation therapy as soon as possible. And it promises to greatly improve the genetic matching of donors and recipients for bone marrow transplantation.

PCR can even diagnose the diseases of the past. Hubert H. Humphrey underwent tests for bladder cancer in 1967. Although the tests were negative, he died of the disease in 1978. In 1994, researchers compared a 1976 tissue sample from his cancer-ridden bladder with his 1967 urine sample. With the help of PCR amplification of the small amount of DNA in the 27-year-old urine, they found identical mutations in the p53 gene, well known for suppressing tumors, in both samples. "Humphrey's examination in 1967 may have revealed the cancerous growth if the techniques of molecular biology were as well understood then as they have become," the researchers said.

Historical medical genetics has gone even further back in time with PCR. After the colour-blind British chemist John Dalton died in 1844, some tissue from his eyes was preserved. Dalton had asked for a posthumous investigation of the reason why he confused scarlet with green and pink with blue. A recent examination of DNA taken from that tissue, carefully amplified by PCR, has shown that Dalton lacked a gene for making one of the three photopigments essential for normal colour vision.

Many of the new genetic tests are the result of the Human Genome Project, the huge international effort to identify and study all human genes. DNA sequencing reveals crucial variations in the nucleotides that constitute genes. These mutational changes produce disease and even death by forcing the genes to produce abnormal proteins, or sometimes no proteins at all. DNA sequencing involves isolating and duplicating DNA segments for nucleotide analysis. Thus PCR is an essential tool for the Human Genome Project because it can quickly and easily generate an unlimited amount of any piece of DNA for this kind of study.

Genetic Fingerprinting

Genetic fingerprinting is a forensic technique used to identify a person by comparing his or her DNA with a given sample, such as blood from a crime scene can be genetically compared to blood from a suspect. The sample may contain only a tiny amount of DNA, obtained from a source such as blood, semen, saliva, hair, or other organic material. Theoretically, just a single strand is needed. First, one breaks the DNA sample into fragments, then amplifies them using PCR. The amplified fragments are then separated using gel electrophoresis. The overall layout of the DNA fragments is called a **DNA fingerprint**. Since there is a very small possibility that two individuals may have the same sequences, the technique is more effective at acquitting a suspect than proving the suspect guilty.

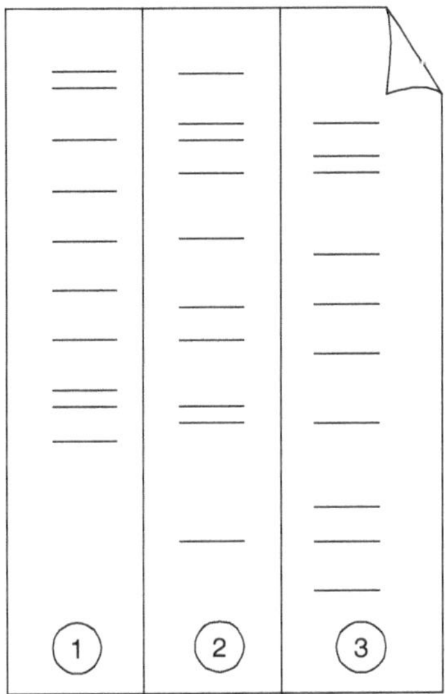

Figure 15.4 Electrophoresis of PCR-amplified DNA fragments. (1) Father (2) Child (3) Mother

Figure 15.4 shows the electrophoresis of PCR-amplified DNA fragments of a father, child and a mother. The child has inherited some, but not all of the fingerprint of each of its parents, giving it a new, unique fingerprint. Although these resulting 'fingerprints' are unique (except for identical twins), genetic relationships, for example, parent–child or siblings, can be determined from two or more genetic fingerprints, which can be used for paternity tests. A variation of this technique can also be used to determine evolutionary relationships between organisms.

Detection of Hereditary Diseases

The detection of hereditary diseases in a given genome is a long and difficult process, which can be shortened significantly by using PCR. Each gene in question can easily be amplified through PCR by using the appropriate primers and then sequenced to detect mutations. Viral diseases can also be detected using PCR through amplification of the viral DNA. This analysis is possible right after infection, which can be from several days to several months before actual symptoms occur. Such early diagnoses give physicians a significant lead in treatment.

Cloning Genes

Cloning a gene, not to be confused with cloning a whole organism, describes the process of isolating a gene from one organism and then inserting it into another organism (now termed a genetically modified organism (GMO)). PCR is often used to amplify the gene, which can then be inserted into a vector (a vector is a piece of DNA which 'carries' the gene into the GMO) such as a plasmid (a circular DNA molecule) (Figure 15.5). The DNA can then be transferred into

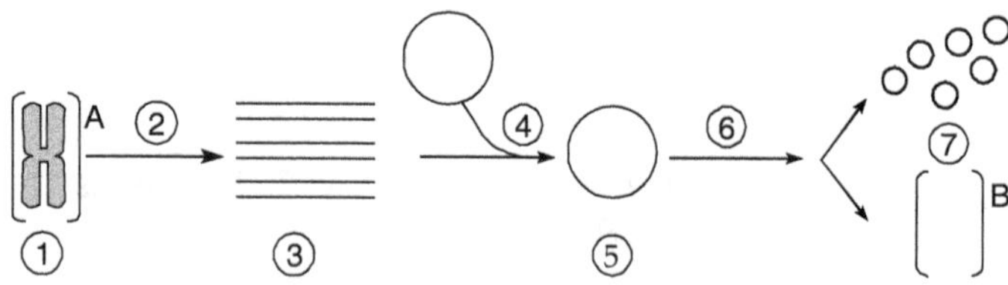

Figure 15.5 Cloning a gene using a plasmid (1) Chromosomal DNA of organism A (2) PCR (3) Multiple copies of a single gene from organism A (4) Insertion of the gene into a plasmid (5) Plasmid with gene from organism A (6) Insertion of the plasmid in organism B (7) Multiplication or expression of the gene, originally from organism A, occurring in organism B

an organism (the GMO) where the gene and its product can be studied more closely. Expressing a cloned gene [when a gene is expressed, the gene product (usually protein or RNA) is produced by the GMO] can also be a way of mass-producing useful proteins, for example medicines or the enzymes in biological washing powders. The incorporation of an affinity tag on a recombinant protein will generate a fusion protein which can be more easily purified by affinity chromatography.

Mutagenesis

Mutagenesis is a way of making changes to the sequence of nucleotides in the DNA. There are situations in which one is interested in mutated (changed) copies of a given DNA strand, for example, when trying to assess the function of a gene or in *in vitro* protein evolution. Mutations can be introduced into copied DNA sequences in two fundamentally different ways in the PCR process. **Site-directed mutagenesis** allows the experimenter to introduce a mutation at a specific location on the DNA strand. Usually, the desired mutation is incorporated in the primers used for the PCR program. **Random mutagenesis,** on the other hand, is based on the use of error-prone polymerases in the PCR process. In the case of random mutagenesis, the location and nature of the mutations cannot be controlled. One application of random mutagenesis is to analyse structure–function relationships of a protein. By randomly altering a DNA sequence, one can compare the resulting protein with the original and determine the function of each part of the protein.

Analysis of Ancient DNA

Using PCR, it becomes possible to analyse DNA that is thousands of years old. PCR techniques have been successfully used on animals such as a forty thousand-year-old mammoth, and also on human DNA, in applications ranging from the analysis of Egyptian mummies to the identification of a Russian tsar.

Genotyping of Specific Mutations

Through the use of allele-specific PCR, one can easily determine which allele of a mutation or polymorphism an individual has. Here, one of the two primers is common, and would anneal a short distance away from the mutation, while the other anneals right on the variation. The 3´ end of the allele-specific primer is modified, to only anneal if it matches one of the alleles. If the mutation of interest is a T or C single nucleotide polymorphism (T/C SNP), one would use two reactions, one containing a primer ending in T, and the other ending in C. The common primer would be the same. Following PCR, these two sets of reactions would be run out on an agarose gel, and the band pattern will tell us if the individual is homozygous T, homozygous C, or heterozygous T/C. This methodology has several applications such as amplifying certain haplotypes (when certain alleles at 2 or more SNPs occur together on the same chromosome [Linkage Disequilibrium]) or detection of recombinant chromosomes and the study of meiotic recombination.

Comparison of Gene Expression

Researchers have used traditional PCR as a way to estimate changes in the amount of a gene's expression. RNA is the molecule into which DNA is transcribed prior to making a protein, and those strands of RNA that hold the instructions for protein sequence are known as messenger RNA (mRNA). Once RNA is isolated it can be reverse transcribed back into DNA (complementary DNA to be precise, known as cDNA), at which point traditional PCR can be applied to amplify the gene. This methodology is called RT-PCR. In most cases, if there is more starting material (mRNA) of a gene, then during PCR more copies of the gene will be generated. When the product of the PCR reaction are run on an agarose gel, a band, corresponding to a gene, will appear larger on the gel. By running samples of amplified cDNA from differently treated organisms one can get a general idea of which sample expressed more of the gene of interest.

REVIEW YOUR LEARNING

1. Explain the applications of the following:
   i. PCR
   ii. Nested PCR
   iii. Inverse PCR
   iv. RT-PCR
   v. Asymmetric PCR
   vi. QRT-PCR
   vii. Touchdown PCR
   viii. TAIL-PCR
2. Explain the principle and working mechanism of PCR. Add a note on its applications.

DNA MICROARRAY

Recent advances in technologies such as microarray analysis are bringing about a revolution in our understanding of the molecular mechanisms underlying normal and dysfunctional biological processes. Microarray technology evolved from Southern blotting, where fragmented DNA is attached to a substrate and then probed with a known gene or fragment. Measuring gene expression using microarrays is relevant to many areas of biology and medicine.

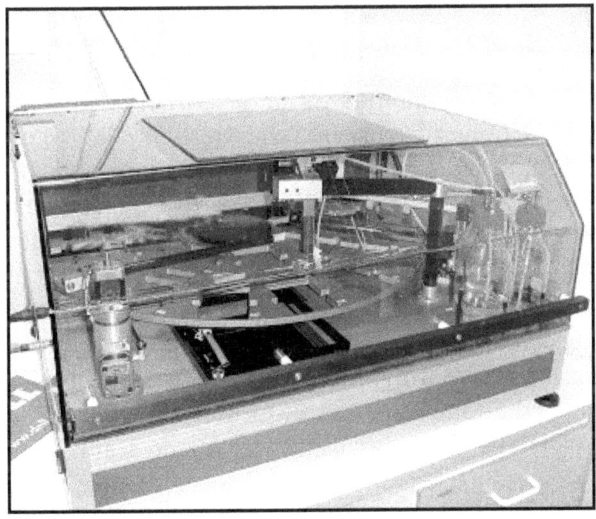

Figure 16.1 Generation of microarray slides using robotic machine

A DNA microarray is a collection of microscopic DNA spots attached to a solid surface such as glass, plastic or silicon chip forming an array. A specialized robotic machine uses super-thin stainless steel needles to dot the slide with the spots. The robot places the spots at precise intervals as it moves over the surface (Figure 16.1). The spots are incredibly minuscule, measured in micrometres (millionths of a metre); they typically range from 20 to 100 microns in diameter. The affixed DNA segments are known as probes. There can be thousands or tens of thousands of these tiny spots on a single slide. A computer keeps track of the gene contained in each spot (Figure 16.2).

Figure 16.2 DNA microarrayer

The most common use of microarrays is to quantify mRNAs transcribed from different genes and which encode different proteins. RNA is extracted from many cells, ideally from a single cell type, then converted to cDNA or cRNA. The copies may be amplified by RT-PCR. Fluorescent tags are enzymatically incorporated into the newly synthesized cDNA/cRNA or can be chemically attached to the new strands of DNA or RNA. A cDNA or cRNA that contains a sequence complementary to one of the single-stranded probe sequences on the array will hybridize, via base-pairing, to the spot at which the complementary sequences are affixed. The spot will then glow when examined using a microarray scanner (Figure 16.3). The scanned images of the spotted slides are analysed using analyser (Figure 16.4).

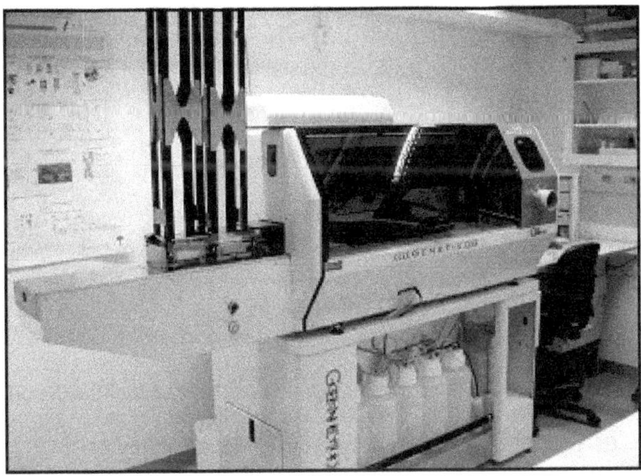

Figure 16.3 Hybridized slides are scanned using the high-throughput DNA microarray scanner

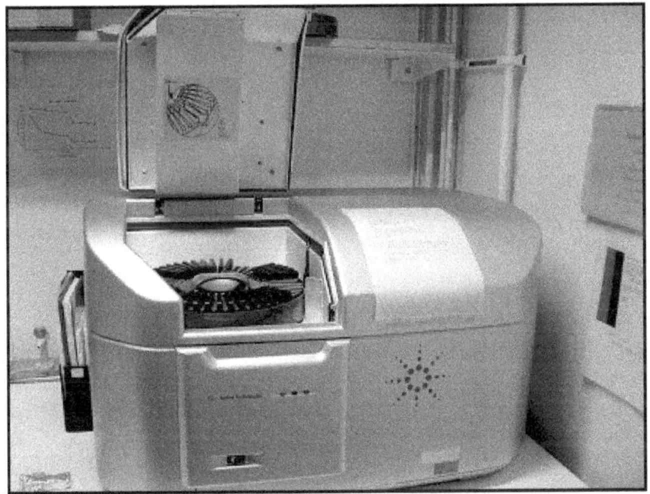

Figure 16.4 Analysis of scanned images of the spotted slides by microarray analyser

Increased or decreased fluorescence intensity indicates that cells in the sample have recently transcribed, or ceased transcription of a gene that contains the probed sequence (Figure 16.5).

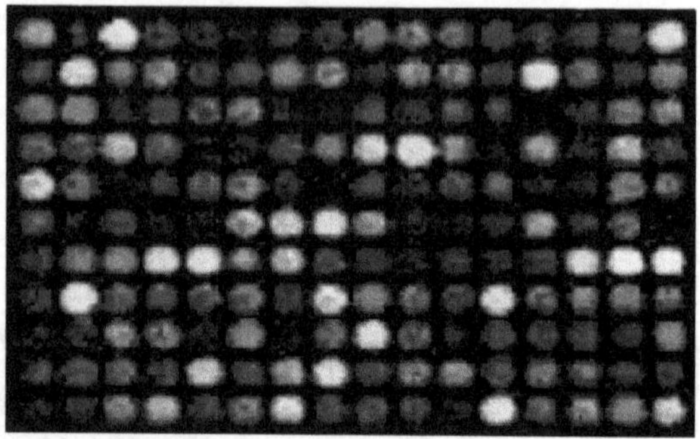

Figure 16.5 Pattern of gene activity on a microarray chip

The intensity of the fluorescence is roughly proportional to the number of copies of a particular mRNA that were present and thus roughly indicates the activity or expression level of that gene. Arrays can paint a picture or profile of which genes in the genome are active in a particular cell type and under a particular condition. The entire workflow in DNA microarray is shown in Figure 16.6.

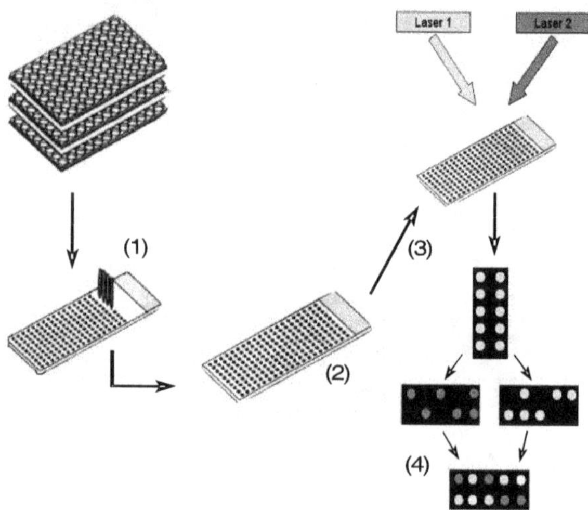

(1) Affixing DNA segment
(2) Synthesis of cDNA or cRNA
(3) Fluorescent tags are incorporated into cDNA/cRNA
(4) Hybridization

Figure 16.6 Microarray workflow

APPLICATIONS

Because many proteins have unknown functions, and because many genes are active all the time in all kinds of cells, researchers usually use microarrays to make comparisons between similar cell types. For example, an RNA sample from brain tumour cells, might be compared to a sample from healthy neurons or glia. Probes that bind RNA in the tumour sample but not in the healthy one may indicate genes that are uniquely associated with the disease. Typically in such a test, the two sample cDNAs are tagged with two distinct colours, enabling comparison on a single chip. Researchers hope to find molecules that can be targeted for treatment with drugs among the various proteins encoded by disease-associated genes.

Although the chips detect RNAs that may or may not be translated into active proteins, scientists refer to these kinds of analyses as expression analysis or expression profiling. Since there are hundreds or thousands of distinct probes on an array, each microarray experiment can accomplish the equivalent of thousands of genetic tests in parallel. Arrays have therefore dramatically accelerated many types of investigations.

Microarrays are also being used to identify genetic variation in individuals and across populations. Short oligonucleotide arrays can be used to identify the single nucleotide polymorphisms (SNPs) that are thought to be responsible for genetic variation and the source of susceptibility to genetically caused diseases. Generally termed 'genotyping' applications, chips may be used in this fashion for forensic applications, rapidly discovering or measuring genetic predisposition to disease, or identifying DNA-based drug candidates.

These SNP microarrays are also being used to profile somatic mutations in cancer, specifically loss of heterozygosity events and amplifications and deletions of regions of DNA. Amplifications and deletions can also be detected using comparative genomic hybridization in conjunction with microarrays.

Resequencing arrays have also been developed to sequence portions of the genome in individuals. These arrays may be used to evaluate germ-line mutations in individuals, or somatic mutations in cancer.

Genome tiling arrays include overlapping oligonucleotides designed to blanket an entire genomic region of interest. Many companies have successfully designed tiling arrays that cover whole human chromosomes.

Microarray studies and other genomic techniques are also stimulating the discovery of new targets for the treatment of disease, which is aiding drug development, immunotherapeutics and gene therapy.

Gene chips allow scientists to examine so many genes at once; hence, they have greatly reduced the time it takes to do experiments. Studies that once took months or even years to perform can now be done in a matter of days or even a few hours.

REVIEW YOUR LEARNING

1. Explain the process of DNA microarray analysis.
2. Explain the applications of DNA microarray.

PROTEIN SEQUENCING

Proteins are found in every cell and are essential for every biological process. Determination of protein structure involves protein sequencing—determining the amino acid sequences of its constituent peptides and also determining what conformation it adopts and whether it is complexed with any non-peptide molecules. Discovering the structure and function of proteins in living organisms is an important tool for understanding cellular processes, and allows drugs that target specific metabolic pathways to be invented more easily.

The two major direct methods of protein sequencing are mass spectrometry and the Edman degradation reaction. It is also possible to generate an amino acid sequence from the DNA or mRNA sequence encoding the protein, if it is known. However, there are a number of other reactions, which can be used to gain more limited information about protein sequences and can be used as preliminaries to the aforementioned methods of sequencing or to overcome specific inadequacies within them.

DETERMINING AMINO ACID COMPOSITION

It is often desirable to know the unordered amino acid composition of a protein prior to attempting to find the ordered sequence, as this knowledge can be used to facilitate the discovery of errors in the sequencing process or to distinguish between ambiguous results. Knowledge of the frequency of certain amino acids may also be used to choose which protease to be used for digestion of the protein. A generalized method for doing this is as follows:

1. Hydrolyse a known quantity of protein into its constituent amino acids.
2. Separate the amino acids by ion-exchange chromatography.
3. Determine the respective quantities of the amino acids.

Once the amino acids have been separated, their respective quantities are determined by adding a reagent that will form a coloured derivative. If the amounts of amino acids are in excess of 10 nmol, ninhydrin can be used for this—it gives a yellow colour when reacted with proline, and a vivid blue with other amino acids. The concentration of amino acid is proportional to the absorbance of the resulting solution. With very small quantities, down to 10 pmol, fluoresceamine can be used as a marker; this forms a fluorescent derivative on reacting with an amino acid.

N-terminal Amino Acid Analysis

A generalized method for N-terminal amino acid analysis is as follows:

1. React the peptide with a reagent, which will selectively label the terminal amino acid. There are many different reagents, which can be used to label terminal amino acids. They all react with amine groups and will therefore also bind to amine groups in the side chains of amino acids such as lysine—for this reason it is necessary to be careful in interpreting chromatograms to ensure that the right spot is chosen. Two of the more common reagents are Sanger's reagent (2,4-dinitrofluorobenzene) and dansyl derivatives such as dansyl chloride. Phenylisothiocyanate, the reagent for the Edman degradation, can also be used.
2. Hydrolyse the protein.
3. Determine the amino acid by comparison with standards or using thin layer chromatography or high performance liquid chromatography (HPLC).

C-terminal Amino Acid Analysis

The most common method for C-terminal amino acid analysis is to add carboxypeptidases to a solution of the protein. Determine the terminal amino acid in the samples taken at regular intervals, by analysing a plot of amino acid concentrations against time.

EDMAN DEGRADATION

The Edman degradation is a very important reaction for protein sequencing, because it allows the ordered amino acid composition of a protein to be discovered. Automated Edman sequencers are now in widespread use, and are able to sequence peptides up to approximately 50 amino acids long. A reaction scheme for sequencing a protein by the Edman degradation is as follows :

1. Break any disulphide bridge in the protein by oxidizing with performic acid.
2. Separate and purify the individual chains of the protein complex, if there is more than one.
3. Determine the amino acid composition of each chain.
4. Determine the terminal amino acids of each chain.
5. Break each chain into fragments under 50 amino acids long.
6. Separate and purify the fragments.
7. Determine the sequence of each fragment.
8. Repeat with a different pattern of cleavage.
9. Construct the sequence of the overall protein.

Peptides longer than about 50–70 amino acids cannot be sequenced reliably by the Edman degradation. Because of this, long protein chains need to be broken up into small fragments, which can then be sequenced individually. Digestion is done either by endopeptidases such as trypsin or pepsin or by chemical reagents such as cyanogen bromide. Different enzymes give different cleavage patterns, and the overlap between fragments can be used to construct an overall sequence.

The Edman Degradation Reaction

The peptide to be sequenced is adsorbed onto a solid surface—one common substrate is glass fibre coated with polybrene, a cationic polymer. The Edman reagent, phenylisothiocyanate (PITC), is added to the adsorbed peptide, together with a basic buffer solution of 12% trimethylamine. This reacts with the amine group of the N-terminal amino acid. The terminal amino acid derivative can then be selectively detached by the addition of anhydrous acid. The derivative then isomerizes to give a substituted phenylthiohydantoin (PTH), which can be washed off and identified by chromatography, and the cycle can be repeated. The efficiency of each step is about 98%, which allows about 50 amino acids to be reliably determined.

Limitations of the Edman Degradation

Because the Edman degradation proceeds from the N-terminus of the protein, it will not work if the N-terminal amino acid has been chemically modified or if it is concealed within the body of the protein.

MASS SPECTROSCOPY

The other major direct method by which the sequence of a protein can be determined is mass spectrometry. This method has been gaining popularity in recent years as new techniques and increasing computing power have facilitated it. Mass spectrometry can, in principle, sequence any size of protein, but the problem becomes computationally more difficult as the size increases. Peptides are also easier to prepare for mass spectrometry than whole proteins, because they are more soluble. One method of delivering the peptides to the spectrometer is electrospray ionization, invented by Kurt Wuthrich, J. B. Fenn and K. Tanaka. They were awarded Nobel Prize in chemistry in 2002 for this invention. The protein is digested by an endoprotease, and the resulting solution is passed through a high-pressure liquid chromatography column. At the end of this column, the solution is sprayed out of a narrow nozzle charged to a high positive potential into the mass spectrometer. The charge on the droplets causes them to fragment until only single ion remain. The mass spectrum is analysed by computer and often compared against a database of previously sequenced proteins in order to determine the sequences of the fragments. This process is then repeated with a different digestion enzyme, and the overlaps in the sequences are used to construct a sequence for the protein.

PREDICTING PROTEIN SEQUENCE FROM DNA/RNA SEQUENCES

The amino acid sequence of a protein can also be determined indirectly from the mRNA or, in organisms such as prokaryotes that do not have introns, the DNA that codes for the protein. If the sequence of the gene is already known, then this is very easy. However, it is rare that the DNA sequence of a newly isolated protein will be known, and so if this method is to be used, it has to be found in some way. One way that this can be done is to sequence a short section, perhaps 15 amino acids long, of the protein by one of the above methods, and then use this sequence to generate a complementary marker for the protein's RNA. This can then be used to isolate the mRNA coding for the protein, which can then be replicated in a polymerase chain reaction to yield a significant amount of DNA, which can then be sequenced relatively easily. The amino acid sequence of the protein can then be deduced from this. However, it is necessary to take into account the possibility of amino acids being removed after the mRNA has been translated.

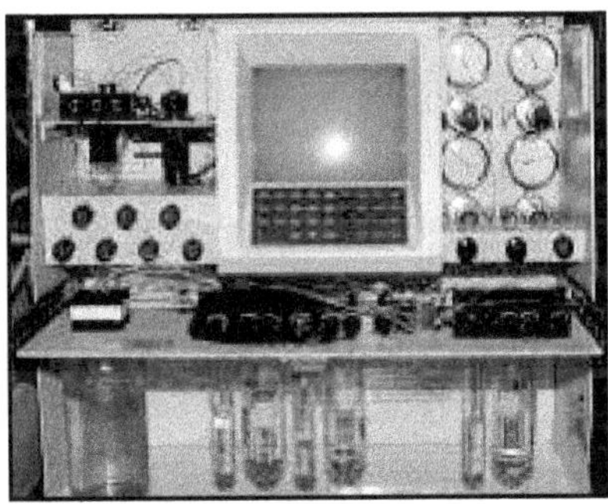

Figure 17.1 Protein sequencing system

SEQUENATORS

Protein sequencing provides information vital in gene cloning strategies involving synthetic oligodeoxynucleotides, for identifying amino acids at the active sites of enzymes, for primary structure characterization of peptides and proteins, for elucidating post-translational processing pathways, for characterizing products of peptide synthesis, for studies of domain structure in proteins, and for many other purposes.

Protein Sequenators (Figure 17.1) are now available for automated amino-terminal sequence analysis of protein and peptides. The instrument stores the PTH derivative of each amino acid residue of a polypeptide chain in separate tubes for many cycles. The PTH derivatives are identified by high performance liquid chromatography (HPLC). A number of instruments are used to sequence peptides using the Edman technique.

Solid-phase Liquid-pulse Sequenator

In solid-phase liquid-pulse sequenator a thin polymer membrane is held in the reaction vessel to which the protein molecules are covalently attached through its C-terminal. The reagents are added to the upper chamber of the reaction vessel in which the reactions occur and phenylthiohydantoin derivatives are automatically collected and analysed. In gas phase sequenator, the reagents are added in gaseous state so that the losses are minimized and the analysis is more rapid.

Pulsed-liquid Sequenator

Automated protein sequencing is done using on-line PTH-amino acid analysis. Samples can be provided as liquids or solids, or blotted onto solid substrates. A minimum of 10–100 picomoles of material is required, however lower limits may be obtainable depending on the sample. All samples are sequenced a minimum of five cycles to determine whether, and at what levels, sequence information is being produced. In case no sequences are observed, the background levels of amino acid PTH derivatives produced during this short initial run provide useful information about whether the sample was blocked or simply was present in insufficient amount.

Gas–liquid Solid Phase Peptide and Protein Sequenator

RM Hewick, MW Hunkapiller, LE Hood and WJ Dreyer have constructed a miniaturized protein and peptide sequenator in 1981, which uses gas phase reagents at the coupling and cleavage steps of the Edman degradation. The sample is embedded in a matrix of polybrene dried onto a porous glass fibre disc located in a small cartridge-style reaction cell. The protein or peptide, though not covalently attached to the support, is essentially immobile throughout the degradative cycle, since only relatively apolar, liquid-phase solvents pass through the cell. This instrument can give useful sequence data on as little as 5 pmol of protein, can perform extended sequence runs (greater than 30 residues) on subnanomole quantities of proteins purified by sodium dodecyl sulphate-polyacrylamide gel electrophoresis, and can sequence hydrophobic peptides to completion. The sequenator is characterized by a high repetitive yield during the degradation, low reagent consumption, low maintenance requirements, and a degradative cycle time of only 50 min using a complete double-cleavage program.

REVIEW YOUR LEARNING

1. Explain protein sequencing.
2. Write short notes on the following:
   i. Edman degradation
   ii. Protein sequenators

BIOINFORMATICS

Bioinformatics is a newly emerging interdisciplinary research area, which is playing an increasingly important and central role in biology research. Computers and the World Wide Web (www) are dramatically changing the face of biological research. Research that used to start in the laboratory now starts at the computer, as scientists search databases for information that might suggest new hypotheses. Bioinformatics integrates the many disparate bodies of data from different fields. Bioinformatics tools and databases are powerful, valuable, and essential for life science research in the Genomic Age.

Over the last decade, biologists have handled a number of genome research projects that include DNA sequencing, proteomics, expression studies and metabolomics, and the methodology used by them has been supported by automation. The development of the rapid DNA sequencing technology tremendously increased the amount of biological information available to biologists. The human genome project sequenced more than 100,000 genes encoded by 3×10^9 bases. Other genome projects such as *Haemophilus influenzae* (microbial genome), *Saccharomyces cerevisiae*, *Drosophila melanogaster*, *Schistosoma* sp. and *Plasmodium* sp. have also increased this information repertory. The completely sequenced genomes of mammals, primates, rodents, insects, bacteria and other life forms have been classified and stored in various public databases, and if the current shift in investigation continues, the data will increase manifold.

The results from genomic studies are immensely important in biological and medical research. Hence, the vast amount of biological information needs to be stored, organized, and indexed so that the information can be retrieved and used. Biological data is highly complex and interrelated. For example, while working with genome sequences, one has to pay attention to the structure, function and amino acid sequences of proteins encoded by the genome. The identification of these proteins is further complicated by several factors. In order to understand and organize the variety of information, information systems have to be created. Today's biologists must, therefore, be adequately equipped through proper training to cope with the management of this sophisticated computer-aided biological information systems. The first significant macromolecular sequence database was created by M. Dayhoff in the form of an atlas of protein sequence and structure, which has generated a number of databases all over the world. These databases include GenBank at National Institute of Health (NIH) in the USA, EMBL at the European Bioinformatics Institute (EBI) at Cambridge, PIR which is the US protein information resource, Swiss-Prot, OMIM, Prosite, Medline, Pubmed, etc., which store sequence information on nucleic acids, proteins, restriction enzymes, cloning vectors and transcription factors, besides storing information on cell lines and genetic disorders in humans. In future, biologists would need the information in digital form for correct and meaningful interpretation. Genomics and proteomics will need the skills and tools for organizing, interpreting, and storing biological information. There are many levels at which this biological information can be used, such as genomics, proteomics, metabolomics, etc. so that predictive methods can be developed to study the functioning of the phenotype of an organism. The future of genomics and proteomics is promising, as these fields have to play an important role in the treatment of genetic diseases.

The data from the human genome project is likely to be of significant assistance in medical genetics, including diagnosis of diseases. A genetic disease is the one that is caused by a defective or abnormal gene inherited from the parents. Detection of all the defective genes is always not possible because all of them may not be manifested in the phenotype. Medical genetics has made huge strides in diagnosing genetic diseases by devising DNA probes for screening diseases even when they are not functional.

Bioinformatics can significantly influence the solutions to the following types of problems:

1. Prediction of 3-D structure based on linear genomic information, i.e., the study of structural genomics.
2. Gene expression analysis, prediction of gene function, and establishment of gene libraries (functional genomics).
3. The ability to use genome sequences to identify proteins and their functions, protein interactions, modifications and functions, i.e., the field of proteomics.
4. Elucidating the function of a molecule based on its structure.
5. Simulating metabolism from the biochemical functions of an organism.
6. Molecular modeling and molecular dynamics which are the methods to predict structure from function. Several methods of machine learning are used to predict function from sequence structure.
7. Medical science would need to know the pathways to know which genomic changes could give rise to each known inherited disease, i.e., identification of the gene causing disease, and also genetic therapies that can reverse disease phenotypes.
8. Data obtained from functional genomics and proteomics could be used in drug designing and discovery.

The importance of databases and the increasing sophisticated communication network in biological and biomedical research is tremendous. The ability to use the different online accessible software in molecular biology is becoming mandatory for all biomedical scientists. The current quest is to sequence all genes, and to make the information available in databases, such that all biological investigations must start with browsing the data banks, making computer literacy compulsory for all biologists. The knowledge base arising from the genomics and proteomics research calls for a well-planned bioinformatics program.

The history of bioinformatics can possibly be traced back to the 1920s. During that period, A.J. Lotka and V. Volterra introduced mathematical models representing host/prey interaction. This was perhaps the first attempt to mathematically represent a model to achieve a cyclic balance in mean population density, i.e., to attain a dynamic equilibrium. It was however only recently that the development of powerful computers, and the availability of experimental data that can be readily treated by computation (for example, DNA or amino acid sequences and 3-D structures of proteins) launched bioinformatics as an independent field. It is now a rapidly evolving field with multidisciplinary applications.

APPLICATIONS OF BIOINFORMATICS

Applications in bioinformatics can be looked at the following levels:

1. At the basic level, bioinformatics is used to organize biological data, to help the researchers to access information, add new information arising out of experiments and modify existing information. Fundamentally, there are three types of data sets: genome sequences, macromolecular structures, and data from functional genomics experiments. Besides these data sets, bioinformatics analysis is also applied to other kinds of data, such as phylogenetic trees, metabolic pathways, the text of scientific papers, and medical information. For example, the Protein Data Bank (http://www.rcsb.org/pdb/) (Figure 18.1) is the single worldwide repository for the processing and distribution of 3-D biological macromolecular structure data.
2. While the first level is aligned more to the management of large quantities of data, the second level is to develop tools and resources that aid in the analysis of data. Just the compilation and maintenance of data is a simple, but highly

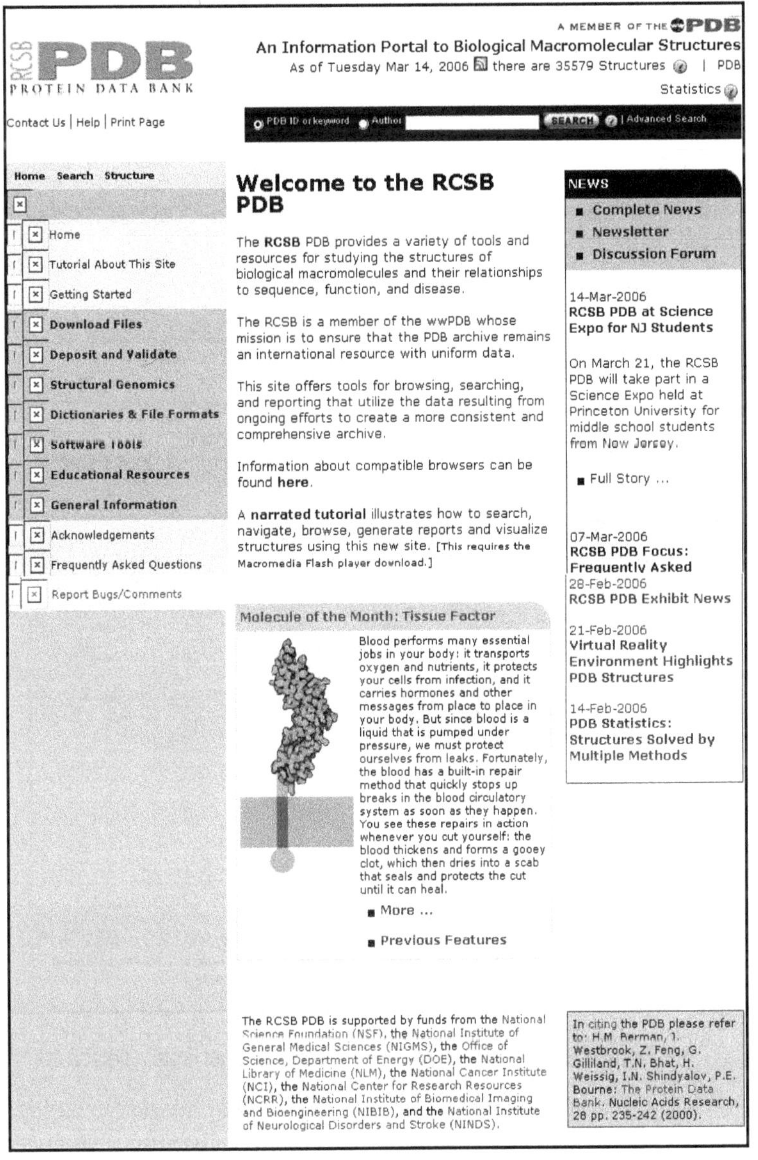

Figure 18.1 RCSB Protein Data Bank Home page

labour-intensive task and its importance cannot be overemphasized. Curated databases are a very useful compendium of biological knowledge.

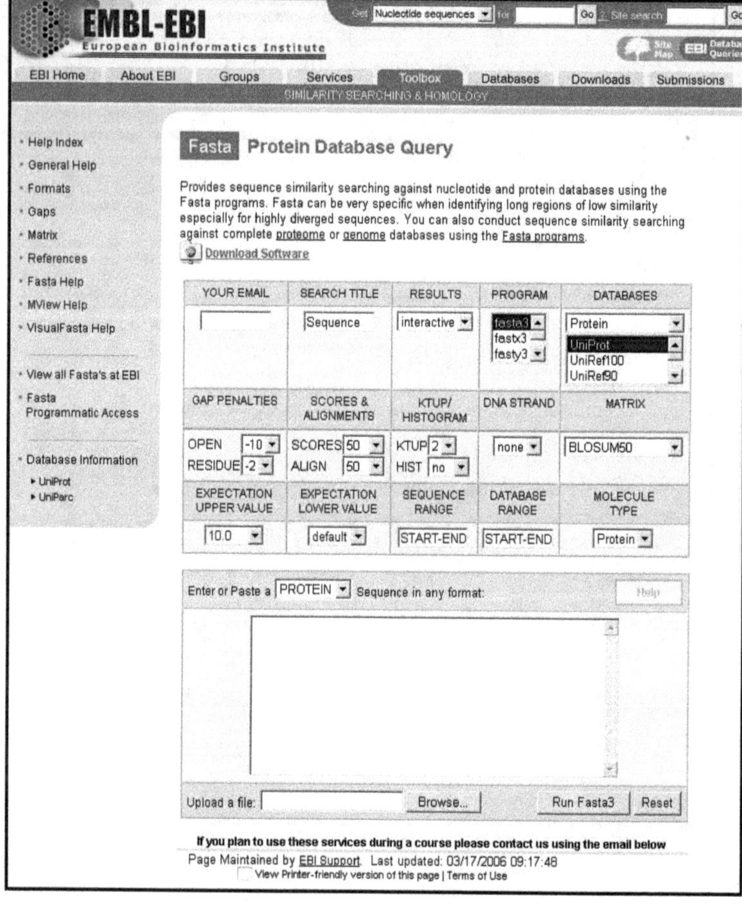

Figure 18.2 Fast a similarity searching against protein databases

It is hence important and of great interest to biologists to analyse the data. For example, having sequenced a particular protein, it is of interest to the biologist to compare it with previously characterized sequences. Sequence analysis is perhaps the most popular of the applications of bioinformatics and it is not just a straightforward database search. Programs such as **Fasta** (http://www.ebi.ac.uk/fasta33/) (Figure 18.2) provide sequence similarity and homology searching against nucleotide and protein databases. Fasta can be very specific when identifying long regions of low similarity especially for highly diverged sequences. Sequence similarity and homology searching against complete proteome or genome databases can be done using the Fasta programs.

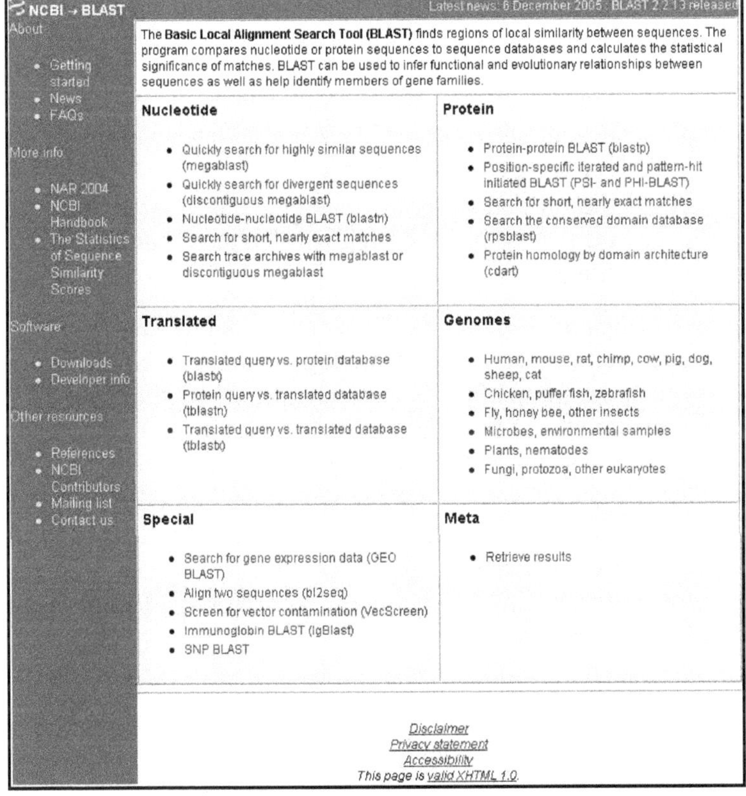

Figure 18.3 NCBI—Basic Local Alignment Search Tool (BLAST)

BLAST® (http://www.ncbi.nlm.nih.gov/BLAST) (Figure 18.3) (Basic Local Alignment Search Tool) is a set of similarity search programs designed to explore all of the available sequence databases regardless of whether the query is a protein or DNA. The BLAST programs have been designed for speed, with a minimal sacrifice of sensitivity to distant sequence relationships. The scores assigned in a BLAST search have a well-designed statistical interpretation, making real matches easier to distinguish from random background hits. BLAST uses a heuristic algorithm that seeks local as opposed to global alignments and is therefore able to detect relationships among sequences that share only isolated regions of similarity.

3. The third level of bioinformatics applications is to use these tools to analyse the data and interpret the results in a biologically meaningful manner. Using bioinformatics, one can conduct global analysis of all the available data with the aim of uncovering common principles that apply across many systems and highlight features that are unique to some.

Comparative genomics is one of such large-scale applications using multiple set of data involving the analysis and comparison of genomes from different species. It can be used to gain a better understanding of the evolution of species and to determine the function of genes and non-coding regions of the genome. A great deal about the function of human genes has been deduced by examining their counterpart genes in simpler organisms such as the mouse. Comparative genomics involves the use of computer programs that can line up multiple genomes and look for regions of similarity among them.

Bioinformatics can impact the traditional wet-lab approaches in several significant fields. The major impact is on the data acquisition, storage, analysis and interpretation of biological data.

INFORMATION SEARCH AND RETRIEVAL

Information search and retrieval is one of the most powerful applications of bioinformatics. For example, PubMed (http://www.ncbi.nlm.nih.gov/PubMed/) (Figure 18.4) which is a service of the National Library of Medicine, provides access to over 12 million MEDLINE citations back to the mid-1960s and additional life science journals. PubMed includes links to many sites providing full text articles and other related resources.

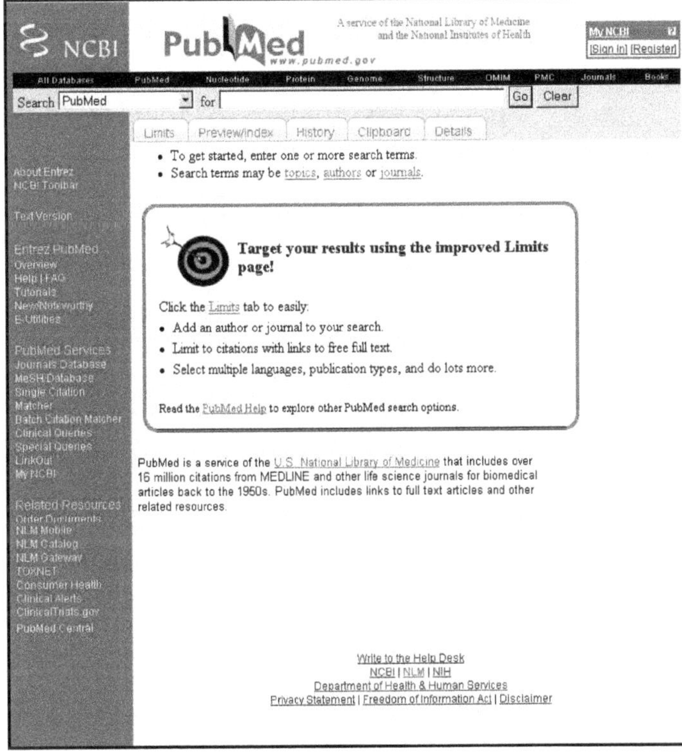

Figure 18.4 PubMed—a service of National Library of Medicine provides access to life science journals and books

eTBLAST (http://chaos.swmed.edu/etblast/index.shtml) is an application that is used to compare a query set of sentences with a database of other text to identify the text in the database that is most similar to the query. A number of different algorithms can be used in eTBLAST. For example, eTBLAST can be used to compare an abstract in a paper, with every abstract in MEDLINE to identify papers and information of interest.

Genetics-Related Applications

There are three types of computational problems in genetics:

1. Analysis of a single sequence to assess similarity with known genes.
2. Identification of typical features such as binding sites, or derivation of evolutionary relationships through phylogenetic trees.
3. Complete genome analysis to identify members of gene families, determination of the chromosomal location of the gene, etc.

One of the most useful and popular applications for the biologists is the sequence comparison or similarity search. The most popular tools for similarity search are BLAST and FASTA. These tools can perform pair-wise comparison of sequences. There are several variants of BLAST: PSI-BLAST, PHI-BLAST, MEGABLAST, RPS-BLAST, etc.

Geneological research and linkage analysis involves the analysis of a large amount of data. Linkage analysis is used to identify the chromosomal location of genes—this has important implications in disease identification. There are several tools that can be used for linkage analysis. Many of these programs are given at http://linkage.rockefeller.edu/.

Phylogenetic Analysis

Phylogenetic analysis is also known as molecular taxonomy. It uses the representation of evolutionary information in the form of phylogenetic trees. There are several methods of conducting phylogenetic analysis. One of the most popular tools is PHYLIP (http://evolution.genetics.washington.edu/phylip.html) (Figure 18.5).

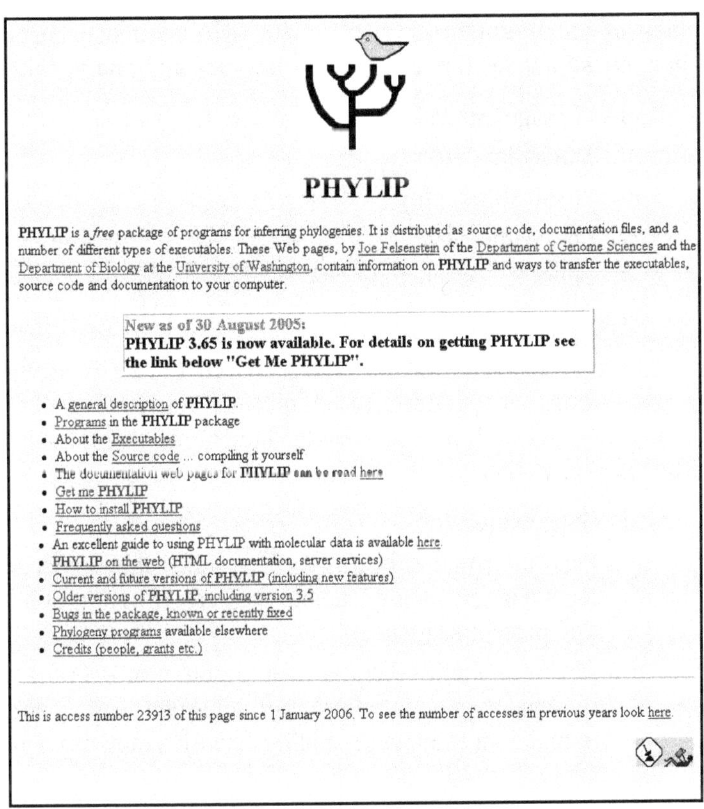

Figure 18.5 PHYLIP Home page

The tree of life (http://tolweb.org/tree/phylogeny.html) is a collaborative Internet project containing information about phylogeny and biodiversity.

Genomics

Genomics is a term coined by **Thomas Roderick** in 1986. It refers to mapping, sequencing, and analysis of genomes. This definition has now enlarged to include the genome function also.

Structural genomics represents the initial phase of genome analysis and leads to the construction of high-resolution genetic, physical, and transcript maps of an organism. The ultimate physical map of an organism is its complete DNA sequence. Structural genomics includes the following techniques and analytical tools:

i. Linkage analysis
ii. Molecular cytogenetics
iii. Physical mapping
iv. EST sequencing
v. Genome sequencing
vi. Genome organization

Functional genomics is characterized by high throughput experimental methodologies combined with statistical and computational analysis of the results. Functional genomics includes the following tools:

i. Gene expression
ii. Forward genetics (begins with a mutant phenotype, defines a gene or genes and leads to the determination of its DNA and protein sequence)
iii. Reverse genetics (begins with a protein or DNA for which there is no genetic information and then works backward to make a mutant gene, ending up with a mutant phenotype gene)
iv. Comparative genomics
v. Proteomics
vi. Metabolomics

Microarrays

The transcriptional profiles of most genes within a genome can be analysed using microarrays and quantitative analysis of microarray data. Transcriptional profile can help to generate gene expression data that can be used to define a cell type or condition. There is a huge amount of data produced in any microarray experiment and hence computational methods are essential to conduct this analysis.

Microarrays utilize the preferential binding of complementary single-stranded nucleic acid (cDNA) sequences. Microarray is built using a glass slide, on to which cDNA molecules are attached at fixed locations. There are a very large number of spots on an array, each containing a huge number of identical DNA molecules, of lengths from twenty to hundreds of nucleotides. Gene expression monitoring and SNP detection are the two most important applications of microarray technology.

ArrayExpress (http://www.ebi.ac.uk/arrayexpress/) (Figure 18.6) is a public repository for microarray data, which is aimed at storing annotated data.

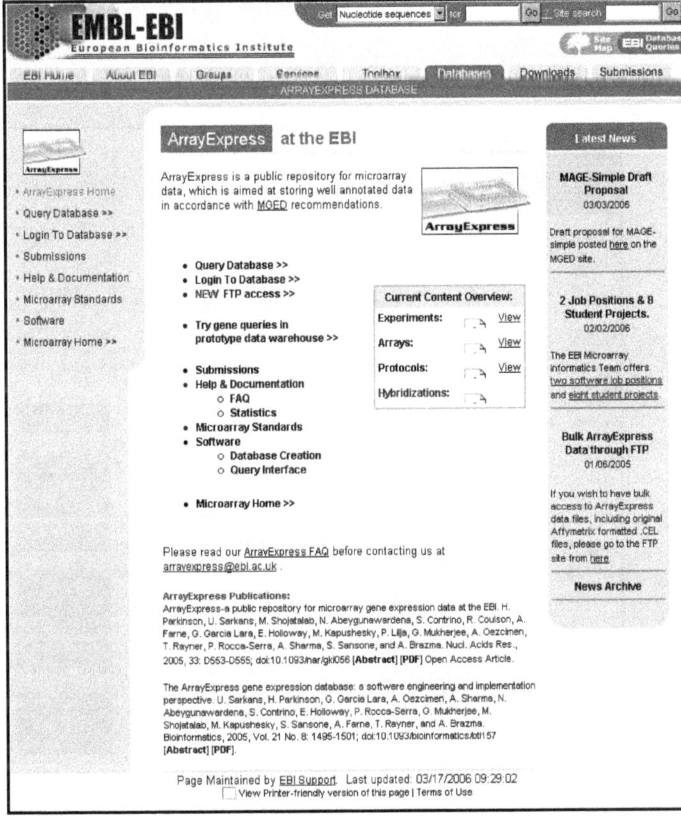

Figure 18.6 EBI Database—Array Express Home page

Sequence Assembly

The Human Genome Project popularized sequence assembly tools. The sequence assembly tools were extensively used to efficiently and accurately assemble large numbers of individual sequence reads. There are several tools for conducting sequence assembly.

The phred/phrap/consed suite of programs (http://www.phrap.org/) is used to do the assembly and finishing of shotgun sequencing information. PolyPhred (http://droog.mbt.washington.edu/PolyPhred.html) (Figure 18.7) is a tool integrated with these above tools and is used for comparison with fluorescence-based sequences across traces obtained from different individuals to identify heterozygous sites for single nucleotide substitutions.

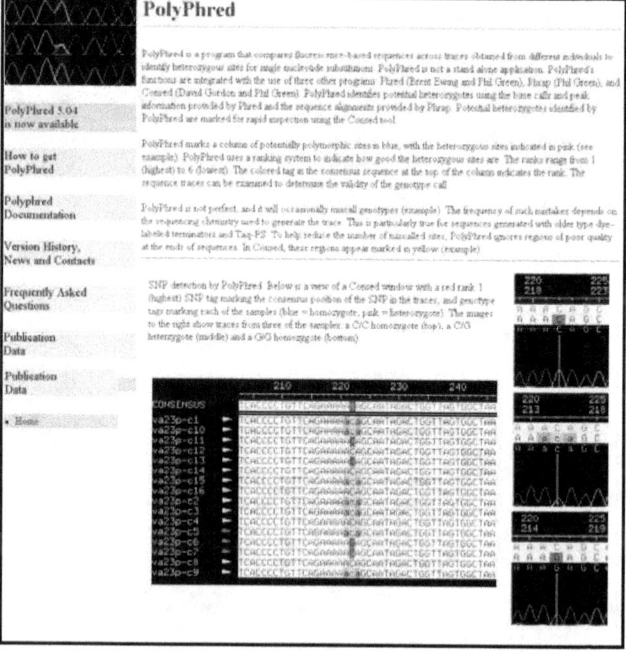

Figure 18.7 PolyPhred—a program that compares fluorescence-based sequences across traces obtained from different individuals to identify heterozygous sites for single nucleotide substitutions

There are various tools available for conducting genome annotation. The Genome Annotation Consortium (http:// compbio.ornl.gov/gac/) is a multi-institution collaboration to assist in the annotation and analysis of genome sequences by bioinformatics. Some of the popular tools for genome annotation are GRAIL, GRAILExp and Pipeline III at Genome Annotation Consortium.

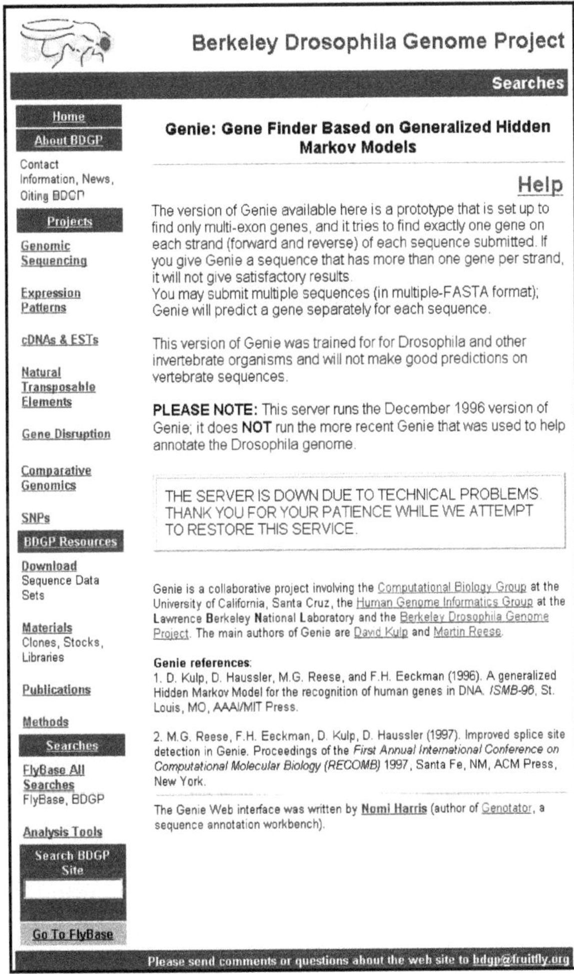

Figure 18.8 Genie: Gene finder based on generalized Hidden Morkov Models

Genie (http://www.fruitfly.org/seq_tools/genie.html) (Figure 18.8) is a tool to locate genes and it uses Hidden Markov Models (HMMs). There are several tools available at (http:// www.softberry.com/) (Figure 18.9).

Figure 18.9 SoftBerry Home page

Proteomics

The term 'proteomics' was coined by Marc Wilkins in 1995. Proteomics is the study of proteomes (proteins expressed by a genome). Proteome refers to all proteins produced by a species and it varies with time.

Proteomics is a rapidly evolving field and there has been a very fast development in proteomics technologies like mass spectrometry for rapid and quantitative measurements of proteins in a complex mixture. This technology has led to the development of protein sequence databases and has immense applications in health care, drug research and diagnostics.

Pharmacogenomics

Pharmacogenomics is the study of how an individual's genetic inheritance affects the body's response to drugs. The term comes from the words pharmacology and genomics and is thus the intersection of pharmaceuticals and genetics. Pharmacogenomics applies the information in molecular biology and genetic tools to drug design, discovery and clinical development.

HapMap (http://www.genome.gov/) (Figure 18.10) is a genetic variation-mapping project launched by an international consortium to help identify genetic contributions to common diseases. It is aimed at "speeding the discovery of genes related to common illnesses such as asthma, cancer, diabetes and heart disease".

DRUG DISCOVERY AND COMPUTER-AIDED DRUG DESIGN

Drug discovery process usually starts with an analysis of binding sites in target proteins, or an identification of structural features common to active compounds. The process ends with the generation of small molecule "leads" suitable for further chemical synthetic work. This process includes analysis of drug–ligand complexes, quantitative assessment of binding interactions and pharmacophore development.

Computer-aided drug design (CADD) is a recent and emerging discipline that uses several bioinformatics tools and allied fields like chemoinformatics and combinational chemistry. This is a very important commercial application and has seen the introduction of tools like Quantitative Structure Activity Relationship (QSAR), structure-based design, combinational library design, etc.

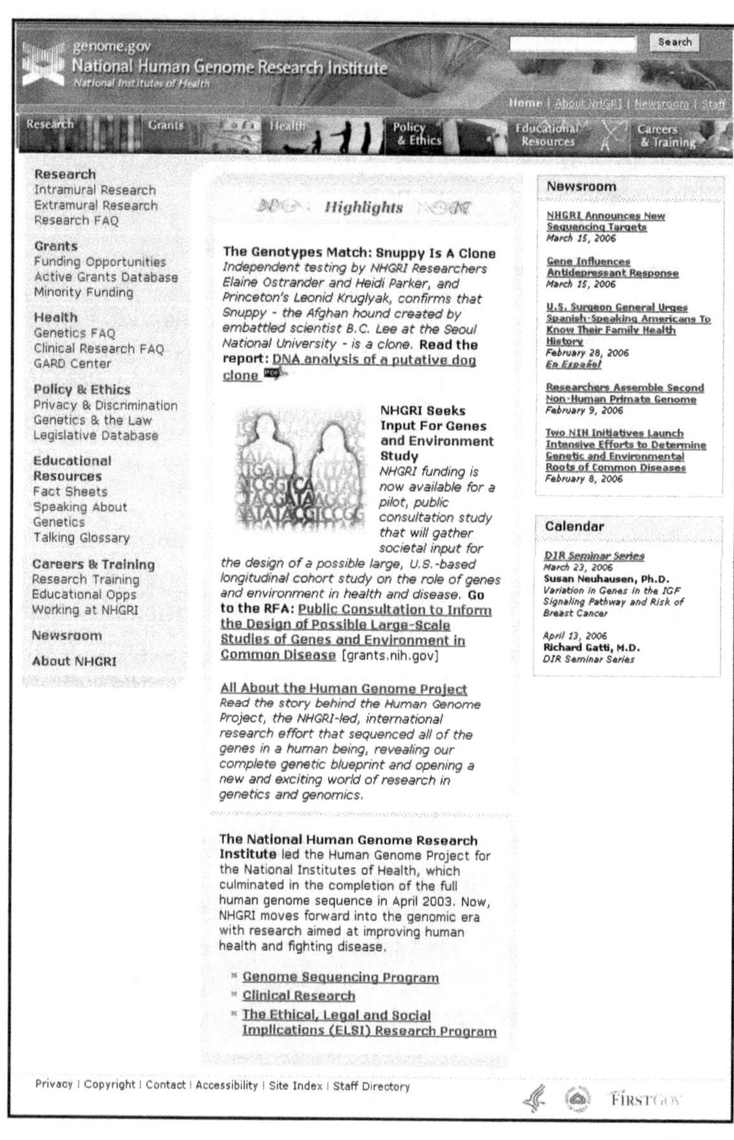

Figure 18.10 Genome.gov—National Human Genome Research Institute—Home page

SYSTEMS BIOLOGY

Systems biology is an integrated multidisciplinary approach in which one can study pathways and networks. Systems biology has applications that encompass all areas of biology, including drug discovery. It is an evolving field that includes dynamic modeling of pathways and even the cell, multivariate analysis, and Bayesian networks.

The goal of Alliance for Cellular Signaling (**http://www.cellularsignaling.org/**) (Figure 18.11) is "to understand as completely as possible the relationships between sets of inputs and outputs in signaling cells that vary both temporally and spatially. The same goal, stated from a slightly different perspective, is to understand fully how cells interpret signals in a context-dependent manner."

Systems biology is highly dependant on experimental design like the e-cell (modeling and simulation environment for biochemical and genetic processes). NRCAM (**http://www.nrcam.uchc.edu/**) (Figure 18.12) has developed a "general computational tool, the Virtual Cell for modeling cellular processes. This new technology associates biochemical and electrophysiological data describing individual reactions with experimental microscopic image data and their subcellular locations".

TOOLS AND DATABASES IN BIOINFORMATICS

A biological database is a large organized body of persistent data, usually associated with computerized software designed to update, query and retrieve components of data stored within the system. Making the biological information available for analysis, and developing applications is the key and there are huge numbers of databases in the public and private domain to do so. Some of the major databases are discussed here.

Protein Databases

Protein sequence databases are categorized as primary and composite or secondary. Primary protein databases contain over

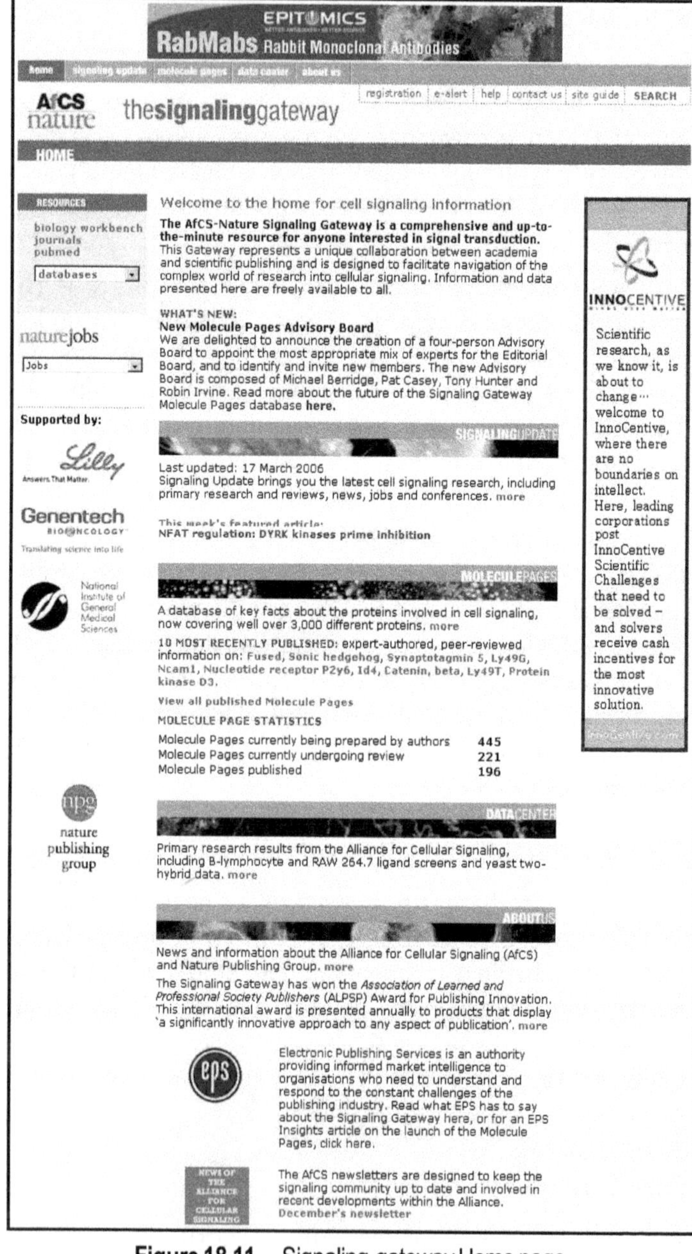

Figure 18.11 Signaling-gateway Home page

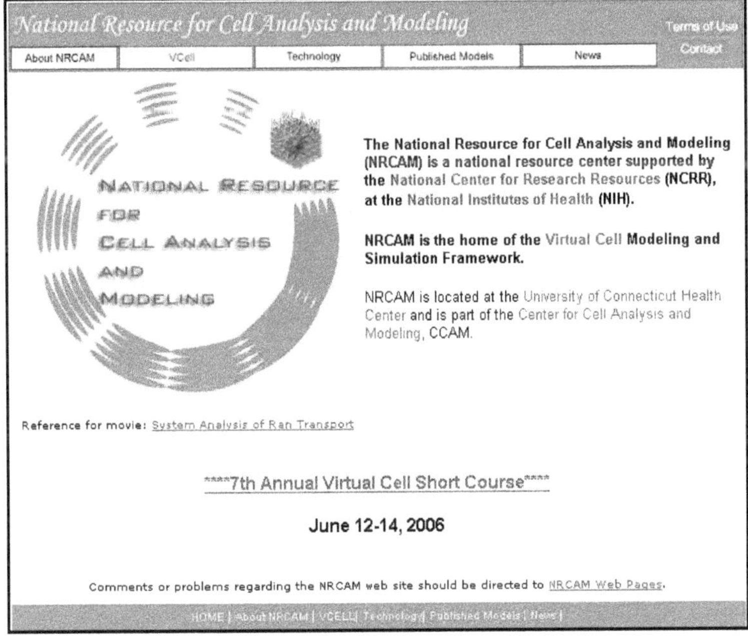

Figure 18.12 National Resource for Cell Analysis and Modeling (NRCAM)

300,000 protein sequences that function as a repository for the raw data. Some of the common repositories such as SWISS-PROT (http://us.expasy.org/sprout/) and Protein Information Resource (PIR) (http://pir.georgetown.edu/) (Figure 18.13) annotate the sequences as well as describe the functions of proteins, its domain structure and post-translational modifications. SWISS-PROT and PIR are both curated: groups of designated scientists have prepared the entries from literature and/or contacts with external experts.

SWISS-PROT is a curated protein sequence database that strives to providc a high level of annotations (such as the description of the function of a protein, its domain structure, post-translational modifications, variants, etc.), a minimal level of redundancy and high level of integration with other databases.

TrEMBL is a computer-annotated supplement of SWISS-PROT that contains all the translations of EMBL nucleotide sequence entries not yet integrated in SWISS-PROT.

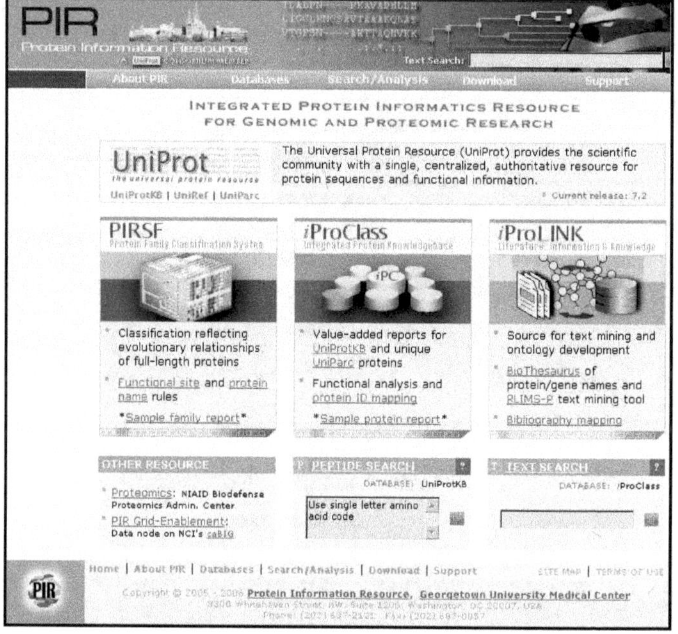

Figure 18.13 Protein Information Resource (PIR)

The databases can be accessed and searched through the SRS system (http://us.expasy.org/srs5/) (Figure 18.14) at ExPASy, or one can download the entire database as one single flat file. The SWISS-PROT database has some legal restrictions—the entries themselves are copyrighted, but freely accessible and usable by researchers.

PIR (pir.georgetown.edu) produces the Protein Sequence Database (PSD) of functionally annotated protein sequences, which grew out of the Atlas of Protein Sequence and Structure (1965–1978) edited by Margaret Dayhoff and has been incorporated into an integrated knowledge base system of value-added databases and analytical tools. Although SWISS-PROT and PIR overlap extensively, there are still many sequences that can be found in only one of them.

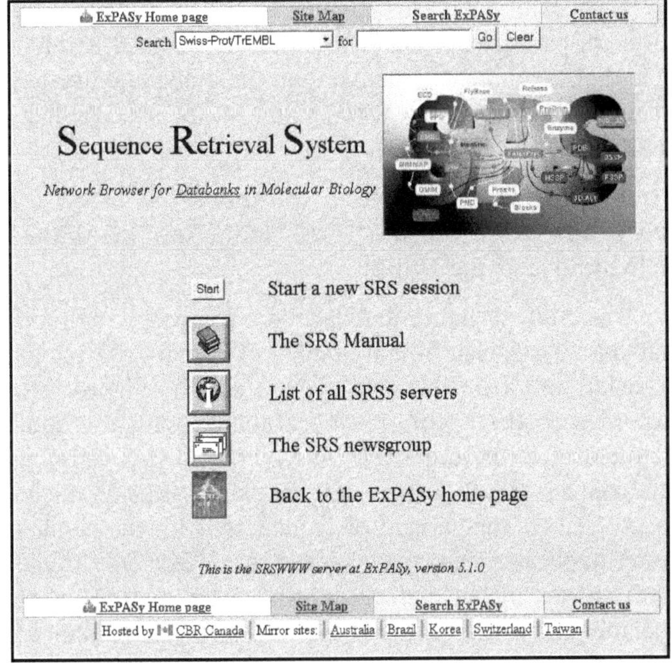

Figure 18.14 Sequence Retrieval System at ExPASy

PIR has collaborated with EBI (European Bioinformatics Institute) and SIB (Swiss Institute of Bioinformatics) to establish the UniProt (United Protein Databases), the central resource of protein sequence and function.

iProClass (http://pir.georgetown.edu/iproclass/) is a central point for exploration of protein information. iProClass provides summary descriptions of protein family, function and structure for PIR-PSD, SWISS-PROT, and TrEMBL sequences, with links to over 50 biological databascs.

PIR-NREF (http://pir.georgetown.edu/iproclass/) is a comprehensive database for sequence searching and protein identification, contains non-redundant protein sequences from PIR-PSD, SWISS-PROT, TrEMBL, RefSeq, GenPept, and PDB.

Composite databases compile and filter sequence data from different primary databases to produce combined non-redundant sets that are more exhaustive than the individual databases and also include protein sequence data from the translated coding regions in DNA sequence databases.

An example of such databases is OWL (http://www.hgmp.mrc.ac.uk/Bioinformatics/Databases/owl-help.html) and the NRDB.

The OWL sequence database is a composite, non-redundant database assembled from a number of primary sources including translations of nucleic acid sequences. The highest priority is accorded to the SWISS-PROT databank, with the addition of sequences extracted from NBRF/PIR (PIRI-3 only) and the Brookhaven PDB 3-D structural database (NRL3-D). Redundancy is avoided by comparison of sequences, with the elimination of exact duplicates and of sequences that differ only trivially. The data entries include references and other textual information, including cross-references to the PDB 3-D structural and PRINTS databases.

PROSITE (www.expasy.ch/prosite/) (Figure 18.15) is a database of short sequence patterns and profiles that characterize biologically significant sites in proteins. PROSITE is a database of protein families and domains. It consists of biologically significant sites, patterns and profiles that help to reliably identify a new sequence with the known protein family. PROSITE has been extended to contain also some profiles, which can be described as probability patterns for specific protein sequence families.

Pfam (www.sanger.ac.uk/Software/Pfam/) (Figure 18.16), www.cgr.ki.se/Pfam/) is a database of protein families defined as domains (contiguous segments of entire protein sequences). For each domain, it contains a multiple alignment of a set of defining sequences and the other sequences in SWISS-PROT and TrEMBL that can be matched to that alignment.

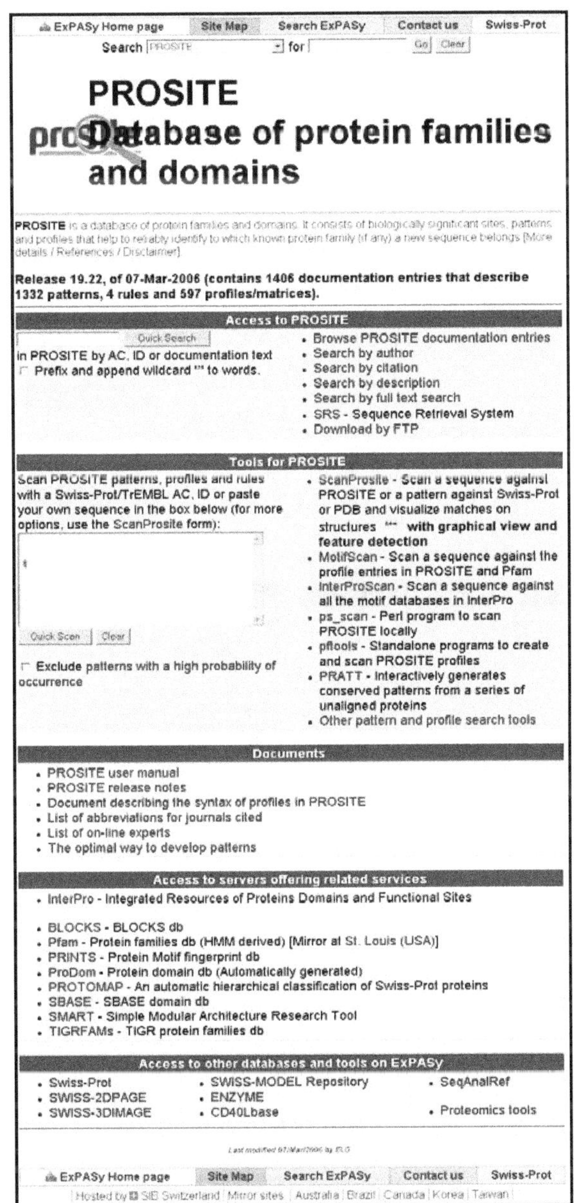

Figure 18.15 ExPASy–PROSITE Database of protein families and domains

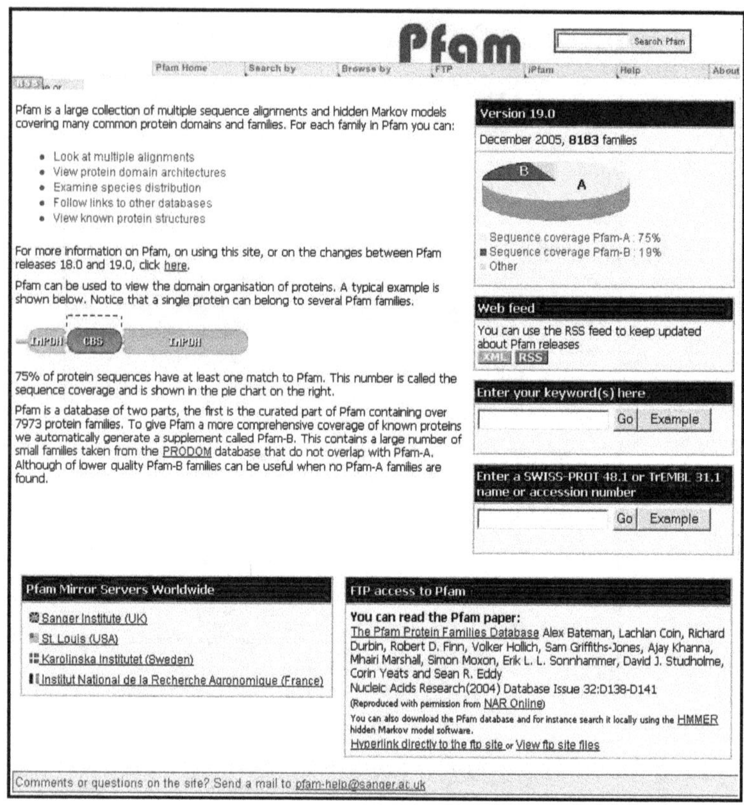

Figure 18.16 Pfam Home page

The Pfam database can be searched, or used to identify domains in a sequence, or downloaded from the websites above. It is licensed under the GNU General Public License, which basically makes it available to anyone, but imposes the restriction that derivative works (new databases, modifications) must be made available in source form.

Structural Databases

Structural databases are related to macromolecular structures. PDB (www.rcsb.org/pdb/) is the main primary database for 3-D structures of macromolecules determined by X-ray crystallography and NMR. The PDB entries contain the atomic coordinates, and some structural parameters connected with the atoms, or computed from the structures (secondary structure). The PDB entries contain some annotation, but it is not as comprehensive as in SWISS-PROT. PDB provides a primary archive of all 3-D structures for macromolecules such as proteins, RNA, DNA and various complexes.

As the information provided in individual PDB entries can be difficult to extract, PDBsum (**http://www.biochem.ucl.ac.uk/bsm/pdbsum/doc/**) provides summary information and derived data on entries in the PDB. The summary information gives an overview of the contents of each PDB entry in terms of numbers of protein, ligands, metal ions, etc. The derived data include PROMOTIF analyses, summary PROCHECK statistics, links to CATH and other databases, etc.

There are two major databases that classify proteins based on the structure in order to identify structural and phylogenetic relationships. They are CATH (**www.biochem.ucl.ac.uk/bsm/cath/**) (Figure 18.17) and SCOP (**scop.mrc-lmb.cam.ac.uk/scop/**) (Figure 18.18) databases.

The CATH database is a hierarchical classification of protein domain structures, which form clusters at four major structural levels that include Class (C), Architecture (A), Topology (T) and Homologous superfamily (H).

Class, derived from secondary structure content, is assigned for more than 90% of protein structures automatically. Architecture, which describes the gross orientation of secondary structures independent of connectivity, is currently assigned manually. The topology level clusters the structures according to their topological connections and numbers of secondary structures. The homologous superfamilies cluster proteins with highly similar structures and functions. The assignments of structures to topology families and homologous superfamilies are made by sequence and structure comparisons.

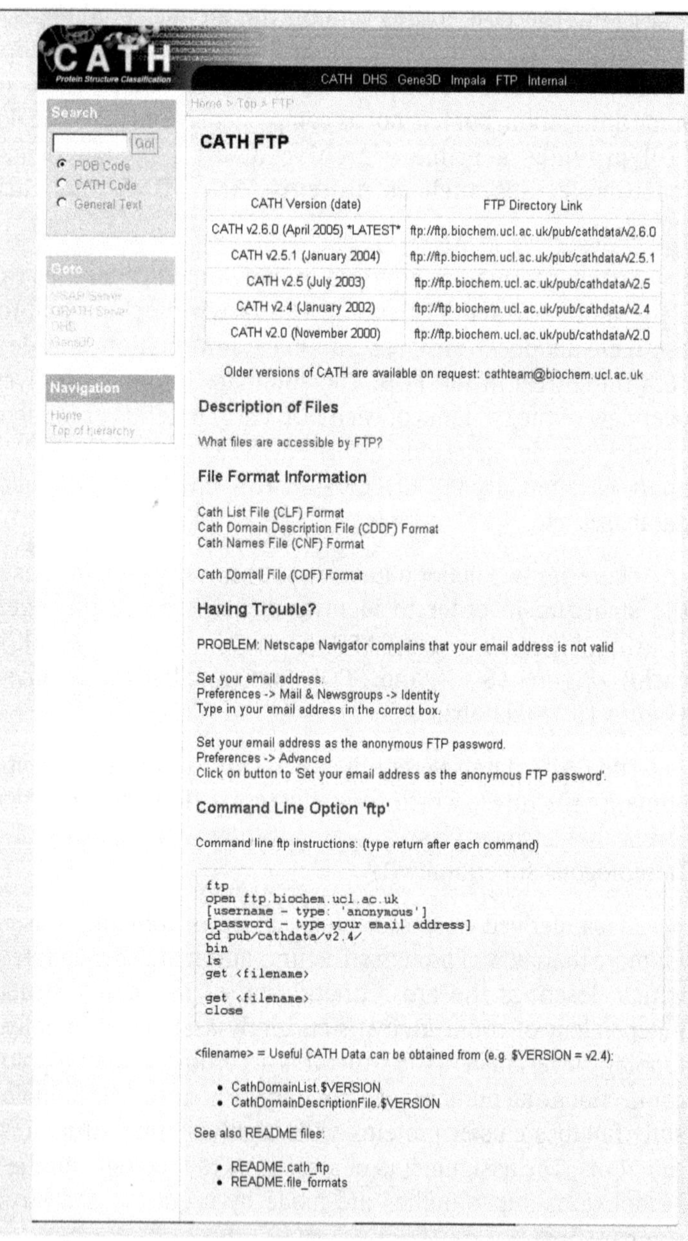

Figure 18.17 CATH Protein structure classification Database

Structural Classification of Proteins

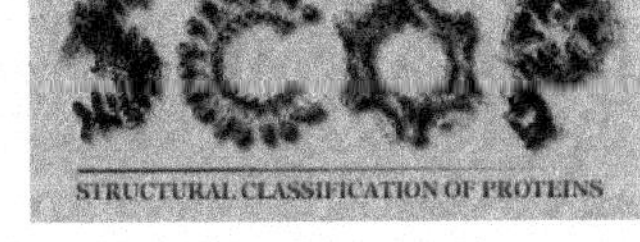

Welcome to SCOP: Structural Classification of Proteins.
1.69 release (July 2005)

25973 PDB Entries. 1 Literature Reference. 70859 Domains.
(excluding nucleic acids and theoretical models).
Folds, superfamilies, and families statistics here.
New folds superfamilies families.
List of obsolete entries and their replacements.

Authors. Alexey G. Murzin, John-Marc Chandonia, Antonina Andreeva, Dave Howorth, Loredana Lo Conte, Bartlett G. Ailey, Steven E. Brenner, Tim J. P. Hubbard, and Cyrus Chothia. scop@mrc-lmb.cam.ac.uk
Reference: Murzin A. G., Brenner S. E., Hubbard T., Chothia C. (1995). SCOP: a structural classification of proteins database for the investigation of sequences and structures. *J. Mol. Biol.* 247, 536-540. [PDF]
Recent changes are described in: Lo Conte L., Brenner S. E., Hubbard T.J.P., Chothia C., Murzin A. (2002). SCOP database in 2002: refinements accommodate structural genomics. *Nucl. Acid Res.* 30(1), 264-267. [PDF] and
Andreeva A., Howorth D., Brenner S.E., Hubbard T.J.P., Chothia C., Murzin A.G. (2004). SCOP database in 2004: refinements integrate structure and sequence family data. *Nucl. Acid Res.* 32:D226-D229. [PDF].

Access methods

- **Enter SCOP at the top of the hierarchy**
- Keyword search of SCOP entries
- SCOP parseable files
- All SCOP releases and reclassified entry history
- SCOP domain sequences and pdb-style coordinate files (ASTRAL)

> NEW **ASTRAL survey - please help our partners!**
>
> The ASTRAL authors are currently seeking NIH funding for further maintenance and development of the ASTRAL compendium, which has been critical for SCOP production since its creation.
>
> Please take the survey on the ASTRAL website, as it will greatly increase the chances of future ASTRAL funding.

- Hidden Markov Model library for SCOP superfamilies (SUPERFAMILY)
- Online resources of potential interest to SCOP users

SCOP **mirrors** around the world may speed your access.

News

- SCOP has been updated to include all PDB entries released up to 1 October 2004. See folds, superfamilies, and families statistics.
- The process of SCOP production is gradually being changed and has resulted in some internal and external database information being incorporated in new names and comments. This information is primarily intended for use in the ongoing SCOP redevelopment but it has been displayed on the web pages as it may be of some help to users.
- Links to some external databases are still being calculated, and may be updated when this process is completed.
- This release is similar in appearance to the previous release, so the generic release notes from that release still apply. Please read the notes; they contain more detailed explanations and examples of SCOP features.

- Previous releases' news.

Synopsis

Nearly all proteins have structural similarities with other proteins and, in some of these cases, share a common evolutionary origin. The **SCOP** database, created by manual inspection and abetted by a battery of automated methods, aims to provide a detailed and comprehensive description of the structural and evolutionary relationships between all proteins whose structure is known. As such, it provides a broad survey of all known protein folds, detailed information about the close relatives of any particular protein, and a framework for future research and classification.

A more detailed description of the database is available. Help on using the database may be obtained on any screen by pressing the question mark button.

Online resources of potential interest to scop users

- Structural similarity search of SCOP using SSM
- Combinatorial Extension (CE) method for structural comparison
- PALI pairwise and multiple alignments of SCOP families
- SUPFAM structure/sequence relationships
- Stuctural similarity search of SCOP using 3dSearch
- Stuctural alignment of SCOP sequences (database + server)
- PINTS - Patterns In Non-homologous Tertiary Structures

- Sequence similarity search of SCOP using FPS

- CATH structural classification
- Dali structural comparison and FSSP structural classification
- PDB at a Glance
- 3Dee Protein Domain Definitions

- Protein Data Bank (PDB)
- Macromolecular Structure Database (EBI)
- Nucleic Acid Database (NDB)

- Swiss-Model
- Macromolecular Motions Database

- The PRESAGE Database for Structural Genomics
- Genome Census
- Function assignment and metabolic models

- Licensing information for commercial users (MRC site)
- Glossary of terms used in the fold classification
- References related to fold classification methods and fold definitions

Figure 18.18 SCOP: Structural Classification of Proteins

The SCOP (Structural Classification of Proteins) database created by manual inspection and abetted by a battery of automated methods, aims to provide a detailed and comprehensive description of the structural and evolutionary relationships between all proteins whose structures are known. As such, it provides a broad survey of all known protein folds, detailed information about the close relatives of any particular protein, and a framework for future research and classification.

Nucleotide and Genome Sequences

GenBank (www.ncbi.nlm.nih.giv/Genbank/) is the NIH genetic sequence database, an annotated collection of all publicly available DNA sequences. A new release is made every two months. GenBank is part of the International Nucleotide Sequence Database Collaboration, which is comprised of the DNA DataBank of Japan

(DDBJ), the European Molecular Biology Laboratory (EMBL), and GenBank at the National Center for Biotechnology Information. These three organizations exchange data on a daily basis.

Each GenBank entry includes a concise description of the sequence, scientific nomenclature and taxonomy of the source organism, and a table of features that identifies coding regions and other sites of biological significance, such as transcription units, sites of mutations or modifications, and repeats. Protein translations for coding regions are included in the feature table. Bibliographic references are included along with a link to the Medline unique identifier for all published sequences.

The European Bioinformatics Institute (EBI) in Hinxton, Cambridge, UK maintains The European Molecular Biology Laboratory (EMBL) (www.ebi.ac.uk/embl/) (Figure 18.19) nucleotide sequence database. It can be accessed and searched through the SRS system at EBI, or one can download the entire database as flat files.

UniGene (www.ncbi.nlm.nih.gov/UniGene/) system attempts to process the GenBank sequence data into a non-redundant set of gene-oriented clusters. Each UniGene cluster contains sequences that represent a unique gene, as well as related information such as the tissue types in which the gene has been expressed and map location.

SGD (genome-www.stanford.edu/Saccharomyces/) *Saccharomyc*es Genome Database is a scientific database of the molecular biology and genetics of the yeast *Saccharomyces cerevisiae.*

EBI Genomes (www.ebi.ac.uk/genomes/) web site provides access and statistics for the completed genomes, and information about ongoing projects.

Ensembl (www.ensembl.org) is a joint project between EMBL-EBI and the Sanger Centre to develop a software system that produces and maintains automatic annotation on eukaryotic genomes.

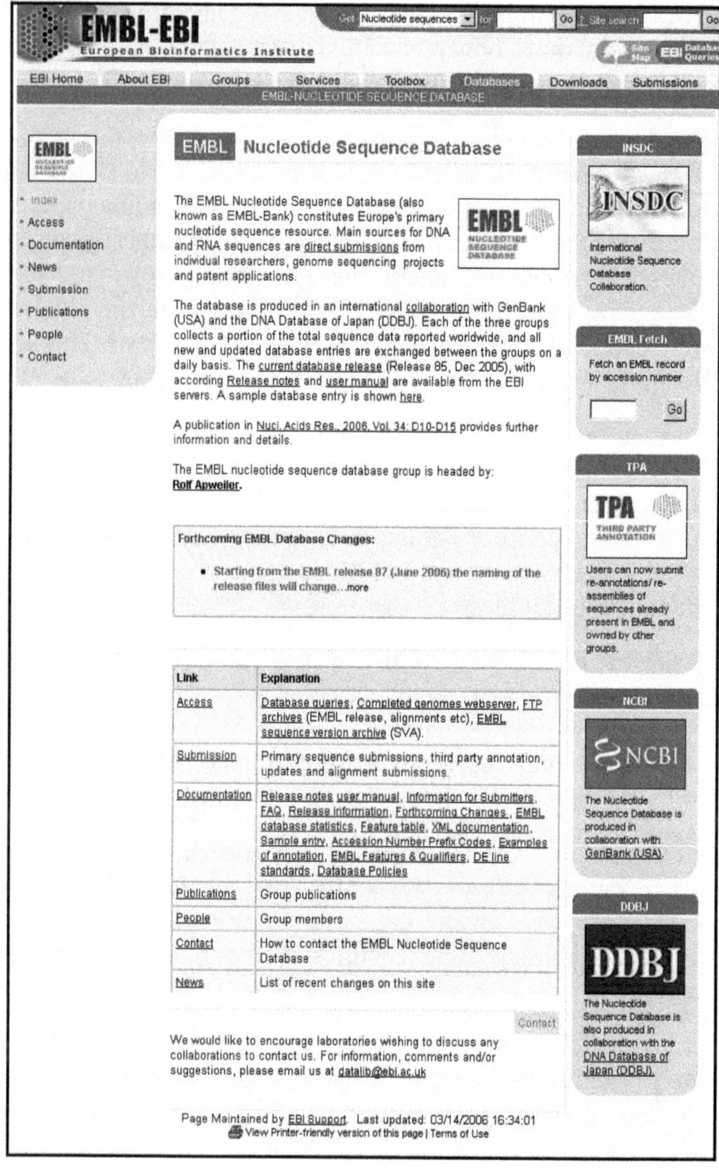

EMBL-EBI
European Bioinformatics Institute

Get Nucleotide sequences for Go Site search Go

Site Map | EBI Database Queries

EBI Home | About EBI | Groups | Services | Toolbox | Databases | Downloads | Submissions

EMBL-NUCLEOTIDE SEQUENCE DATABASE

- Index
- Access
- Documentation
- News
- Submission
- Publications
- People
- Contact

EMBL Nucleotide Sequence Database

The EMBL Nucleotide Sequence Database (also known as EMBL-Bank) constitutes Europe's primary nucleotide sequence resource. Main sources for DNA and RNA sequences are direct submissions from individual researchers, genome sequencing projects and patent applications.

The database is produced in an international collaboration with GenBank (USA) and the DNA Database of Japan (DDBJ). Each of the three groups collects a portion of the total sequence data reported worldwide, and all new and updated database entries are exchanged between the groups on a daily basis. The current database release (Release 85, Dec 2005), with according Release notes and user manual are available from the EBI servers. A sample database entry is shown here.

A publication in Nucl. Acids Res., 2006, Vol. 34: D10-D15 provides further information and details.

The EMBL nucleotide sequence database group is headed by: **Rolf Apweiler.**

Forthcoming EMBL Database Changes:

- Starting from the EMBL release 87 (June 2006) the naming of the release files will change...more

Link	Explanation
Access	Database queries, Completed genomes webserver, FTP archives (EMBL release, alignments etc), EMBL sequence version archive (SVA).
Submission	Primary sequence submissions, third party annotation, updates and alignment submissions.
Documentation	Release notes user manual, Information for Submitters, FAQ, Release Information, Forthcoming Changes, EMBL database statistics, Feature table, XML documentation, Sample entry, Accession Number Prefix Codes, Examples of annotation, EMBL Features & Qualifiers, DE line standards, Database Policies
Publications	Group publications
People	Group members
Contact	How to contact the EMBL Nucleotide Sequence Database
News	List of recent changes on this site

Contact

We would like to encourage laboratories wishing to discuss any collaborations to contact us. For information, comments and/or suggestions, please email us at datalib@ebi.ac.uk

Page Maintained by EBI Support. Last updated: 03/14/2006 16:34:01
View Printer-friendly version of this page | Terms of Use

INSDC
International Nucleotide Sequence Database Collaboration.

EMBL Fetch
Fetch an EMBL record by accession number
Go

TPA
TPA THIRD PARTY ANNOTATION
Users can now submit re-annotations/ re-assemblies of sequences already present in EMBL and owned by other groups.

NCBI
The Nucleotide Sequence Database is produced in collaboration with GenBank (USA).

DDBJ
The Nucleotide Sequence Database is also produced in collaboration with the DNA Database of Japan (DDBJ).

Figure 18.19 THE EMBL Nucleotide Sequence Database

The Entrez (http://www.ncbi.nlm.nih.gov/Entrez/) (Figure 18.20) is a retrieval system for searching several linked databases. It provides access to several databases.

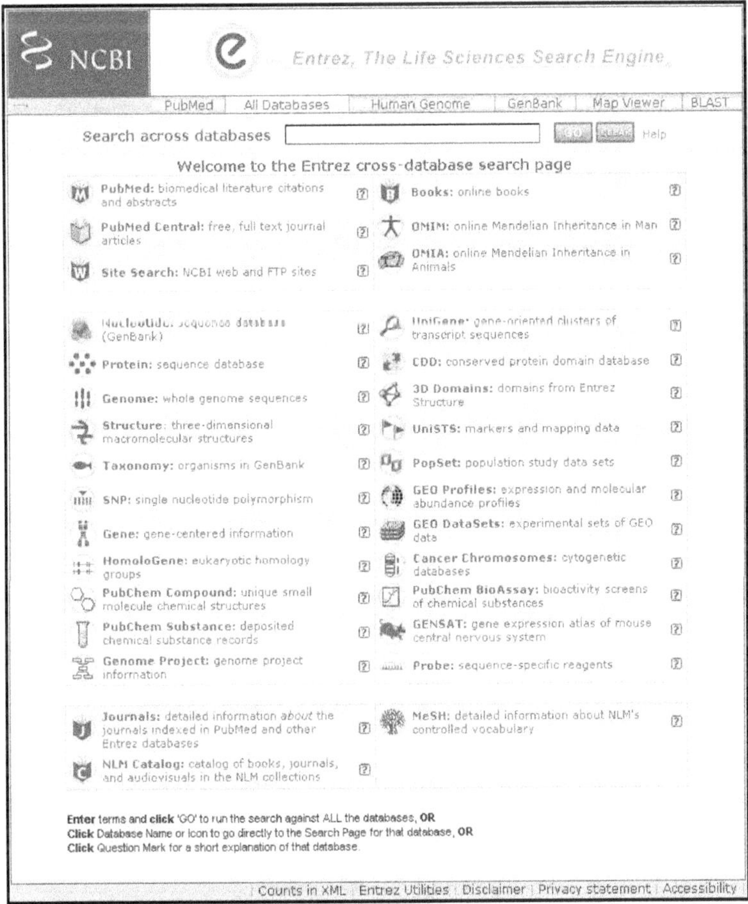

Figure 18.20 Entrez cross-database search

GeneCensus (http://bioinfo.mbb.yale.edu/genome/) provides an entry point for genome analysis with an interactive whole-genome comparison from an evolutionary perspective. The database allows building of phylogenetic trees based on different

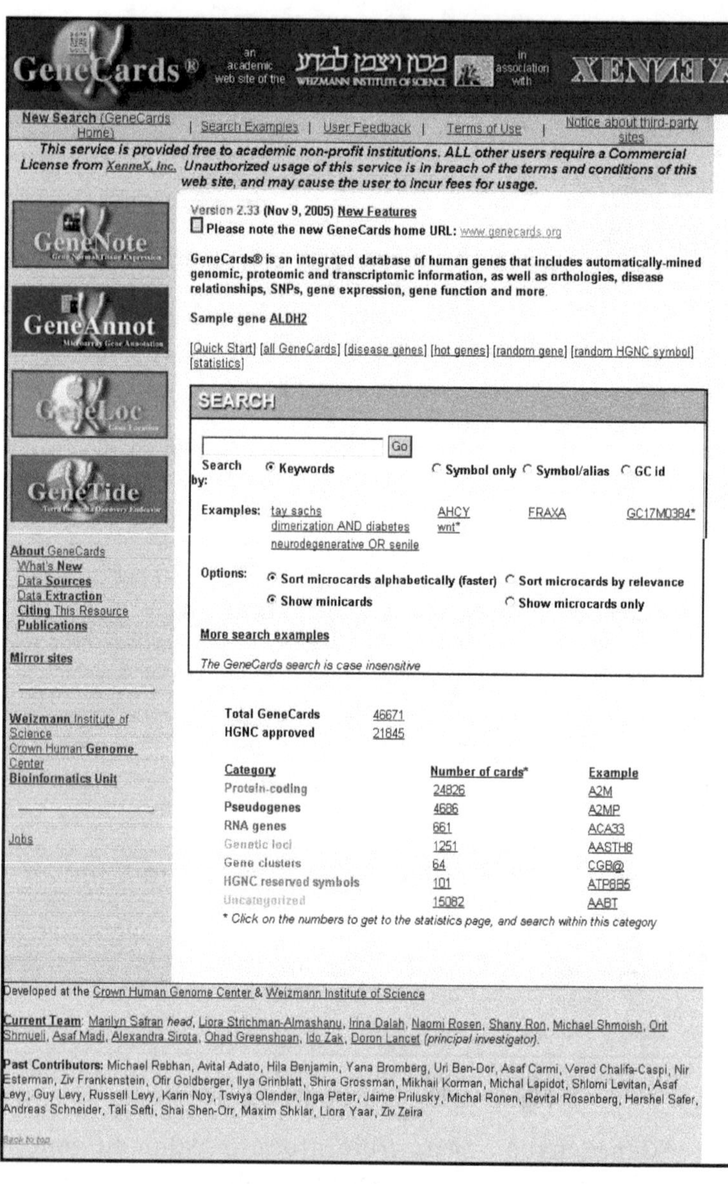

Figure 18.21 GeneCards Home page

criteria such a ribosomal RNA or protein fold occurrence. The site also enables multiple genome comparisons, analysis of single genomes and retrieval of information for individual genes.

Gene Expression Data

Three main technologies are used in gene expression data analysis: the cDNA microarray, Affymetrix GeneChip and SAGE methods. The first method measures relative levels of mRNA abundance between different samples, while the other two measure absolute levels. Most of the effort in gene expression analysis has concentrated on the yeast and human genomes.

Molecular Profiling project (**http://llmpp.nih.gov/lymphoma/**) provides data from microarray experiments on human cancer cells.

There are many more databases, some of which provide very specialized information. GeneCards (**bioinformatics.weizmann.ac.il/cards/**) (Figure 18.21) is a database of human genes, their products and their involvement in diseases. It offers concise information about the functions of all human genes that have an approved symbol, as well as selected others. It is a secondary database, which contains many links to other databases, and attempts to consolidate the information that is available for a specific class or entity, in this case human genes.

KEGG (**www.genome.ad.jp/kegg/**) stands for the Kyoto Encyclopedia of Genes and Genomes (KEGG). It is an effort to computerize current knowledge of molecular and cellular biology in terms of the information pathways that consist of interacting molecules or genes and to provide links from the gene catalogues produced by genome sequencing projects.

REVIEW YOUR LEARNING

1. Define/explain the following:
   i. Bioinformatics
   ii. Genomics

   iii. Microarrays
   iv. Proteomics
   v. Pharmacogenomics
2. What are the various types of protein databases?
3. What are the various types of nucleotide and genome sequence databases?
4. Explain the tools and databases in Bioinformatics.
5. Describe the applications of Bioinformatics.

Appendix I

UNITS OF MEASUREMENTS

Weight

1 kg	=	1000 g
1 gm	–	1000 mg
1 mg	=	1000 μg (microgram)
1 μg	=	1000 ng (nanogram)
1 mg	=	10^{-3} g
1 μg	=	10^{-6} g
1mg	=	10^{-9} g

Volume

1 L	=	1000 ml
1 ml	=	1000 μl (microlitre)
1 ml	=	10^{-3} L
1 μl	=	10^{-6} L

Note: 1 cc = 1.0004 ml (approx.). But for practical purposes 1 ml = 1cc.

Length

1 m	=	100 cm
1 cm	=	10 mm
1 mm	=	1000 μm (micrometre)
1 μ (1 micron)	=	1000 mμ (millimicrons)

1 mμ	=	1 nm (nanometre)
1 nm	=	10 Å (Angstrom units)

Strength of solution

1M	=	1000 mM (millimolar)
0.1M	=	100 mM
0.01M	=	10 mM
0.001M	=	1 mM
1 mM	=	1000 μM (micromoles)
1Eq/L	=	1000 mEq/L
1 mEq/L	=	(mg/dl $\times$ 10) $\div$ Eq.wt.

Appendix II

PREFIXES USED IN THE INTERNATIONAL SYSTEM OF UNITS

Prefix	Multiple	Abbreviation
Exa	10^{18}	E
Peta	10^{15}	P
Tera	10^{12}	T
Giga	10^{9}	G
Mega	10^{6}	M
Kilo	10^{3}	k
deci	10^{-1}	d
centi	10^{-2}	c
milli	10^{-3}	m
micro	10^{-6}	m
nano	10^{-9}	n
pico	10^{-12}	p
femto	10^{-15}	f
atto	10^{-18}	a

Appendix III

GREEK ALPHABETS

Α	α	alpha	Ο	ο	omicron
Β	β	beta	Π	π	pi
Γ	γ	gamma	Ρ	ρ	rho
Δ	δ	delta	Σ	σ	sigma
Ε	ε	epsilon	Τ	τ	tau
Ζ	ζ	zeta	Υ	υ	upsilon
Η	η	eta	Φ	φ	phi
Θ	θ	theta	Χ	χ	chi
Ι	ι	iota	Ψ	ψ	psi
Κ	κ	kappa	Ω	ω	omega
Λ	λ	lambda	Μ	μ	mu
Ν	ν	nu	Ξ	ξ	xi

Appendix IV

AMINO ACID REPRESENTATION

Amino acid	Three letter representation	Single letter representation
Alanine	Ala	A
Valine	Val	V
Leucine	Leu	L
Isoleucine	Ile	I
Proline	Pro	P
Phenylalanine	Phe	F
Tryptophan	Trp	W
Methionine	Met	M
Glycine	Gly	G
Serine	Ser	S
Threonine	Thr	T
Cysteine	Cys	C
Tyrosine	Tyr	Y
Asparagine	Asn	N
Glutamine	Gln	Q
Aspartic acid	Asp	D

(Contd.)

Amino acid	Three letter representation	Single letter representation
Glutamic acid	Glu	E
Lysine	Lys	K
Arginine	Arg	R
Histidine	His	H

Appendix V

PREPARATION OF BUFFERS

1. ACETATE BUFFER

Stock solution

A 0.2M solution of acetic acid (11.55 ml in 1000 ml)

B 0.2M solution of sodium acetate (16.4 g of C_2H_2Na or 27.2 g of $C_2H_3O_2Na \cdot 3H_2O$ in 1000 ml).

x ml of A, y ml of B diluted to a total of 100 ml.

x	y	pH
46.3	3.7	3.6
44.0	6.0	3.8
41.0	9.0	4.0
36.8	13.2	4.2
30.5	19.5	4.4
25.5	24.5	4.6
20.0	30.0	4.8
14.8	35.2	5.0
10.5	39.5	5.2
8.8	41.2	5.4
4.8	54.2	5.6

2. BORIC ACID–BORAX BUFFER

Stock solution

A 0.2M solution of boric acid (12.4 g in 1000 ml)

B 0.05 solution of borax (19.05 g in 1000 ml; 0.2M in terms of sodium borate).

50 ml of A, x ml of B, diluted to a total of 200 ml.

x	pH
2.0	7.6
3.1	7.8
4.9	8.0
7.3	8.2
11.5	8.4
17.5	8.6
22.5	8.7
30.0	8.8
42.5	8.9
59.0	9.0
83.0	9.1
115.0	9.2

3. CARBONATE–BICARBONATE BUFFER

Stock solution

A 0.2M solution of anhudrous sodium carbonate (21.2 g in 1000 ml)

B 0.2M solution of sodium bicarbonate (16.8 g in 1000 ml)

x ml of A, y ml of B, diluted to a total of 200 ml.

x	y	pH
4.0	46.0	9.2
7.5	42.5	9.3
9.5	40.5	9.4
13.0	37.0	9.5
16.0	34.0	9.6
19.5	30.5	9.7
22.0	28.0	9.8
25.0	25.0	9.9
27.5	22.5	10.0
30.0	20.0	10.1
33.0	17.0	10.2
35.5	14.5	10.3
38.5	11.5	10.4
40.5	9.5	10.5
42.5	7.5	10.6
45.0	5.0	10.7

4. CITRATE BUFFER

Stock solution

A 0.1M solution of citric acid (21.01 g in 1000 ml)

B 0.1M solution of sodium citrate (29.41 g of $C_6H_5O_7Na_3 \cdot 2H_2O$ in 1000 ml)

x ml of A, y ml of B diluted to a total of 100 ml.

x	y	pH	x	y	pH	x	y	pH
46.5	3.5	3.0	33.0	17.0	4.0	18.0	32.0	5.2
43.7	6.3	3.2	31.5	18.5	4.2	16.0	34.0	5.4
40.0	10.0	3.4	28.0	22.0	4.4	13.7	36.3	5.6
37.0	13.0	3.6	25.5	24.5	4.6	11.8	38.2	5.8
35.0	15.0	3.8	23.0	27.0	4.8	9.5	41.5	6.0
			20.5	29.5	5.0	7.2	42.8	6.2

5. GLYCINE–HCl BUFFER

Stock solution

A 0.2M Glycine (15.01 g in 1000 ml)

B 0.2N HCl

25 ml of A, x ml of B diluted to a total of 100 ml.

x	pH
22.0	2.2
16.2	2.4
12.1	2.6
8.4	2.8
5.7	3.0
4.1	3.2
3.2	3.4
2.5	3.6

6. PHOSPHATE BUFFER

Stock solution

A 0.2M solution of monobasic sodium phosphate (27.8 g in 1000 ml)

B 0.2M solution of di basic sodium phosphate (53.66 g of $Na_2HPO_4 \cdot 7H_2O$ or 71·7 g if $Na_2HPO_4 \cdot 12H_2O$ in 1000 ml).

x ml of A, y ml of B diluted to a total of 200 ml.

x	y	pH	x	y	pH
93.5	6.5	5.7	45.0	55.0	6.9
92.0	8.0	5.8	39.0	61.0	7.0
90.0	10.0	5.9	33.0	67.0	7.1
87.7	12.3	6.0	28.0	72.0	7.2
85.0	15.0	6.1	23.0	77.0	7.3
81.5	18.5	6.2	19.0	81.0	7.4
77.5	22.5	6.3	16.0	84.0	7.5
73.5	26.5	6.4	13.0	87.0	7.6
68.5	31.5	6.5	10.5	89.5	7.7
62.5	37.5	6.6	8.5	91.5	7.8
56.5	43.5	6.7	7.0	93.0	7.9
51.0	49.0	6.8	5.3	94.7	8.0

7. TRIS (HYDROXYMETHYL) AMINOMETHANE (TRIS-HCl) BUFFER

Stock solution

A 0.2M solution fo Tris (hydroxymethyl) aminomethane (24.2 g in 1000 ml)

B 0.2N HCl

50 ml of A, x ml of B diluted to a total of 200 ml.

x	pH
5.0	9.0
8.1	8.8
12.2	8.6
16.5	8.4
21.9	8.2
26.8	8.0
32.5	7.8
38.4	7.6
41.4	7.4
44.2	7.2

8. GLYCINE (NaOH) BUFFER

Stock solutions

A 0.2M solution of Glycine (15.01 g in 1000 ml)

B 0.2M NaOH

50 ml of A, x ml of B diluted to a total of 200 ml.

x	pH
4.0	8.6
6.0	8.8
8.8	9.0
12.0	9.2
16.8	9.4
22.4	9.6
27.2	9.8
32.0	10.0
38.6	10.4
45.5	10.6

Appendix VI

COMMONLY USED PROTEIN MARKERS IN GEL FILTRATION CHROMATOGRAPHY

Proteins	Molecular weight
Cytochrome C	12,400
Carbonic anhydrase	29,000
Bovine serum albumin	66,000
Alcohol dehydrogenase	150,000
β-amylase	200,000
Apoferritin	443,000
Dextran blue	2,000,000

Appendix VII

COMMONLY USED PROTEIN MARKERS IN SDS-PAGE

Proteins	Molecular weight
α-Lactalbumin	14,200
Trypsin inhibitor	20,100
Carbonic anhydrase	29,000
Egg albumin	45,000
Bovine serum albumin	66,000
β-galactosidase	116,000
β_2-macroglobulin	180,000
Urease	272,000

Appendix VIII

CONVERSION TABLE OF PERCENT TRANSMISSION TO OPTICAL DENSITY

% Transmission	Optical Density
1	2.000
10	1.301
20	0.699
30	0.523
40	0.398
50	0.301
60	0.222
70	0.155
80	0.097
90	0.046
100	0.000

Appendix IX

ABSORPTION MAXIMA AND EXTINCTION COEFFICIENTS OF SOME IMPORTANT BIOCHEMICAL COMPOUNDS

Compound	λ_{max} (mμ)	Molar extinction coefficient $a_m \times 10^{-3}$
Adenine	260.5	13.3
Cytosine	267.0	6.1
Guanine	275.5	8.1
	246.0	10.7
Thymine	264.0	7.9
Uracil	259.5	8.2
NAD^+, $NADP^+$	259.0	18.0
NADH, NADPH	339.0	6.2
FAD	450.0	11.3
	260.0	37.0
FMN	450.0	12.2
Pyridoxal phosphate	388.0	4.9
Tryptophan (0.1N HCl)	278.0	5.6
Tyrosine (0.1N HCl)	274.5	1.3

Extinction coefficients are given for 1 cm light path

Appendix X

SOME COMMONLY USED RADIOISOTOPES AND THEIR PROPERTIES

Radioisotope	Type of radiation	Half-life
^{3}H	β^-	121.1 yr
^{14}C	β^-	5700 yr
^{22}Na	β^+, γ	2.6 yr
^{32}P	β^-	14.3 days
^{35}S	β^-	87.1 days
^{45}Ca	β^-	180 days
^{54}Mn	γ	310 days
^{59}Fe	β^-, γ	47 days
^{60}Co	β^-, γ	5.3 yr
^{65}Zn	β^+, γ	250 days
^{82}Br	β^-, γ	34 yrs
^{131}I	β^-, γ	8 days

Appendix XI

UNITS COMMONLY USED TO DESCRIBE RADIOACTIVITY

Unit	Abbreviation	Definition
Counts per minute	c.p.m.	The recorded rate of decay
Counts per second	c.p.s.	
Disintegrations per minute	d.p.m.	The actual rate of decay
Disintegrations per second	d.p.s.	
Curie	Ci	The number of d.p.s. equivalent to 1 g of radium (3.7×10^{10} d.p.s.)
Millicurie	mCi	Ci $\times 10^{-3}$ or 2.22×10^{9} d.p.m.
Microcurie	mCi	Ci $\times 10^{-6}$ or 2.22×10^{6} d.p.m.
Becquerel (SI unit)	Bq	1 d.p.s.
Terabecquerel (SI unit)	TBq	10^{12} Bq or 27.027 Ci
Gigabecquerel (SI unit)	GBq	10^{9}Bq or 27.027 mCi
Megabecquerel (SI unit)	MBq	10^{6} Bq or 27.027 mCi

(Contd.)

Unit	Abbreviation	Definition
Electron volt	eV	The energy attained by an electron accelerated through a potential difference of 1 volt. Equivalent to 1.6×10^{-19} J
Roentgen	R	The amount of radiation that produces 1.61×10^{15} ion-pairs kg^{-1}
Rad	rad	The dose that gives an energy absorption of 0.01 J k^{-1}
Gray	Gy	The dose that gives an energy absorption of 1 J k^{-1}. Thus 1 Gy = 100 rad.
Rem	rem	The amount of radiation that gives a dose in humans equivalent to 1 rad of X-rays
Sievert	Sv	The amount of radiation that gives a dose in humans equivalent to 1 Gy of X-rays. Thus 1 Sv = 100 rem.

Glossary

Absorption spectrum The pattern of energy absorption by a solution of any substance when light of different wavelengths passes through it. This pattern is known as absorption spectrum. The absorption spectrum of a substance is established by measuring either optical density or transmittance of a particular concentration of the substance at different wavelengths.

Acid A proton donor.

Acid error Values registered by the glass electrode that tend to be somewhat high when the pH is less than 0.5.

Adsorption A process that occurs when a liquid or gas called adsorbate accumulates on the surface of a solid or liquid (adsorbent) by physical forces (dispersive, polar or ionic), forming a molecular or atomic film.

Adsorption chromatography A technique in which separation of components of a mixture takes place by adsorption efficiency of the substances to the solid stationary phase. The most strongly adsorbed component forms the topmost band while the least strongly adsorbed material forms the lowermost band on the adsorbent column.

Affinity chromatography A type of chromatography which makes use of a specific affinity between a substance to be isolated and a molecule that can specifically bind (a ligand). This technique is useful for the isolation of proteins, polysac-charides, nucleic acids and other classes of naturally occurring compounds.

Alkaline error An ordinary glass electrode that becomes somewhat sensitive to alkali metal ions and gives low readings at pH values greater than 9.

Analytical centrifugation The centrifugation involving measurement of the physical properties of the sedimenting particles such as sedimentation coefficient or molecular weight. Optical methods are used in analytical ultracentrifugation.

Annular disc (aperture stop) A major component of the phase contrast microscope. It is a disc with a transparent circular ring (annulus). It is placed just below the condenser. Its function is to transmit the hollow beam of light or to produce an annulus of light (a ring of light). It avoids hitting of bright light rays on the object.

Atomic force microscope (AFM) A microscope that measures the interaction force between the tip and surface. The tip may be dragged across the surface, or may vibrate as it moves. The interaction force will depend on the nature of the sample, the probe tip and the distance between them. It is used to obtain images of conductive surfaces at an atomic scale 2×10^{-10} m or 0.2 nanometre.

Atomic number The number of protons in the nucleus of an atom.

Atomic weight The total number of protons and neutrons in the atom of an element.

Autoradiography A technique in which the presence of radioactive isotopes in tissues is detected by covering tissue sections with photographic emulsion. At sites of radioactive material, the radioactive emission acts on the silver halide in the emulsion. Subsequent development and fixation turn radiated silver halide into black grains.

Base A proton acceptor.

Beer's law This law states that when a parallel beam of monochromatic light passes through an isotopic, light-absorbing medium, the amount of light that is absorbed is directly proportional to the number of light-absorbing molecules in that medium. In other words, it is the concentration of the substance in that medium.

Bioinformatics Conceptualizing biology in terms of molecules (in the sense of physical chemistry) and applying "informatics techniques" (derived from disciplines such as applied mathematics, computer science and statistics) to understand and organize the information associated with these molecules on a large scale. In short, bioinformatics is a management information system for molecular biology and has many practical applications.

Biosensor A compact analytical device incorporating a biological or biologically derived sensing element either integrated within or intimately associated with a physico-chemical transducer. The aim of a biosensor is to produce either discrete or continuous digital electronic signals, which are proportional to a single analyte or a related group of analytes.

Buffer A solution which resists a change in pH when an acid or base is added to it.

Centrifugal elutriation A technique by which the separation and purification of a large variety of cells from different tissues and species can be achieved by gentle washing action, using an elutriator rotor. The technique is based upon differences in the size, equilibrium and set-up in the separation chamber of the rotor.

Centrifugation A process used to separate or concentrate materials suspended in a liquid medium. The theoretical basis of this technique is the effect of gravity on particles (including macromolecules) in suspension. Two particles of different masses will settle in a tube at different rates in response to gravity.

Centrifuge A device that spins liquid samples at high speeds and thus creates a strong centripetal force causing the denser materials to travel towards the bottom of the centrifuge tube more rapidly than they would under the force of normal gravity.

Chromatic aberration Visible light is made of different colours. When visible light passes through a glass lens or a prism, it gets dispersed, or split into many colours. A lens focuses each colour at a different point, causing a fringe of colour to appear around bright objects.

Chromatography A technique in which the components of a mixture are separated based upon the rates at which they are carried through a stationary phase by a liquid or gaseous mobile phase. It is widely used for the separation, identification and estimation of chemical components present in a complex mixture.

Circular dichroism The difference in absorption between left- and right-handed circularly polarized light in chiral molecules.

Colorimetry A form of photometry which deals with the measurement of light absorption by coloured substances in solutions. The instrument which measures the intensity of the colour is known as colorimeter. This is based on the principle that when a beam of incident light passes through a coloured solution, the coloured substances in the solution absorb a part of the light and hence, the intensity of the transmitted light is always less than that of the incident light. As the number of light-absorbing molecules increases, the intensity of light coming out of the medium decreases exponentially and vice-versa. The difference in intensities between the incident and transmitted light, in turn, reflects the number of absorbing molecules or in other words, the concentration of the absorbing molecules in that solution.

Column chromatography A separation process involving uniform percolation of a liquid through a column packed with finely divided material. The separation in the column is effected either by direct interaction between the solute components and the surface of the stationary phase or by adsorption of solute by the stationary phase.

Computational biology The development and application of data-analytical and theoretical methods, mathematical modeling and computational simulation techniques to the study of biological, behavioural, and social systems.

Confocal microscopy An imaging technique used to increase micrograph contrast and/or to reconstruct three-dimensional images by using a spatial pinhole to eliminate out-of-focus light or flare in specimens that are thicker than the focal plane. This technique has been gaining popularity in the scientific and industrial communities. Typical applications include life sciences and semiconductor inspection.

Continuous flow electrophoresis A type of electrophoresis used for separation of particles in large-scale productions. Electrophoresis takes place continuously as the separating material is carried upwards by flow of carrier buffer through annular space between two vertical concentric cylinders. The outer cylinder is rotated to maintain a stable laminar flow of the buffer solution. An electrical field is applied between the two cylinders, causing the sample material to separate radially as it is carried upwards by the buffer flow.

Dark-field Illumination An optical technique where the specimen is seen as a bright object against a dark background.

Dark-field microscope A microscope which permits light rays which are heavily retarded to form the image. Thus only objects which retard light strongly are viewed.

Density gradient centrifugation A type of centrifugation in which separation depends upon the buoyant densities of the particle. This method gives a much better separation than differential centrifugation.

Differential centrifugation The type of centrifugation used for fractionation of the contents of a homogenate. This method is based upon the differences in the sedimentation rate of particles of different size and density, e.g. separation of subcellular organelles.

Differential interference contrast An excellent mechanism for rendering contrast in transparent specimens. Differential interference contrast (DIC) microscopy is a beam-shearing interference system in which the reference beam is sheared by a minuscule amount, generally somewhat less than the diameter of an airy disc. The technique produces a monochromatic shadow-cast image that effectively displays the gradient of optical paths for both high and low spatial frequencies present in the specimen. Those regions of the specimen where the optical paths increase along a reference direction appear brighter (or darker), while regions where the path differences decrease appear in reverse contrast. As the gradient of optical path difference grows steeper, image contrast is dramatically increased.

DNA footprinting A technique for identifying exactly where a protein binds to DNA. Knowing where a protein binds to DNA often aids in understanding how gene expression is regulated.

DNA microarray A collection of microscopic DNA spots attached to a solid surface such as glass, plastic or silicon chip forming an array for the purpose of expression profiling, monitoring expression levels for thousands of genes simultaneously and is commonly known as gene chip, DNA chip, or biochip.

DNA sequencing The determination of all or part of the nucleotide sequence of a specific deoxyribonucleic acid (DNA) molecule.

Electrochemical scanning tunnelling microscope (EC-STM) A microscope which works on the general working principle of a scanning tunnelling microscope, with the difference that the EC-STM is designed to run in an electrochemical cell.

Electron spin resonance When the molecules of a solid exhibit paramagnetism as a result of unpaired electron spins, transitions can be induced between spin states by applying a magnetic field and then supplying electromagnetic energy, usually in the microwave range of frequencies. The resulting absorption spectra are described as electron spin resonance (ESR) or electron paramagnetic resonance (EPR).

Electrophoresis An important analytical tool used to separate different compounds in a mixture with different electrophoretic mobility. Mixture of amino acids, proteins, and nucleic acids can be separated by electrophoresis.

Electrophoretic mobility Defined as the distance travelled by the charged particles in an electric field, in one second under the potential gradient of one volt per centimetre.

Elution The removal of the sample from the solid matrix using a suitable solvent.

Embank (www.ncbi.nlm.nih.giv/Genbank/) The NIH genetic sequence database, which is an annotated collection of all publicly available DNA sequences.

Eyepiece diaphragm The part inside the eyepiece, which holds gracticules and reticules. It also defines the round field of view that is seen through the microscope.

Eyepiece Also known as ocular, it produces the second stage of magnification enlarging the image magnified by the objective lens.

Flame photometry A technique based on the measurement of intensity of the light emitted when a metal is introduced into a flame. The wavelength of the colour indicates the nature of the element and the intensity of the colour indicates the quantity of the element present.

Fluorescence A luminescence that is mostly found as an optical phenomenon in cold bodies, in which the molecular absorption of a photon triggers the emission of a lower-energy photon with a longer wavelength. The energy difference between the absorbed and emitted photons ends up as molecular vibrations or heat. Usually the absorbed photon is in the ultraviolet, and the emitted light is in the visible range, but this depends on the absorbance curve and Stokes' shift of the particular fluorophore. Fluorescence is named after the mineral fluorite, composed of calcium fluoride, which exhibits this phenomenon.

Fluorescence microscope A light microscope used to study properties of organic or inorganic substances using the phenomena of fluorescence and phosphorescence instead of, or in addition to, reflection and absorption.

Fluorimeter An instrument used to measure the fluorescence.

Fluorophore A component of a molecule which causes the molecule to be fluorescent. It is a functional group in a molecule which will absorb energy of a specific wavelength and re-emit energy at a different (but equally specific) wavelength. The amount and wavelength of the emitted energy depends on both the fluorophore and the chemical environment of the fluorophore. This technology has particular importance in the field of biochemistry and protein studies, e.g. in immunofluorescence and immunohistochemistry. Fluorescein isothiocyanate is an example of a fluorophore which can be chemically attached to a different, non-fluorescent molecule to create a new and fluorescent molecule.

Fourier transform infrared spectroscopy A technique for obtaining high quality infrared spectra by mathematical conversion of an interference pattern into a spectrum.

Functional genomics A field of molecular biology that is attempting to make use of the vast wealth of data produced by genome sequencing projects to describe genome function. Functional genomics uses techniques like DNA microarrays, proteomics, metabolomics and mutation analysis to describe the function and interactions of genes.

Gas chromatography (GC) See gas liquid chromatography.

Gas liquid chromatography (GLC) A type of partition chromatography in which the mobile phase is a carrier gas, usually an inert gas such as helium or nitrogen, and the stationary phase is a microscopic layer of liquid on an inert solid support. A gas chromatograph uses a thin capillary fibre known as the column, through which different chemicals pass at different rates depending on various chemical and physical properties. As the chemicals exit the end of the column, they are detected and identified electronically.

Gel filtration chromatography See gel permeation chromatography.

Gel permeation chromatography A special type of partition chromatography in which separation of macromolecules is based only on molecular size. A column is prepared of tiny particles of an inert substance that contains small pores. If a solution containing substances with different molecular weights is passed through the column, molecules larger than the pores move only in the space between the particles and hence are not retarded by the column material. However, molecules smaller than the pores diffuse in and out of particles. Thus the smaller molecules are slowed in their movement down the column. Hence, molecules are eluted from the column in order of decreasing size or decreasing molecular weight.

Genetic fingerprinting A technique to distinguish between individuals of the same species using only samples of their DNA.

Genomics Any attempt to analyse or compare the entire genetic complement of species. It is, of course, possible to compare genomes by comparing more or less representative subsets of genes within genomes.

Huygenian eyepiece A type of eyepiece that has two planoconvex lens elements with the eyepiece diaphragm between them. Usually with a narrow field of view.

Immersion objective The most common immersion objective is the 100× oil objective. Oil is placed on the cover glass of the slide (and sometimes on the top element of the condenser) which produces a high magnification and high resolving power of the objective when immersed in oil. This produces the full NA of objective lens.

Immunoelectrophoresis A technique based on the electrophoretic mobility and precipitin. In precipitin reaction, an antigen combines with its specific antibody to form antigen–antibody complex. Since most of the antigen–antibody complexes are insoluble, they can be seen with the naked eye. Thus, immunoelectrophoresis exploits the specificity of reaction between an antigen and antibody and molecular sieving of the gel in which this reaction takes place.

Immunofluorescence The labelling of antibodies or antigens with fluorescent dyes. Immunofluorescently labelled tissue sections are studied using a fluorescence microscope or by confocal microscopy.

Infrared spectrophotometer A technique based upon the simple fact that a chemical substance shows marked selective absorption in the infrared region. After absorption of IR radiations, the molecules of a chemical substance vibrate at many rates of vibration, giving rise to close-packed absorption bands, called an IR absorption spectrum, which may extend over a wide wavelength range. Various bands will be present in IR spectrum that will correspond to the characteristic functional groups and bonds present in a chemical substance. Thus, an IR spectrum of a substance is a fingerprint for its identification.

Ion-exchange chromatography A type of chromatography based on the principle of the attraction between oppositely charged particles. If a mixture of charged and uncharged molecules is passed through a column of ion-exchanger, charged molecules adsorb to ion exchangers reversibly. The column is eluted with buffers of different pH or ionic strength. The components of the eluting buffer compete with the bound material for the binding sites and eventually displace the charged particle originally present in the mixture.

Iris diaphragm The diaphragm usually mounted under the condenser and which controls the amount of light converging on the specimen by opening or closing the leaf diaphragm.

Isoelectric focusing The method of separating proteins according to their isoelectric points in a pH gradient.

Isoelectric point A particular pH where molecules are electrophoretically immobile since they carry no net charge.

Iso-osmolar solution Two solutions producing the same osmotic effect.

Isopycnic centrifugation A type of centrifugation which depends only upon the buoyant density of the particle. In isopycnic centrifugation, the maximum density of the gradient always exceeds the density of the densest particle. A continuous density gradient is always used. During centrifugation, sedimentation of the particle occurs until the buoyant density of the particle and the density of the gradient are equal. At this point of isodensity, no further sedimentation occurs, irrespective of how long centrifugation continues. This technique is used to separate particles of similar size but of differing density. Subcellular organelles such as Golgi apparatus, mitochondria and peroxisomes can be effectively separated by this method.

Isotopes Atoms with the same atomic number but different mass numbers.

Kohler Illumination A technique for uniformly illuminating a field from non-uniform light such as a coiled filament lamp.

Lambert's law This law states that when a parallel beam of monochromatic light passes through an isotopic, light-absorbing medium, the amount of light that is absorbed is directly proportional to the length of the medium through which the light passes.

Liquid–liquid chromatography A type of chromatography in which the stationary phase is polar and the mobile phase is relatively non-polar. The mechanism of separation exploits the ability of the analyte to displace molecules of the mobile phase adsorbed as a monolayer on the stationary phase, as well as the ability of the analytes to compete with mobile phase molecules in the formation of a bilayer on the stationery phase surface. The order of elution of analytes is such that the least polar is eluted first and the most polar last.

Magnification The enlargement of an object through the lens system. This is determined by multiplying the magnifying power of the objective by the eyepiece.

Magnifying power The number of times the image seen through the microscope is larger than the item appears to the unaided eye.

Manometer An instrument used to measure pressure. Measurements involving gases are governed by three parameters, namely volume, pressure and temperature. At constant temperature, any change in the amount of a gas can be measured by a change in its volume if the pressure is kept the same or by a change in its pressure if its volume is kept constant.

Mass spectrometry A technique in which the compound under investigation is bombarded with a beam of electrons, which produce an ionic molecule, or ionic fragments of the original species. The resulting assortment of charged particles is then separated according to their masses. The spectrum produced is known as mass spectrum, which can provide information concerning the molecular structure of organic and inorganic compounds.

Matrix The material supporting the stationary phase, in column chromatography.

Mechanical stage A device on the platform for holding and moving the slide (or specimen) on an X or Y axis.

Micrometer disc A glass disc with a scale or grid mounted to the eyepiece diaphragm used for measurement.

Microphotography The process by which miniature photographs of comparatively large objects are taken, as in document copying.

Microscope (Greek: *micron* = small and *scopos* = aim) An instrument for viewing objects that are too small to be seen by the naked or unaided eye. The science of investigating small objects using such an instrument is called microscopy, and the term microscopic means minute or very small, not easily visible with the unaided eye; in other words, requiring a microscope to examine.

Molal solution A solution which is prepared by dissolving one gram-molecular weight of a substance in one thousand grams of its solvent.

Molar solution A solution in which a litre solution contains 1 g molecular weight of the substance.

Molecular sieve chromatography See gel permeation chromatography.

Mole fraction The ratio of the number of moles of a particular substance to the total number of moles of all substances present in the solution.

NCBI The US National Center for Biotechnology Information.

Nephelometry When light is allowed to pass directly through a solution having suspended particles, the amount of radiation scattered by the particles is measured at an angle (usually 90°) to the incident beam. The measurement of the intensity of the scattered light as a function of the concentration of the dispersed phase forms the basis of nephelometric analysis.

NIH The US National Institute of Health.

NMR spectroscopy A type of spectroscopy involving transition of a nucleus from one spin state to another with the resultant absorption of electromagnetic radiation by spin active nuclei (having nuclear spin not equal to zero) when they are placed in a magnetic field.

Normal solution A solution which contains one gram equivalent weight of solute dissolved in one litre of its solution.

Northern blotting A transfer technique for RNA from agarose gel onto nitrocellulose membrane.

Nuclear magnetic resonance A technique for detecting atoms which have nuclei that possess a magnetic moment.

Numerical aperture (NA) The angle included by a cone of

light accepted by the objective of a microscope. The higher the NA, the greater the resolving power.

Objective lens The lens forming the primary image of the microscope, which is seen through the eyepiece. The markings on the objective lens are the magnifying power (such as 10×), followed by the NA (0.25) and the tube length. Other numbers which appear on the objective lens may refer to the manufacturer's catalog number of the particular item.

Optical rotatory dispersion If a linearly polarized light wave passes through an optically active substance, the direction of polarization will change. This change is wavelength-dependent. This phenomenon is called **optical rotatory dispersion (ORD)**.

Osmolar concentration The molar concentration of a substance, multiplied by the number of particles or ions produced by that substance when in solution, e.g. for sucrose or glucose, a non-dissociating solute, a 1M solution is one osmolar. But for sodium chloride, which dissociates into two particles, viz. Na^+ and Cl^-, a 1 M solution is equal to 2 osmolar.

Osmometer A device for measuring the osmotic strength of a solution, colloid or compound.

Osmosis The net diffusion of water molecules from a diluter solution or pure water itself to a more concentrated solution across a semipermeable membrane, which permits the diffusion of water but not of a solute.

Osmotic pressure Hydrostatic pressure that balances the osmotic influx of water from pure water to the concentrated solution.

Parts per million (ppm) 1 mg of any substance in one litre.

PDB (Protein data bank) A database and format of files that describe the 3D structure of a protein or nucleic acid as determined by X-ray crystallography or nuclear magnetic resonance (NMR) imaging.

Per cent by weight (W/W) A solution is said to be 1% W/W, when it contains one gram of solute plus 99 g of solvent.

Percent solution A solution which contains a known weight of the substance in a specified volume of its solvent.

pH The negative logarithm of hydrogen ion concentration.

pH Meter A potentiometer which measures the voltage between two electrodes placed in a solution. pH meter is used for accurate pH measurements of solutions.

Pharmacogenetics The study of how the action of and reactions to drugs vary with the patient's genes. All individuals respond differently to drug treatments; some positively, others with little obvious change in their conditions and yet others with side effects or allergic reactions. Much of this variation is known to have a genetic basis. Pharmacogenetics is a subset of pharmacogenomics which uses genomic/bioinformatic methods to identify genomic correlates, for example SNPs (Single Nucleotide Polymorphisms), characteristic of particular patient-response profiles and uses those markers to inform the administration and development of therapies. Strikingly, such approaches have been used to "resurrect" drugs thought previously to be ineffective, but subsequently found to work within the subset of patients or in optimizing the doses of chemotherapy for particular patients.

Pharmacogenomics The application of genomic approaches and technologies to the identification of drug targets. In short, pharmacogenomics is using genetic information to predict whether a drug will help make a patient well or sick. It studies how genes influence the response of humans to drugs, from the population level to the molecular level.

Phase contrast An optical technique used to view the structure of transparent objects whose varying but invisible differences in thickness result in differences in the phase of transmitted light. This is done when the transmitted light changes its optical path by about 1/3 wavelength.

Phase contrast microscope The microscope widely used for examining such specimens as biological tissues. It is a type of light microscopy that enhances contrasts of transparent and colourless objects by influencing the optical path of light. The phase contrast microscope is able to show components in a cell or bacteria, which would be very difficult to see in an ordinary light microscope.

Phosphorescence A specific type of photoluminescence, related to fluorescence. However unlike fluorescence, a phosphorescent material does not immediately discharge the radiation it absorbs. Phosphorescence is a process in which energy stored in a substance is released very slowly and continuously in the form of glowing light.

Phosphorimeter An instrument used to measure phosphorescence radiation.

Photomicrography The technique concerned with the photography of enlarged images of small objects.

Planachromat lens An achromat lens which has been corrected for a flat field.

Planoconcave mirror A two-sided mirror 50 mm in diameter, with one side flat (plano) and the other curved (concave). The concave side is used for low NA when no condenser is used; the plano surface is used with a substage condenser.

Polarizer A transparent material which can absorb all vibrations of light passing through it except those in a single plane.

Polymerase chain reaction (PCR) A technique of creating copies of specific fragments of DNA. PCR rapidly amplifies a single DNA molecule into many billions of molecules using a heat-stable polymerase, two "primers" that flag the beginning and end of the DNA stretch to be copied and a pile of DNA building blocks that the polymerase needs to make that copy.

PROSITE (www.expasy.ch/prosite/) A database of short sequence patterns and profiles that characterize biologically significant sites in proteins. PROSITE is a database of protein families and domains. It consists of biologically significant sites, patterns and profiles that help to reliably identify a new sequence with the known protein family.

Proteomics The study of proteins—their location, structure and function. It is the identification, characterization and quantification of all proteins involved in a particular pathway, organelle, cell, tissue, organ or organism that can be studied to provide accurate and comprehensive data about that system.

Rack and pinion The term used to describe the gear system for lowering and raising the stage or barrel when focusing. The coarse adjustment control (knob) usually moves the barrel or stage.

Radioactivity The spontaneous disintegration of the nuclei of some of the isotopes of certain elements, with the emission of alpha (α) particles, beta (β) particles or gamma (γ) rays.

Radiodating Determination of the age of rocks, fossils or sediments using radioisotopes.

Radioimmunoassay One of the most important techniques used in the clinical and biochemical fields for the quantitative analysis of hormones, steroids and drugs. It combines the specificity of the immune reaction with the sensitive radioisotope techniques.

Raman effect The discovery made by Sir C.V. Raman that when a beam of monochromatic light was allowed to pass through a substance in the solid, liquid or gaseous state, the scattered light contains some additional frequencies over and above that of the incident frequency.

Ramsden or Kellner eyepiece Positive type eyepieces in which the primary image lies beneath the field lens rather than between the field and eye lenses. Such lenses are useful for measurement and comparison purposes since the size of the primary image is not affected by the field lens.

Rate-zonal centrifugation technique A type of density gradient centrifugation in which particle separation is based upon (i) differences in the size, shape and density of the particle (ii) the density and viscosity of the medium and (iii) the applied centrifugal field.

Refractive Index (RI) The ratio of the speed of light in vacuum to its speed in some other medium. This will determine how much light rays are bent. When using immersion objectives, it is important to keep the values as close together as possible.

Relative centrifugal force (RCF) The ratio of the weight of the particle in the centrifugal field to the weight of the same particle when acted by gravity alone which is commonly referred to as the "number times g".

Hence,

$$RCF = \frac{4\pi^2(\text{rev min}^{-1})^2 r}{3600 \times 981}$$

$$RCF = (1.118 \times 10^{-5})(\text{rev min}^{-1})^2 r$$

RCF depends upon the rpm and the radius of rotation, r. If r is a constant for a given rotor, then variations in rpm alone determine the variations in RCF.

Resolution A measure of the ability of a lens to image closely spaced objects so that they are recognized as separate objects.

Resolving power The capacity of any optical system to distinguish and separate details in a specimen.

Restriction fragment length polymorphism (RFLP) The identification of pattern difference between the DNA fragment sizes in individual organisms using specific restriction enzymes. Differences in fragment length result from base substitutions, additions, deletions or sequence rearrangements within restriction enzyme recognition sequences.

Reversed-phase liquid chromatography A form of liquid chromatography in which the stationary phase is non-polar and the mobile phase is relatively polar. Chromatographic separation of analytes is determined principally by the characteristics of mobile phase and probably involves a combination of adsorption and partition mechanisms.

Rocket electrophoresis A type of electrophoresis in which antigen samples are placed in a gel containing monospecific antibodies, and when an electrical field is applied, antigens tend to migrate towards anode. During this process, antigen molecules meet the corresponding antibodies and form a rocket-like precipitin peak. The height of the rocket varies depending on the concentration of antigen.

Saturated solution A solution which holds as much solute as it can.

Scanning electron microscope (SEM) A microscope used to provide a three-dimensional image of cells and micro-organisms. In scanning electron microscopy the electron beam does not pass through the specimen. Instead, the surface of the cell is coated with a heavy metal, and a beam of electrons is used to scan across the specimen. Electrons that are scattered or emitted from the sample surface are collected to generate a three-dimensional image as the electron beam moves across the cell.

Scanning probe microscopy (SPM) A technology that covers several related technologies for imaging and measuring surfaces on a fine scale, down to the level of molecules and groups of atoms. SPM technologies share the concept of scanning an extremely sharp tip (3–50 nm radius of curvature) across the object surface. The tip is mounted on a flexible cantilever, allowing the tip to follow the surface profile. When the tip moves in proximity to the investigated object, forces of interaction between the tip and the surface influence the movement of the cantilever. These movements are detected by selective sensors. Various interactions can be studied depending on the mechanics of the probe.

Scanning tunnelling microscopy (STM) A non-optical microscope which employs principles of quantum mechanics. An atomically sharp probe (the tip) is moved over the surface of the material under study, and a voltage is applied between the probe and the surface. Depending on the voltage, electrons will tunnel (this is a quantum-mechanical effect) or jump from the tip to the surface (or vice-versa depending on the polarity), resulting in a weak electric current. The STM measures a weak electrical current flowing between the tip and sample as they are held a great distance apart. The scanning tunnelling microscope is one of the most important tools for surface physics and surface chemistry, where it shows the structure of the topmost layer of atoms or molecules.

Sedimentation coefficient The sedimentation rate or velocity (v) of a particle expressed in terms of its sedimentation rate per unit of centrifugal field (commonly referred to as its sedimentation coefficient,S). Since sedimentation rate studies are performed using a wide variety of solvent–solute systems or at different temperatures, the value is affected by temperature, solution viscosity and density. Therefore, these values are corrected to the sedimentation constant theoretically obtainable in water at 20°C. The equation for standard sedimentation coefficient is

$$S_{20w} = \frac{S_{obs}(1 - \bar{v}\,\rho_{20w})}{(1-\bar{v}\rho_T)} \times \frac{\eta_T}{\eta_{20}} \times \frac{\eta}{\eta_{20}}$$

where,

S_{20w} is the standard sedimentation coefficient,

S_{obs} is the experimentally measured sedimentation coefficient,

η_T / η_{20} is the relative viscosity of water at temperature T compared with that at 20°C,

η / η_{20} is the relative viscosity of the solvent to that of water,

ρ_{20w} is the density of water at 20°C,

ρ_T is the density of the solvent at temperature T(°C) and

$\bar{v}$ is the partial specific volume of the solute.

The basic unit of sedimentation coefficient is 1×10^{-13} sec. This is also termed as one Svedberg unit (S).

Sedimentation rate The sedimentation rate of a given particle depends upon a number of factors such as density of the particle (ρ_p), radius of the particle (r_p), the density of the medium (ρ_m) and the viscosity (η) of the suspending medium. A mathematical expression relating to all these factors for sedimentation of a rigid spherical particle is given below

$$v = \frac{2}{9} \times \frac{r_p^2\,(\rho_p - \rho_m)}{\eta} \times g$$

where,

v = rate of sedimentation

g = gravitational field

2/9 = shape factor constant for a spherical particle.

Single nucleotide polymorphism (SNP) A change in DNA sequence at a single residue.

Southern blot The first blotting technique described by Southern (1975) for the transfer of DNA.

Spectroscopy The branch of science dealing with the study of interaction of electromagnetic radiation with matter. The ways in which the measurements of radiation frequency (emitted or absorbed) are made experi-mentally and the energy levels deduced from these comprise the practice of spectroscopy.

Spherical aberration The blurring of an image that occurs when light from the margin of a lens or mirror with a spherical surface comes to a shorter focus than light from the central portion. The changing focal length is caused by deviations in the lens or mirror surface from a true sphere.

Stokes' shift The difference (in wavelength or frequency units) between positions of the band maxima of the absorption and luminescence spectra of the same electronic transition. When a molecule or atom absorbs light, it enters an excited electronic state. Stokes' shift occurs because the molecule loses a small amount of the absorbed energy before re-releasing the rest of the energy as luminescence. This energy is often lost as thermal energy. It is named after the Irish physicist George G. Stokes. Stokes noted that fluorescence emission always occurred at a longer wavelength than that of the excitation light. This shift towards longer wavelength is known as Stokes' shift.

Structural genomics Refers to the analysis of macromolecular structures particularly proteins, using computational tools and theoretical frameworks. One of the goals of structural genomics is the extension of the idea of genomics, to obtain accurate three-dimensional structural models for all known protein families, protein domains or protein folds. Structural alignment is a tool of structural genomics. This field is also known as structural bioinformatics.

Svedberg unit (symbol S, sometimes Sv) A non-SI physical unit used in ultracentrifugation to express sedimentation coefficient. It is named after the Swedish physicist and chemist Theodor Svedberg.

SWISS-PROT A curated protein sequence database that strives to provide a high level of annotations such as the description of the function of a protein, its domain structure, post-translational modifications, variants, etc.

Transmission electron microscope (TEM) A microscope where the principle of observation of stained cells as used in the bright-field light microscope is used. Specimens are fixed and stained with salts of heavy metals, which provide contrast by scattering electrons. A beam of electrons is then passed through the specimen and focused to form an image on a fluorescent screen. Electrons that encounter a heavy metal ion as they pass through the sample are deflected and do not contribute to the final image, so stained areas of the specimen appear dark. Owing to the high resolving power of the electron microscope, it is possible to take negatives at very high magnification.

TrEMBL A computer-annotated supplement of SWISS-PROT that contains all the translations of EMBL nucleotide sequence entries not yet integrated in SWISS PROT.

Turbidimetry When light is allowed to pass through a suspension, the part of the incident radiant energy is dissipated by absorption, reflection, and refraction while the remainder is transmitted. Measurement of the intensity of the transmitted light as a function of the concentration of the suspended particles forms the basis of turbidimetric analysis.

Two-photon excitation microscopy An alternative to confocal microscopy that can be applied to living cells. The specimen is illuminated with a wavelength of light such that excitation of the fluorescent dye requires the simultaneous absorption of two photons. The probability of two photons simultaneously exciting the fluorescent dye is only significant at the point in the specimen upon which the input laser beam is focused, so fluorescence is only emitted from the plane of focus of the input light. This highly localized excitation automatically provides three-dimensional resolution, without the need for passing the emitted light through a pinhole aperture, as in confocal microscopy. Moreover, the localization of excitation minimizes damage to the specimen, allowing three-dimensional imaging of living cells.

Vant Hoff's laws of osmotic pressure

Law 1 The osmotic pressure (π) of a solution is directly proportional to the molar concentration (C) of the solute, as long as the temperature is maintained constant.

$$\pi = k_1 C_2$$

where k_1 is constant.

Law 2 The osmotic pressure of a solution is directly proportional to the absolute temperature (T) as long as its concentration remains constant.

$$\pi = k_2 T$$

where k_2 is constant.

Vapour pressure The pressure of a vapour in equilibrium with its non-vapour phases.

Video imaging Optical images produced in the microscope can be captured using either traditional film techniques, digitally with electronic detectors such as a charge-coupled device (CCD), or with a tube-type video camera. The primary function of video is to produce an electrical signal by scanning an optical image of a dynamic scene, then transmitting the information in real time to a receiver housed in a remote location for viewing or recording of the original event. The signal may be processed or recorded, then sent to an analyser or converted back into a two-dimensional image on a video monitor.

Wavelength Electromagnetic radiation is composed of an electric vector and magnetic vector, which oscillate in planes at right angles to each other and both at right angles to the direction of propagation. The distance along the direction of propagation for one complete cycle is known as wavelength (λ).

Weight/Volume per cent (W/V) A solution is said to be 1% W/V, when it contains one gram of solute in 100 ml solution.

Western blotting A blot transfer technique for proteins.

X-ray diffraction A technique utilized to study atomic structure of crystalline substances by noting the patterns produced by X-rays shot through the crystal.

References

Anbalagan, A. 1999. *An Introduction to Electrophoresis.* The Electrophoresis Institute, Salem.

Chatwal, G. and Anand, S. 1989. *Instrumental Methods of Chemical Analysis*. Himalaya Publishing House, Delhi.

Cooper, G.M. 2000. *The cell: A molecular approach,* 2nd Edn. ASM Press, Washington, DC.

Cynthia, G. and Jambeck, P. 2003. *Developing Bioinformatics Computer Skills*. L.Lejeune, (ed.). Shroff Publishers and Distributors Pvt. Ltd. New Delhi.

Gurumani, N. 2006. *Research Methodology for Biological Sciences*. MJP Publishers, Chennai.

Jayaraman, J. 1981. *Laboratory Manual in Biochemistry.* Wiley Eastern Limited, New Delhi.

Maniatis, T., Fritsch, E.F. and Sambrook, J. 1982. Molecular Cloning—A Laboratory Manual. Cold Spring Harbor Laboratory, Cold Spring Harbor, New York.

Mohan, J. 2001. *Organic Spectroscopy: Principles and Applications.* Narosa Publishing House, New Delhi.

Mukherjee, L. 2004. *Medical Laboratory Technology.* Volume 1.Tata McGraw-Hill Publishing Company Limited, New Delhi.

Plummer, D.T. 1988. *An Introduction to Practical Biochemistry.* Tata McGraw-Hill Publishing Company Limited, New Delhi.

Rastogi, S.C., Mendiratta, N. and Rastogi, P. 2004. *Bioinformatics Methods and Applications: Genomics, Proteomics and Drug Discovery.* Prentice-Hall of India Private Limited, New Delhi.

Robinson, J.W. (ed.). 1991. *Practical Handbook of Spectroscopy.* CRC Press, Boston.

Veerakumari, L. 2004. *Biochemistry.* MJP Publishers, Chennai.

Webster, J.G. (ed.). 2005. *Bioinstrumentation.* John Wiley & Sons Inc., Singapore.

Wilson, K. and Walker, J. (eds.). 2004. *Practical Biochemistry,* 5th edn. Cambridge University Press, UK.

Westermeier, R. 2005. *Electrophoresis in Practice: A Guide to Methods and Applications of DNA and Protein Separations.* Wiley-VCH Verlag MBH & Co. KgaA,Germany.

WEBSITES

http://abacus.bates.edu/~ganderso/biology/resources/index.html

http://books.google.com/

http://dept.kent.edu/projects/cell/index.html

http://en.wikipedia.org/wiki/

http://hplc.chem.shu.edu/NEW/HPLC_Book/

http://las.perkinelmer.com/Catalog/default.htm?CategoryID=Chromatography

http://protist.biology.washington.edu/fingerprint/dnaintro.html

http://protist.biology.washington.edu/fingerprint/restriction.html

http://pubs.acs.org/journals/chromatography/index.html

http://ull.chemistry.uakron.edu/analytical/Chromatography/

http://users.rcn.com/jkimball.ma.ultranet/BiologyPages/D/DNAsequencing.html

http://users.rcn.com/jkimball.ma.ultranet/BiologyPages/N/Nucleotides.html#nucleosides

http://users.rcn.com/jkimball.ma.ultranet/BiologyPages/W/Welcome.html

http://users.ugent.be/~avierstr/index.html

http://www.ai.mit.edu/people/minsky/papers/confocal.microscope.txt

http://www.biochem.mcw.edu/home.html

http://www.biolbull.org/content/vol201/issue2/images/large/1n0510235001.jpeg

http://www.biolbull.org/misc/terms.shtml

http://www.biologie.uni-hamburg.de/b-online/e44/hydrodi.htm#fluor

http://www.chromatography-online.org/

http://www.cranfield.ac.uk/ibst/

http://www.efm.leeds.ac.uk/CIVE/CIVE1400/course.html

http://www.encyclopedia4u.com/

http://www.healthatoz.com/healthatoz/Atoz/default.jsp

http://www.hitl.washington.edu/scivw/EVE/I.D.1.e.OtherDevices.html

http://www.interscience.wiley.com/jpages

http://www.iup.edu/biology/

http://www.lplc.com/

http://www.lsbu.ac.uk/biology/enztech/biosensors.html#fig6_1

http://www.ncbi.nlm.nih.gov/entrez/query.fcgi?db=PubMed

http://www.ornl.gov/divisions/casd/casd_home.htm

http://www.sciencedirect.com/science/journal

http://www.shu.ac.uk/schools/sci/chem/tutorials/

http://www.shu.ac.uk/schools/sci/chem/tutorials/chrom/chrom1.htm

http://www.shu.ac.uk/schools/sci/chem/tutorials/chrom/gaschrm.htm

http://www.unixl.com/google/index.php

http://www.wpi.edu/Academics/Depts/Chemistry/Courses/General/infraredfig5.html

http://www-structure.llnl.gov/cd/cdtutorial.htm

Index

C

D

E

www.ingramcontent.com/pod-product-compliance
Lightning Source LLC
LaVergne TN
LVHW021811240826
846425LV00002B/7